Other volumes in this series:

Volume 1	**Power Circuit Breaker Theory and Design** C.H. Flurscheim (Editor)
Volume 4	**Industrial Microwave Heating** A.C. Metaxas and R.J. Meredith
Volume 7	**Insulators for High Voltages** J.S.T. Looms
Volume 8	**Variable Frequency AC Motor Drive Systems** D. Finney
Volume 10	**SF_6 Switchgear** H.M. Ryan and G.R. Jones
Volume 11	**Conduction and Induction Heating** E.J. Davies
Volume 13	**Statistical Techniques for High Voltage Engineering** W. Hauschild and W. Mosch
Volume 14	**Uninterruptible Power Supplies** J. Platts and J.D. St Aubyn (Editors)
Volume 15	**Digital Protection for Power Systems** A.T. Johns and S.K. Salman
Volume 16	**Electricity Economics and Planning** T.W. Berrie
Volume 18	**Vacuum Switchgear** A. Greenwood
Volume 19	**Electrical Safety: A guide to causes and prevention of hazards** J. Maxwell Adams
Volume 21	**Electricity Distribution Network Design, 2nd Edition** E. Lakervi and E.J. Holmes
Volume 22	**Artificial Intelligence Techniques in Power Systems** K. Warwick, A.O. Ekwue and R. Aggarwal (Editors)
Volume 24	**Power System Commissioning and Maintenance Practice** K. Harker
Volume 25	**Engineers' Handbook of Industrial Microwave Heating** R.J. Meredith
Volume 26	**Small Electric Motors** H. Moczala *et al.*
Volume 27	**AC–DC Power System Analysis** J. Arrillaga and B.C. Smith
Volume 29	**High Voltage Direct Current Transmission, 2nd Edition** J. Arrillaga
Volume 30	**Flexible AC Transmission Systems (FACTS)** Y.-H. Song (Editor)
Volume 31	**Embedded Generation** N. Jenkins *et al.*
Volume 32	**High Voltage Engineering and Testing, 2nd Edition** H.M. Ryan (Editor)
Volume 33	**Overvoltage Protection of Low-Voltage Systems, Revised Edition** P. Hasse
Volume 36	**Voltage Quality in Electrical Power Systems** J. Schlabbach *et al.*
Volume 37	**Electrical Steels for Rotating Machines** P. Beckley
Volume 38	**The Electric Car: Development and future of battery, hybrid and fuel-cell cars** M. Westbrook
Volume 39	**Power Systems Electromagnetic Transients Simulation** J. Arrillaga and N. Watson
Volume 40	**Advances in High Voltage Engineering** M. Haddad and D. Warne
Volume 41	**Electrical Operation of Electrostatic Precipitators** K. Parker
Volume 43	**Thermal Power Plant Simulation and Control** D. Flynn
Volume 44	**Economic Evaluation of Projects in the Electricity Supply Industry** H. Khatib
Volume 45	**Propulsion Systems for Hybrid Vehicles** J. Miller
Volume 46	**Distribution Switchgear** S. Stewart
Volume 47	**Protection of Electricity Distribution Networks, 2nd Edition** J. Gers and E. Holmes
Volume 48	**Wood Pole Overhead Lines** B. Wareing
Volume 49	**Electric Fuses, 3rd Edition** A. Wright and G. Newbery
Volume 50	**Wind Power Integration: Connection and system operational aspects** B. Fox *et al.*
Volume 51	**Short Circuit Currents** J. Schlabbach
Volume 52	**Nuclear Power** J. Wood
Volume 53	**Condition Assessment of High Voltage Insulation in Power System Equipment** R.E. James and Q. Su
Volume 55	**Local Energy: Distributed generation of heat and power** J. Wood
Volume 56	**Condition Monitoring of Rotating Electrical Machines** P. Tavner, L. Ran, J. Penman and H. Sedding
Volume 57	**The Control Techniques Drives and Controls Handbook, 2nd Edition** B. Drury
Volume 58	**Lightning Protection** V. Cooray (Editor)
Volume 59	**Ultracapacitor Applications** J.M. Miller
Volume 62	**Lightning Electromagnetics** V. Cooray
Volume 63	**Energy Storage for Power Systems, 2nd Edition** A. Ter-Gazarian
Volume 65	**Protection of Electricity Distribution Networks, 3rd Edition** J. Gers
Volume 66	**High Voltage Engineering Testing, 3rd Edition** H. Ryan (Editor)
Volume 67	**Multicore Simulation of Power System Transients** F.M. Uriate
Volume 68	**Distribution System Analysis and Automation** J. Gers
Volume 69	**The Lightening Flash, 2nd Edition** V. Cooray (Editor)
Volume 70	**Economic Evaluation of Projects in the Electricity Supply Industry, 3rd Edition** H. Khatib
Volume 72	**Control Circuits in Power Electronics: Practical issues in design and implementation** M. Castilla (Editor)
Volume 73	**Wide Area Monitoring, Protection and Control Systems: The enabler for Smarter Grids** A. Vaccaro and A. Zobaa (Editors)
Volume 74	**Power Electronic Converters and Systems: Frontiers and applications** A.M. Trzynadlowski (Editor)
Volume 75	**Power Distribution Automation** B. Das (Editor)
Volume 76	**Power System Stability: Modelling, analysis and control** B. Om P. Malik
Volume 78	**Numerical Analysis of Power System Transients and Dynamics** A. Ametani (Editor)
Volume 79	**Vehicle-to-Grid: Linking electric vehicles to the smart grid** J. Lu and J. Hossain (Editors)
Volume 81	**Cyber-Physical-Social Systems and Constructs in Electric Power Engineering** Siddharth Suryanarayanan, Robin Roche and Timothy M. Hansen (Editors)
Volume 82	**Periodic Control of Power Electronic Converters** F. Blaabjerg, K.Zhou, D. Wang and Y. Yang
Volume 86	**Advances in Power System Modelling, Control and Stability Analysis** F. Milano (Editor)
Volume 87	**Cogeneration: Technologies, optimisation and implementation** C.A. Frangopoulos (Editor)
Volume 88	**Smarter Energy: From smart metering to the smart grid** H. Sun, N. Hatziargyriou, H. V. Poor, L. Carpanini and M.A. Sánchez Fornié (Editors)
Volume 89	**Hydrogen Production, Separation and Purification for Energy** A. Basile, F. Dalena, J. Tong and T.N.Veziroğlu (Editors)
Volume 90	**Clean Energy Microgrids** S. Obara and J. Morel (Editors)
Volume 91	**Fuzzy Logic Control in Energy Systems with Design Applications in MATLAB®/Simulink®** İ.H. Altaş
Volume 92	**Power Quality in Future Electrical Power Systems** A.F. Zobaa and S.H.E.A. Aleem (Editors)
Volume 93	**Cogeneration and District Energy Systems: Modelling, analysis and optimization** M.A. Rosen and S. Koohi-Fayegh
Volume 94	**Introduction to the Smart Grid: Concepts, technologies and evolution** Salman K. Salman
Volume 95	**Communication, Control and Security Challenges for the Smart Grid** S.M. Muyeen and S. Rahman (Editors)
Volume 97	**Synchronized Phasor Measurements for Smart Grids** M.J.B. Reddy and D.K. Mohanta (Editors)
Volume 98	**Large Scale Grid Integration of Renewable Energy Sources** Antonio Moreno-Munoz (Editor)
Volume 100	**Modeling and Dynamic Behaviour of Hydropower Plants** N. Kishor and J. Fraile-Ardanuy (Editors)
Volume 101	**Methane and Hydrogen for Energy Storage** R. Carriveau and David S.-K. Ting
Volume 104	**Power Transformer Condition Monitoring and Diagnosis** A. Abu-Siada (Editor)
Volume 108	**Fault Diagnosis of Induction Motors** Jawad Faiz, Vahid Ghorbanian and Gojko Joksimović
Volume 110	**High Voltage Power Network Construction** K. Harker
Volume 124	**Power Market Transformation** B. Murray
Volume 130	**Wind and Solar Based Energy Systems for Communities** Rupp Carriveau and David S.-K. Ting (Editors)
Volume 131	**Metaheuristic Optimization in Power Engineering** J. Radosavljević
Volume 905	**Power system protection, 4 volumes**

IET ENERGY ENGINEERING 119

Thermal Power Plant Control and Instrumentation

Thermal Power Plant Control and Instrumentation

The control of boilers and HRSGs

2nd Edition

David Lindsley, John Grist and Don Parker

The Institution of Engineering and Technology

Published by The Institution of Engineering and Technology, London, United Kingdom

The Institution of Engineering and Technology is registered as a Charity in England & Wales (no. 211014) and Scotland (no. SC038698).

First published 2000

Second edition 2018

The Institution of Engineering and Technology
Michael Faraday House
Six Hills Way, Stevenage
Herts, SG1 2AY, United Kingdom

www.theiet.org

British Library Cataloguing in Publication Data
A catalogue record for this product is available from the British Library

ISBN 978-1-78561-419-4 (hardback)
ISBN 978-1-78561-420-0 (PDF)

Typeset in India by MPS Limited

Contents

Preface to the second edition

At the middle of the twentieth century, power plant control systems were pneumatically operated – simple, reliable, easy to mend if they went wrong. They would have been totally immune to interference or the efforts of any hackers who may have existed at the time.

Basically, the situation was this: in the power generation industry, the burgeoning demand for electricity had led to the development of ever bigger, more efficient generators, turbines and boilers together with their ancillary equipment. But plant that was large in terms of output was also large physically. They required the use of long transmission lengths for pneumatic signals and the distance velocity lags inherent when sending pneumatic signals through long copper pipes led to instability.

When the first edition of *Power Plant Control and Instrumentation* was published, power station design and operation was determined almost entirely by engineers. Grid systems operated – as they do now – on the basis of allocating loads to power stations on a merit basis: that is, plants that were cheaper to run were brought on stream before more expensive ones. But that merit basis was at that time relatively uncomplicated: the amortised capital cost of constructing a plant was combined with the operating costs (fuel, staffing and maintenance) to produce a figure representing the generation cost; and load was allocated accordingly. Nuclear plants – expensive to build, slow to start but cheap to run – became base-load stations.

Today, although this principle is still followed, environmental and economic considerations have brought in the need to apply new factors: these include subsidies to encourage the use of 'environmentally friendly' energy sources, and the costs of decommissioning plant at the end of its operating life are added in to the equation.

All of these factors have bearings on plant design – right down to control and instrumentation systems. As a result, the control engineer is faced with handling new plant concepts such as Lo-NOx burners and carbon-capture and recycling.

As ever, the control system designer has to have a good understanding of the plant and its method of operation before he or she can produce an effective design.

Aiding that understanding is the primary objective of this book.

Control technology itself changes and changes again over time: the principles of thermodynamics don't.

Therefore, the concepts behind the control of a mid-twentieth century pneumatically operated feedwater system are basically the same as those employed by today's digital systems.

As the author of the first edition of this book, I recognise that technology has moved on and I recognise that I am no longer sufficiently familiar with it. To bring the book up to date, and to keep it relevant, I have therefore called on the assistance of two of my younger colleagues. Both John Grist and Don Parker have a thorough understanding of the subject and share a wonderful ability to explain complex issues to the uninitiated.

I hope that this latest version of *Power Plant Control and Instrumentation* proves to be as useful and relevant as the first addition apparently was.

David Lindsley
Kingston upon Thames
December 2017

Acknowledgements

We would like to thank the many individuals and organisations who made contributions to the first and second edition of this book, either with direct contributions of diagrams and technical articles, or by the provision of information. In thanking the following, we do not wish to ignore all the others who have helped us with this work:

Company	Web address	Products shown in the book
ABB	www.abb.com	Control room photographs, DCS system drawings, electric actuators, pneumatic positioners, state controller with observer diagrams
Advanced actuators	www.advanced-actuators.com	Hydraulic actuators
AGL	www.agl.com.au	Control room photographs
Ametek Land	www.landinst.com	Mill CO detection
Beck	www.haroldbeck.com	Electric actuators
Air Monitor Corporation	www.airmonitor.com	Multiple averaging Pitots for air flow measurement
Doosan Babcock Ltd	www.doosanbabcock.com	Burners, boiler elevations, spiral furnace tube photographs, over-fire air technology
Doosan Lentjes	www.doosanlentjes.com	Circulating fluidised bed boilers
Durag	www.durag.com	Flame monitors
Emerson	www.emerson.com	Control room HMI, DCS system drawings Coriolis
FCI	www.fluidcomponents.com	Thermal mass flow
FS Cables	www.fscables.com	Cables
Hayward Tyler	www.haywardtyler.com	Boiler circulation pumps
Howden Group	www.howden.com	Air heaters and fans
Intergen	www.intergen.com	Control room photographs
Lias Industrial Ltd.	www.liasindustrial.co.uk	Flame monitors
Loesche	www.loesche.com	Vertical spindle mills
Metso	www.metso.com	Tube mills
Procon	www.proconeng.com	Acoustic leak detection
Promecon	www.promecon.com	Capacitance/cross-correlation air flow measurement, PF flow measurement and carbon in ash
Provecta Process Automation	www.provecta.com.au	Trends of process data

(Continues)

(*Continued*)

Company	Web address	Products shown in the book
Rosemount Measurement Ltd	www.emerson.com	Electronic drum-level measurement
SPX Flow	www.spxflow.com	Attemperators and control valves
Servomex	www.servomex.com	Oxygen & COe analysers and continuous emissions monitoring systems
SICK	www.sick.com	NOx measuring systems
Siemens	www.siemens.com	Benson Technology
Storm Technologies Inc.	www.stormeng.com	Aerofoil and Venturi flow
TC fluid Control	www.tc-fluidcontrol.com	Drum level gauges
WIKA	www.wika.co.uk	Pressure gauges and temperature measurement
Video Technology Ltd	www.videotechnology.tv	CCTV
xSericon	www.xsericon.com	LOPA description

Thanks to Marco Pereira, Neil Ronan and Graham Welford who helped with reviews, and a special thanks to Kevin Bates, Jean Gonese and Ian White for help above and beyond.

Diagrammatic symbols

In spite of the existence of many recognised standards for instrumentation symbols [1], I have chosen to adopt a simple format which should be sufficient to explain the concepts that I want to communicate to the reader. These symbols would not be comprehensive enough to fully define the requirements within a full-scale control system design task (e.g. the controller symbol does not indicate whether or not auto/manual facilities are required, or the form that these should take). Nevertheless, I believe the diagrams will be easily understood by engineers.

In the context of the controllers themselves, it is worth mentioning that different terms are used in the United States and elsewhere to identify the same function. In particular, the plant parameter that is measured and fed to a controller is, in Europe, called the 'measured value' while in the United States it is referred to as the 'process variable'. Also, when referring to controllers, the term 'reset' is sometimes used in the United States instead of 'integral action'.

Abbreviations and terms used in this book

This book is addressed to people working across two very different disciplines: power plant and control systems. Technical terms and abbreviations that are easily understood by professionals in one field can be bewildering to those who understand the other side, and so the following list is provided in an attempt to help readers understand the abbreviations and some of the terms that are used in the text and elsewhere in the industry.

1oo2	One-out-of-two voting
2oo2	Two-out-of-two voting
2oo3	Two-out-of-three voting
2-DOF	Two degrees of freedom
AC	alternating current
ASCII	American Standard Code for Information Interchange (a standard defining the codes used for communication between computers and between computers and their peripherals)
ADC	Analogue-to-digital converter
AGC	Automatic generation control
A/M	Auto/manual control facility
APC	Advanced process control
BSI	British Standards Institution
BMCR	Boiler Maximum Continuous Rating
BMS	Burner Management System
BOFA	Boosted Over Fire Air
C&I	Control and instrumentation
C-BF	Coordinated Boiler Following
CCR	Central control room
CCGT	Combined-cycle gas–turbine plant
CCTV	Closed circuit television
CEM	Continuous emission monitoring
CFD	Computerised fluid dynamics

CHP	Combined heat and power (a type of plant that burns a fuel to produce electricity and steam that is used either to heat a nearby complex or by an industrial process)
CMR	Continuous maximum rating (also MCR)
CPU	Central processing unit
C-TF	Coordinated turbine following
DAC	Digital-to-analogue converter
DC	Direct current
DCS	Distributed control system
(DMZ)	Demilitarised zone. A small network inserted as a buffer between a DCS network and the internet or external networks.
Deterministic	(Network communications) A network system in which events are dealt with in the exact order in which they occur.
DV	Desired value
DP	Differential pressure
EEPROM	Electrically erasable programmable read-only memory
EMC	Electromagnetic compatibility
EMI	Electromagnetic interference
FAT	Factory acceptance test
FD	Forced draught
FDS	Functional design specification
FMEA	Failure mode and effect analysis
FMECA	Failure mode and effect and criticality analysis
FWR	Feedwater regulator (control valve)
H/A	Hand/automatic control facility
HMI	Human–machine interface (frequently used to describe VDUs)
HP	High pressure (the definition is relative: on major central-station plant it is usually above 100 bar g)
HRSG	Heat-recovery steam generator
IC	Integrated circuit
ID	Induced draught
IEC	International Electro-Technical Commission
IEE	Institution of Electrical Engineers
IEEE	Institute of Electrical and Electronics Engineers
EU	European Union
IP	Intermediate pressure (a relative definition, see HP above)
ISA	International Society of Automation (formerly the Instrument Society of America)

ISO	International Standards Organisation
I/O	Input and output
KKS	Kraftwerk Kennzeichensystem (power station designation system)
LAN	Local area network
LED	Light-emitting diode
load	The flow of steam, in kg/s, that is produced at any given time by the boiler or HRSG (sometimes also the electrical load on the generator, in MW)
LP	Low pressure (a relative definition, see HP above)
machine	Turbo-generator or alternator
MCB	Miniature circuit breaker
MCR	Maximum continuous rating (also CMR), typically, the highest rate of steam flow that a boiler can produce for extended periods.
mill	A device (also known as a pulveriser) that is used to crush coal into fine powder before it is fed to the burners
MFT	master fuel trip
MPC	Model-predictive control
MSP	Main steam pressure
MTBF	Mean time between failure
MTTR	Mean time to repair
MV(PV)	Measured value (also known as 'process variable')
NFPA	National Fire Prevention Association
OFA	Over Fire Air A windbox above the normal burner rows where air (but not fuel) is added to reduce NOx
P&ID	Piping and instrumentation diagram
PCB	Printed circuit board
PA	Primary air
PF	Pulverised fuel (coal)
PLC	Programmable-logic controller
PSU	Power supply unit
Pulveriser	A device (also known as a mill) that is used to crush coal into fine powder before it is fed to the burners
PV	Process variable (also known as 'measured value')
RAM	Random access memory
RCM	Reliability centred maintenance
RDF	Refuse-derived fuel
RDS-PP	Reference designation system for power plant
RFI	Radiofrequency interference

ROM	Read-only memory
RTC	Real-time clock
RTD	Resistance temperature detector
SP	Setpoint
SAT	Site acceptance test
SD	System description
SIL	Safety Integrity Level (referring to functional safety) The SIL is the level of risk reduction applied to a system. There are four levels SIL 4 is the most onerous and is not relevant to boiler controls. SIL 1 reduces the risk between 10 and 100 times, SIL 2 between 100 and 1,000 times and SIL 3 between 1,000 and 10,000 times
SIS	Safety-instrumented system
SCADA	Supervisory, control and data-acquisition system
SCR	Selective catalytic reduction
Stoichiometric ratio (lambda, λ)	The ratio of air to fuel in a furnace compared to that needed for complete combustion in ideal conditions.
TDL	Tuneable diode laser
TS	Technical specification
TUV	Technischer Uberwachungs Verein (German Technical Supervisory Association)
UART	Universal asynchronous receiver/transmitter (an electronic device that controls communication with a peripheral)
UL	Underwriters' laboratories
UPS	Uninterruptible power supply
VDU	Visual display unit (also termed a 'monitor' or 'screen')
VLSI	Very large scale integrated … devices/circuits/systems
WTE	Waste-to-energy (a type of plant where waste is burned to produce electricity or heat for a district or industrial processes

Reference

[1] ANSI/ISA-S5. l. Instrumentation symbols and identification International Society of Automation, Research Triangle Park, North Carolina, USA, 1992.

Chapter 1

The basics of steam generation and use

David Lindsley[1]

1.1 Why an understanding of steam is needed

Steam power is fundamental to what is still by far the largest sector of the electricity-generating industry and without it the face of contemporary society would be dramatically different from its present one. With the growth of large-scale intermittent renewable generation such as wind and solar power comes a growing need for fast-responding storage capability. While more efficient simple-cycle gas turbines and diesel-based tri-generation systems have filled the gap to some extent, the work continues towards developing massive batteries, pumped-hydro, molten salt and other energy storage technologies. However, it is the inherent energy storage and responsiveness of large rotating steam-driven synchronous generators that continue to be a mainstay for a secure power network.

Steam *is* important, and the safety and efficiency of its generation and use depend on the application of control and instrumentation, often simply referred to as C&I. The objective of this book is to provide a bridge between the discipline of power-plant process engineering and those of electronics, instrumentation and control engineering.

I shall start by outlining in this chapter the change of state of water to steam, followed by an overview of the basic principles of steam generation and use. This seemingly simple subject is extremely complex. This will necessarily be an overview: it does not pretend to be a detailed treatise and at times it will simplify matters and gloss over some details which may even cause the thermodynamicist or combustion physicist to shudder, but it should be understood that the aim is to provide the C&I engineer with enough understanding of the subject to deal safely with practical control system design, operational and maintenance problems.

1.2 Boiling: the change of state from water to steam

When water is heated its temperature rises in a way that can be detected (e.g. by a thermometer). The heat gained in this way is called *sensible* because its effects can be sensed, but at some point the water starts to boil.

[1]Retired

But here we need to look even deeper into the subject. Exactly what is meant by the expression 'boiling'? To study this we must consider the three basic states of matter: solids, liquids and gases (supercritical fluids, which we shall refer to in later chapters, display both liquid and gas properties but are not considered a separate state of matter). A plasma, produced when the atoms in a gas become ionised, is often referred to as the fourth state of matter, but for most practical purposes it is sufficient to consider only the three basic states. In its solid state, matter consists of many molecules tightly bound together by attractive forces between them. When the matter absorbs heat, the energy levels of its molecules increase and the mean distance between the molecules increases. As more and more heat is applied these effects increase until the attractive force between the molecules is eventually overcome and the particles become capable of moving about independently of each other. This change of state from solid to liquid is commonly recognised as 'melting'.

As more heat is applied to the liquid, some of the molecules gain enough energy to escape from the surface, a process called evaporation (whereby a pool of liquid spilled on a surface will gradually disappear). As a liquid's temperature rises these escapes occur more rapidly and at a certain point the liquid becomes very agitated, with large quantities of bubbles rising to the surface. It is at this time that the liquid is said to start 'boiling'. It is in the process of changing state to a vapour, which is a fluid in a gaseous state.

Let us consider a quantity of water that is contained in an open vessel. Here, the air that blankets the surface exerts a pressure on the surface of the fluid and, as the temperature of the water is raised, enough energy is eventually gained to overcome the blanketing effect of that pressure and the water starts to change its state into that of a vapour (steam). Further heat added at this stage will not cause any further detectable change in temperature: the energy added is used to change the state of the fluid. Its effect can no longer be sensed by a thermometer, but it is still there. For this reason it is called latent, rather than sensible, heat. The temperature at which this happens is called the 'boiling point'. At normal atmospheric pressure the boiling point of water is 100 °C (212 °F).

If the pressure of the air blanket on top of the water were to be increased, more energy would have to be introduced to the water to enable it to break free. In other words, the temperature must be raised further to make it boil. To illustrate this point, if the pressure is increased by 10% above its normal atmospheric value, the temperature of the water must be raised to just above 102 °C before boiling occurs.

The steam emerging from the boiling liquid is said to be saturated and, for any given pressure, the temperature at which boiling occurs is called the saturation temperature.

The information relating to steam at any combination of temperature, pressure and other factors may be found in steam tables, which are nowadays available in software as well as in the more traditional paper form. These tables were originally published in 1915 by Hugh Longbourne Callendar (1863–1930), a British physicist. Because of advances in knowledge and measurement technology, and as a result of changing units of measurement, many different variants of steam tables are today in existence, but they all enable one to look up, for any pressure, the saturation temperature, the heat per unit mass of fluid, the specific volume, etc.

Understanding steam and the steam tables is essential in many stages of the design of power-plant control systems. For example, if a designer needs to compensate a steam-flow measurement for changes in pressure, or to correct for density errors in a water-level measurement, reference to these tables is essential.

Another term relating to steam defines the quantity of liquid mixed in with the vapour. In the United Kingdom this is called the dryness fraction (in the United States the term used is steam quality). What this means is that if each kilogram of the mixture contains 0.9 kg of vapour and 0.1 kg of water, the dryness fraction is 0.9.

Steam becomes superheated when its temperature is raised above the saturation temperature corresponding to its pressure. This is achieved by collecting it from the vessel in which the boiling is occurring, leading it away from the liquid through a pipe, and then adding more heat to it. This process adds further energy to the fluid, which improves the efficiency of the conversion of heat to electricity.

As stated earlier, heat added once the water has started to boil does not cause any further detectable change in temperature. Instead it changes the state of the fluid. Once the steam has formed, heat added to it contributes to the total heat of the vapour. This is the sensible heat plus the latent heat plus the heat used in increasing the temperature of each kilogram of the fluid through the number of degrees of superheat to which it has been raised.

In a power plant, a major objective is the conversion of energy locked up in the input fuel into either usable heat or electricity. In the interests of economics and the environment it is important to obtain the highest possible level of efficiency in this conversion process. As we have already seen, the greatest efficiency is obtained by maximising the energy level of the steam at the point of delivery to the next stage of the process. When as much energy as possible has been abstracted from the steam, the fluid reverts to the form of cold water, which is then warmed and treated to remove any air which may have become entrained in it before it is finally returned to the boiler for reuse.

1.3 The nature of steam

As stated in the Preface, the boilers and steam generators that are the subject of this book provide steam to users such as industrial plant, or housing and other complexes, or to drive turbines that arc the prime movers for electrical generators. For the purposes of this book, such processes are grouped together under the generic name 'power plant'. In all these applications the steam is produced by applying heat to water until it boils, and before we embark on our study of power-plant C&I we must understand the mechanisms involved in this process and the nature of steam itself.

First, we must pause to consider some basic thermodynamic processes. Two of these are the Carnot and Rankine cycles, and although the C&I engineer may not make use of these directly, it is nevertheless useful to have a basic understanding of what they are and how they operate.

1.3.1 The Carnot cycle

The primary function of a power plant is to convert into electricity the energy locked up in some form of fuel resource. In spite of many attempts, it has not proved possible to generate electricity in large quantities from the direct conversion of the energy contained in a fossil fuel (or even a nuclear fuel) without the use of a medium that acts as an intermediary.

Therefore, if one wishes to obtain large quantities of electricity from a fossil fuel or from a nuclear reaction it remains necessary to first release the energy that is available within that resource and then to transfer it to a generator, and this process necessitates the use of a medium to convey the energy from source to destination. Furthermore, it is necessary to employ a medium that is readily available and which can be used with relative safety and efficiency. On planet Earth, water is, at least in general, a plentiful and cheap medium for effecting such transfers. It also has suitable properties for heat transfer, boiling, superheating and condensing at temperatures and pressures that are readily obtainable with cost-effective materials and equipment.

While other possibilities have been considered, over the last century none of these has reached active use outside small, specialised applications, and steam is universally used in power stations.

The use of water and steam to provide motive power has a long history. In the first-century AD, Hero of Alexandria showed that steam leaving via nozzles attached to a heated container filled with water would cause the vessel to rotate, but in this simple machine (the aeolipile) the steam leaving the vessel was wasted and for continuous operation the process therefore necessitated continually replacing the water. With the nature of Hero's design, it was not a simple task to refill the vessel while it was in operation, but even if a method had been found, using water in a one-way process like this necessitates the provision of endless supplies of that fluid. It was not until 1824 that a French engineer, Sadi Carnot, proposed a way to resolve this problem. He used a cycle, where the transfer medium is part of a closed loop and the medium is returned to its starting conditions after it has done the work required of it.

Carnot framed one of the two laws of thermodynamics. The first, Joule's law, had related mechanical energy to work: Carnot's law defined the temperature relations applying to the conversion of heat energy into mechanical energy. He saw that if this process were to be made reversible heat could be converted into work and then extracted and reused to make a closed loop. In his concept (Figure 1.1), a piston moves freely without encountering any friction inside a cylinder made of some perfectly insulating material. The piston is driven by a 'working fluid'. The cylinder has a head at one end that can be switched at will from being a perfect conductor to being a perfect insulator. Outside the cylinder are two bodies, one of which can deliver heat without its own temperature (T_1) falling, the other being a bottomless cold sink at a temperature (T_2) which is also constant.

The operation of the system is shown graphically in Figure 1.2, which shows the pressure/volume relationship of the fluid in the cylinder over the whole cycle.

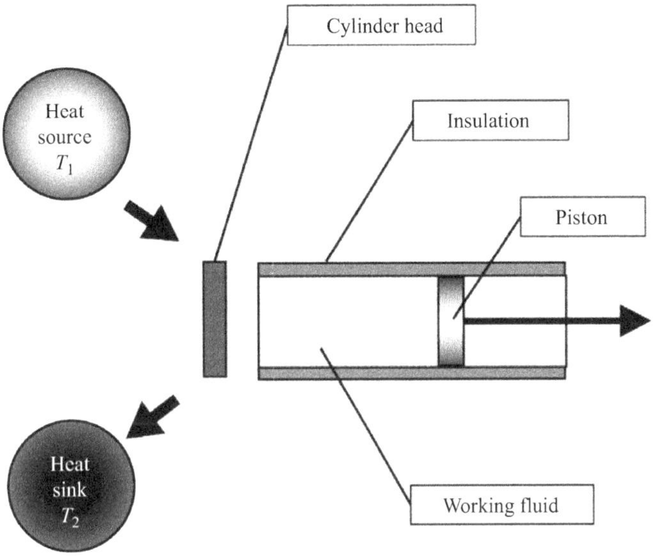

Figure 1.1 Carnot's heat engine

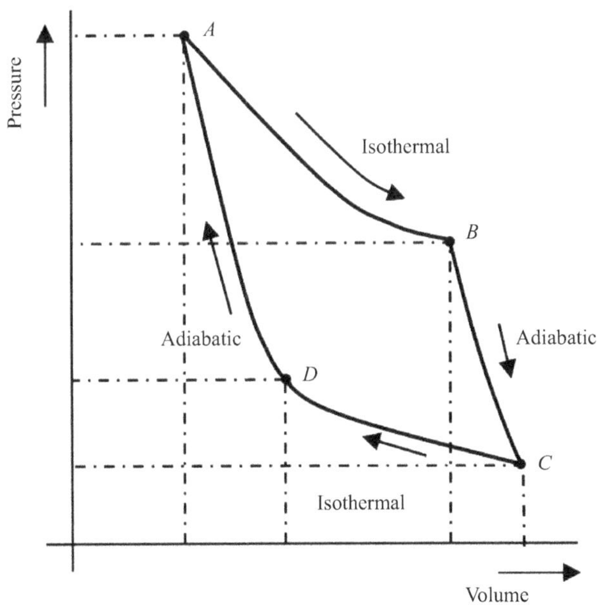

Figure 1.2 The Carnot cycle

As the process is a repeating cycle its operation can be studied from any convenient starting point, and we shall begin at the point A, where the cylinder head (at this time assumed to be a perfect conductor of heat), allows heat from the hot source to enter the cylinder. The result is that the medium begins to expand, and if it is allowed to expand freely, Boyle's law (which states that at any temperature the relationship between pressure and volume is constant) dictates that the temperature will not rise, but will stay at its initial temperature (T_1). This is called isothermal expansion.

When the pressure and volume of the medium have reached the values at point B, the cylinder head is switched from being a perfect conductor to being a perfect insulator and the medium allowed to continue its expansion with no heat being gained or lost. This is known as adiabatic expansion. When the pressure and volume of the medium reach the values at point C, the cylinder head is switched back to being a perfect conductor, but the external heat source is removed and replaced by the heat sink. The piston is driven towards the head, compressing the medium. Heat flows through the head to the heat sink and when the temperature of the medium reaches that of the heat sink (at point D), the cylinder head is once again switched to become a perfect insulator and the medium is compressed until it reaches its starting conditions of pressure and temperature. The cycle is then complete, having taken in and rejected heat while doing external work.

1.3.2 The Rankine cycle

The Carnot cycle postulates a cylinder with perfectly insulating walls and a head which can be switched at will from being a conductor to being an insulator. Even with modifications to enable it to operate in a world where such things are not obtainable, it would have probably remained a scientific concept with no practical application, had not a Scottish professor of engineering, William Rankine, proposed a modification to it at the beginning of the twentieth century [1]. The concepts that Rankine developed form the basis of all thermal power plants in use today. Even today's combined-cycle power plants use his cycle for one of the two phases of their operation.

Figure 1.3 illustrates the principle of the Rankine cycle. Starting at point A again, the source of heat is applied to expand the medium, this time at a constant pressure, to point B, after which adiabatic expansion is again made to occur until the medium reaches the conditions at point C. From here, the volume of the medium is reduced, at a constant pressure, until it reaches point D, when it is compressed back to its initial conditions.

All of this may seem of only theoretical interest, but it takes on a practical form in a thermal power plant, where water is compressed by pumps, then heated until it boils to produce steam which then expands (through a turbine or in some process) until it reverts to water. This operation is shown in Figure 1.4 which this time shows temperature plotted against a quantity called entropy for the processes within the boiler and turbine of a power plant. (Chapter 2 describes in detail the functions of the various items of plant.) Entropy is a measure of the portion of the energy in a

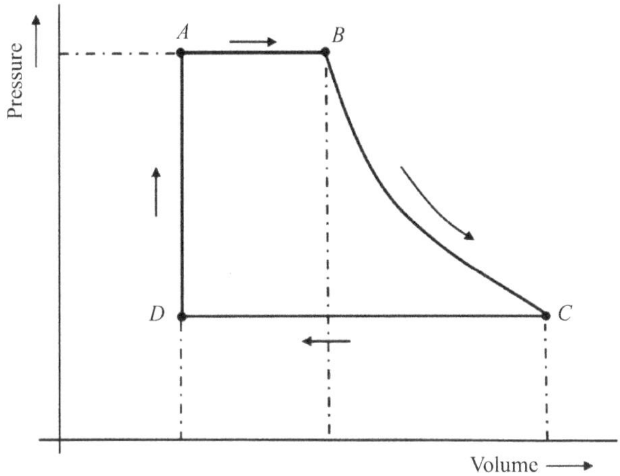

Figure 1.3 The Rankine cycle

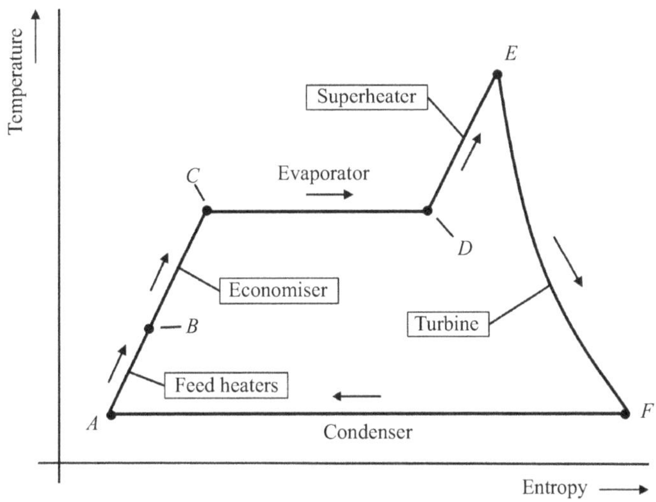

Figure 1.4 The Rankine cycle in a steam-turbine power plant

system that is not available for doing work and it can be used to calculate heat transfer for a reversible process.

In the system shown in Figure 1.4, water is heated in feed heaters (A to B) using steam extracted from the turbine. Within the boiler itself, heat is used to further prewarm the water (in the economiser) before it enters the evaporative stages (C) where it boils. At D, superheat is added until the conditions at E are reached at the turbine inlet. The steam expands in the turbine to the conditions at point F, after which it is condensed and returned to the feedwater heater. The energy in the steam leaving the boiler is converted to mechanical energy in the turbine, which then spins the generator to produce electricity.

The diagram shows that the energy delivered to the turbine is maximised if point E is at the highest possible value and F is at the lowest possible value, and now we begin to see the importance of understanding these cycles if plant operation is to be understood and optimised. It explains why the temperature of the steam leaving the boiler is superheated and why the turbine condenser operates at very low pressures, which correspond with low temperatures.

1.4 Thermal efficiency

The efficiency of a power plant is the measure of its effectiveness in converting fuel into electrical energy or process heat. This factor sets the cost per unit of electricity or heat generated, and in a network of interconnected power stations it is this cost that determines the revenue that will be earned by the plant. Although several steps may be taken to reduce losses, some heat is inevitably lost in the flue gases and in the cooling water that leaves the condenser, and a realistic limit for the efficiency of such a plant has been just over 40%. Although it has long been understood that, for every unit of money put into the operation of the plant, over half was being lost, very little could be done about this situation until developments in materials technology brought forward new opportunities. In recent years such developments have resulted in the construction of very high pressure and temperature power plants referred to as 'ultra-supercritical' (USC) that have achieved around 45% efficiency.

One of the most dramatic power-plant developments of the second half of the twentieth century was the realisation that by employing one cycle in combination with another one heat wasted in one could be used by the other to attain enhanced efficiency; this is the combined cycle.

1.5 The gas turbine and combined-cycle plants

The combined-cycle power station uses gas turbines to increase the efficiency of the power-generation process. Like many other machines that we assume to be products of the twentieth century, the gas turbine is not that new. In fact, Leonardo da Vinci (1452–1519) sketched a machine for extracting mechanical energy from a gas stream. However, no practical implementation of such a machine was considered until the nineteenth century, when George Brayton proposed a cycle that used a combustion chamber exhausting to the atmosphere. In 1872 Germany's F. Stolze patented a machine that anticipated many features of a modern gas turbine engine, although its performance was limited by the constraints of the materials available at the time.

Many other developments across Europe culminated in the development of an efficient gas turbine by Frank Whittle at the British Royal Aircraft Establishment (RAE) in the early 1930s. Subsequent advancements at RAE led to viable axial-flow compressors, which could attain higher efficiencies than the centrifugal

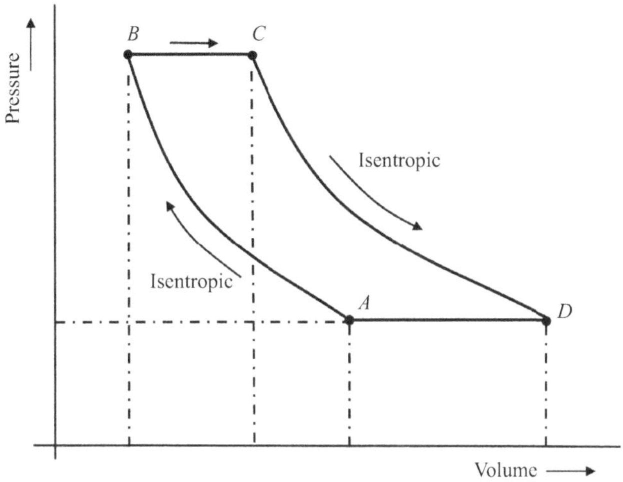

Figure 1.5 The Brayton cycle

counterpart developed by Whittle. All these gas turbines employed the Brayton cycle, whose pressure/volume characteristic is shown in Figure 1.5. Starting at point A in this cycle air is compressed isentropically (A–B) before being fed into a combustion chamber, where fuel is added and burned (B–C). The energy of the expanding air is then converted to mechanical work in a turbine (C–D). From C to D heat is rejected, and in a simple gas-turbine cycle this heat is lost to the atmosphere.

The rotation of the gas turbine can be used to drive a generator (via suitable reduction gearing) but, when used in a simple cycle with no heat recovery, the thermal efficiency of the gas turbine is poor because of the heat lost to the atmosphere. The gases exhausted from the turbine are not only plentiful and hot (400–550 °C), but they also contain substantial amounts of oxygen (in combustion terms, the excess air level for the gas turbine is 200%–300%). These factors point to the possibility of using the hot, oxygen-rich air in a steam-generating plant, whose steam output drives a turbine.

The use of such otherwise wasted heat in a heat-recovery steam generator (HRSG) is the basis of the 'combined-cycle gas-turbine' (CCGT) plant which has been a major development of the past few decades. With the heat used to generate steam in this way, the whole plant becomes a binary unit employing the features of both the Rankine and the Brayton cycles to achieve efficiencies that are simply not possible with either cycle on its own. In fact, the addition of the HRSG yields a thermal efficiency that may be 50% higher than that of the gas turbine operating in simple-cycle mode.

Once again, there is nothing really new about this concept. From the moment when the gas turbine became a practical reality it was very obvious that the hot compressed air it exhausted contained huge amounts of heat. Therefore, the combined cycle was considered in some depth almost as soon as the gas turbine was released from the constraints of military applications. However, because of their use of gases at extremely high temperatures, early machines suffered from limited blade life and they were therefore used only in applications where no other source of power was readily available. With improvements in materials technology this difficulty has been overcome and, nowadays, combined-cycle plants employing gas turbines form the mainstream of modern power-station development.

But whether it is in a combined-cycle plant or a simple-cycle power station, our interest in this chapter is in steam and its use, and this vapour will now be examined in more detail. We shall see that what seems a fairly simple and commonplace thing is, in fact, quite complex.

In spite of its complexities it is important to tackle this subject in some depth because the power-plant C&I engineer will need to deal with the physical parameters of steam through the various stages of designing or using a practical system.

1.6 Summary

In the earlier sections we have looked at the nature of steam and briefly explained how it is derived and used in various parts of the power station. We have also studied simple and combined cycles, and seen that the latter provide an opportunity of achieving higher efficiencies, thereby maximising the revenue earned by the plant.

In the following chapters we shall look at the plant in more detail, starting with the water and steam circuits and then moving on to discuss the combustion process. Once the plant is understood, the principles of its control systems can be better appreciated.

Reference

[1] Rankine, W.J.M. *A manual of the steam engine and other prime movers.* Griffin, London, 1908.

Chapter 2

The steam and water circuits

David Lindsley[1] *and Don Parker*[2]

2.1 Steam generation and use

In a conventional thermal power plant, the heat used for steam generation may be obtained by burning a fossil fuel, or, in the case of a heat-recovery steam generator (HRSG), may be derived from the exhaust of a gas turbine. In a nuclear plant, the heat may be derived from the radioactive decay of a nuclear fuel. In this chapter we shall be examining the water and steam circuits of boilers and HRSGs, as well as the steam turbines and the plant that returns the condensed steam to the boiler.

In the type of plant being considered in this book, the water is contained in tubes lining the walls of a chamber which, in the case of a simple-cycle plant, is called the furnace or combustion chamber. In a combined-cycle plant the tubes form part of the internal components of the HRSG.

The configuration of the steam and water circuits in both fuel-fired boilers and HRSGs falls into two broad groups: drum-type and 'once-through'. While the arrangement and control of the feedwater and steam temperature differ significantly between these types of plant, many of the other systems, including fuel, air and draft in the case of conventional boilers, are very similar. The following sections discuss the main characteristics and operation of the steam and water systems in these two groups of boilers.

2.2 Drum boilers and HRSGs

In drum boilers and conventional HRSGs, the application of the heat causes convection currents to form in the water contained in the tubes, causing it to rise up to a vessel called the drum, in which the steam is separated from the water. In some designs of plant the process of natural circulation is augmented by forced circulation, the water being pumped through the evaporative circuit rather than allowed to circulate by convection. While a conventional boiler has only one drum, an HRSG typically has two or three drums that operate at different pressures.

[1]Retired
[2]Provecta Process Automation LLC, USA

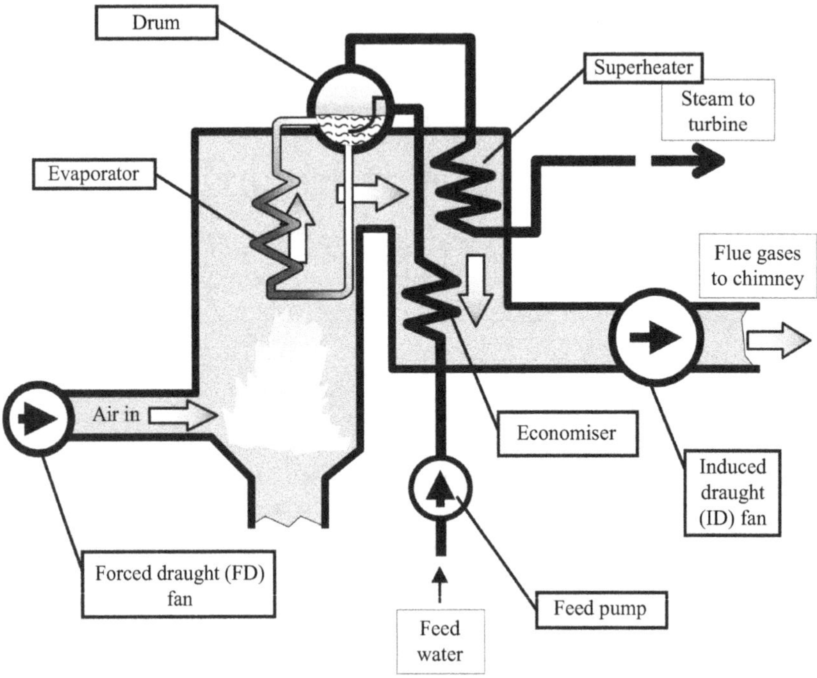

Figure 2.1 Schematic of a drum boiler: air/gas and steam/water paths

Figure 2.1 shows a drum boiler in schematic form. Here, the steam generation occurs in banks of tubes (labelled the Evaporator) that are exposed to the radiant heat of combustion. Of course, with HRSG plant no radiant energy is available (since the combustion process occurs within the gas turbine itself) and the heat of the gas-turbine exhaust is transferred to the evaporator tubes by a mixture of convection and conduction. In this type of plant it is common to have two or more steam/water circuits (see Figure 2.9), each with its own steam drum, and in such plant each of these circuits is as described later.

The steam leaves the drum and enters a bank of tubes where more heat is taken from the gases and added to the steam, superheating it before it is fed to the turbine. In the diagram this part of the plant, the superheater, comprises a single bank of tubes but in many cases multiple stages of superheater tubes are suspended in the gas stream, each abstracting additional heat from the exhaust gases. In boilers (rather than HRSGs), some of these tube banks are exposed to the radiant heat of combustion and are therefore referred to as the radiant superheater. Others, the convection stages, are shielded from the radiant energy but extract heat from the hot gases of combustion.

After the flue gases have left the superheater they pass over a third set of tubes (called the economiser), where almost all of their remaining heat is extracted to prewarm the water before it enters the drum.

Finally the last of the heat in the gases is used to warm the air that is to be used in the process of burning the fuel. (This air heater is not shown in the diagram since it is part of the air and gas plant which is discussed in the next chapter.)

The major moving items of machinery shown in the diagram are the feed pump, which delivers water to the system, and the fan which provides the air needed for combustion of the fuel (in most plants each of these is duplicated). In a combined-cycle plant the place of the combustion-air fan and the fuel firing system is taken by the gas turbine exhaust.

Figure 2.1 shows only the major items associated with the boiler. In a power-generation station, the steam passes to a turbine after which it has to be condensed back to water, thereby necessitating the use of a heat exchanger to extract the last remaining vestiges of heat from the fluid and fully condense it into a liquid. Then, entrained air and gas has to be removed from the condensed fluid before it is returned to the boiler.

2.3 Once-through boilers and OTSGs

In once-through boilers and once-through steam generators (or OTSGs, once-through equivalents of HRSGs), water passes from the liquid to the vapour stage without the use of a steam drum. Unlike a drum boiler, the steam leaving the furnace walls is slightly superheated to ensure there are no remaining water droplets passing to the superheaters.

In a drum boiler, an increase in fuel will result in additional steam production, reducing the drum level and so generating a response from the feedwater control system to increase water inflow. In a once-through boiler there is no ready 'level' measurement inside the tubes to indicate where the steam has fully converted to steam. If the fuel input and feedwater flow are not in balance, tube overheating, or alternatively, carryover of wet steam into the superheater sections may occur. The heat input and feedwater flow must therefore be kept in the correct ratio by monitoring the evaporator's steam temperature or enthalpy.

The removal of the steam drum enables higher pressures to be achieved within the boiler, and with it the opportunity to operate at supercritical pressures and much higher steam temperatures than with drum-type boilers. This results in significantly improved unit efficiency. In a supercritical boiler, the water passes to steam without boiling and expanding, as there is no vaporisation state; instead, as it is heated it transitions smoothly to a state of supercritical fluid which has both liquid and gaseous properties, although by the time it is superheated significantly it can be considered in the same way as steam at subcritical pressures.

During start-up, water flow in the furnace walls is maintained well above the equivalent fuel input to ensure even heating in the wall tubes. The furnace outlet thereby delivers saturated steam, which passes through vertical separators. The water component is collected and returned to the inlet of the furnace walls through a circulation pump, in a manner not unlike a drum boiler. This start-up arrangement is known as 'recirculation', 'wet' or 'bottle' mode. Once the fuel input is raised to the point where the steam is completely dry, the controls pass to 'once-through', 'dry' or 'Benson' mode at which point a different feedwater control regime is activated.

Figure 2.2 shows an outline of the typical water and steam circuit in a once-through boiler, including the start-up recirculation system. The air, furnace and rear

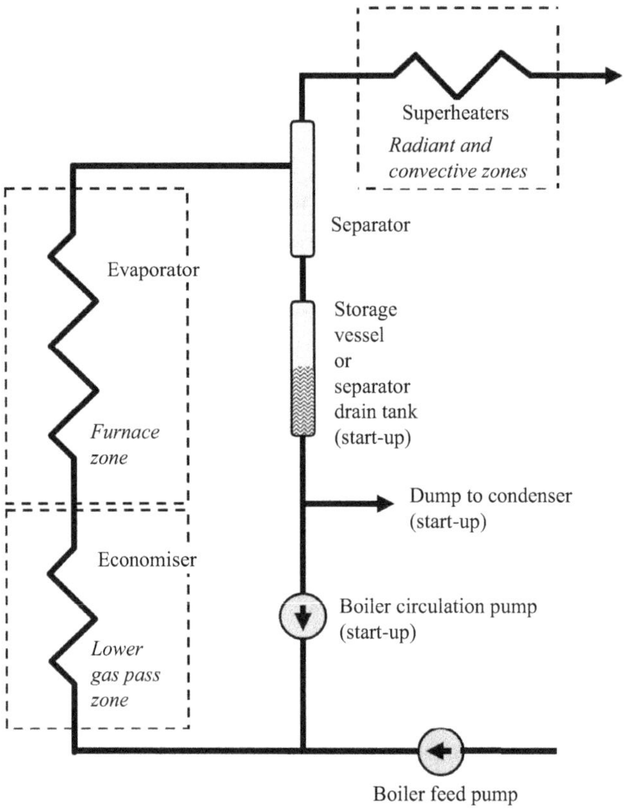

Figure 2.2 Schematic of a once-through boiler

gas pass arrangement shown in Figure 2.1 is generally similar to once-through boilers, with the lack of a drum and the existence of a long, vertical storage vessel receiving drainage from the separators being the most notable physical differences.

Figure 2.3 is a graph of typical relationships between the recirculation, spill, evaporator pass and overall feedwater flows across the boiler's load range. At first, before firing commences, a small flow from the main feed pumps is fed to the furnace inlet.

Since there is not yet any fuel to generate steam, it is all collected by the separator and directed to the storage vessel where it is recirculated to the furnace by the boiler circulation pump (BCP).

The BCP maintains a high level of flow through the furnace walls, well in excess of the water being fed from the main feed pump. In this way the minimum flow through the wall tubes is achieved. However, since the main feed pump flow exceeds the steam production, the drain tank fills and a dump valve to the condenser opens. As fuel flow and steam production rises, the spill to condenser reduces to zero, while the BCP continues to maintain a recirculation flow.

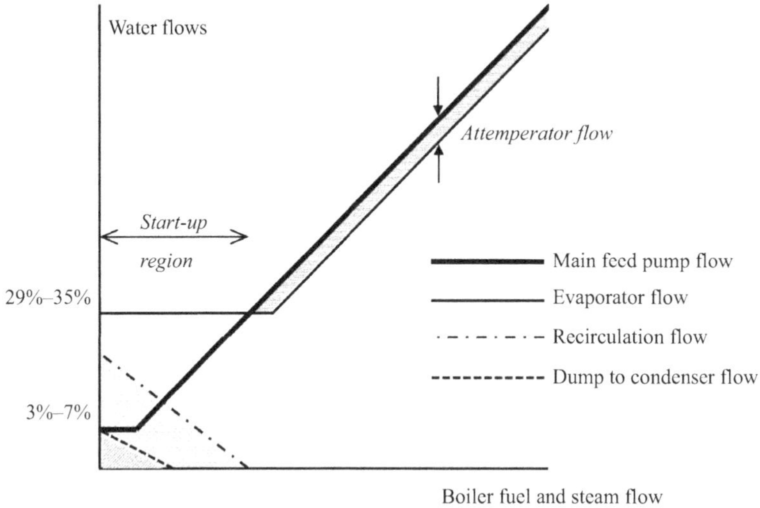

Figure 2.3 Water flows versus boiler load for a once-through boiler

With increasing fuel input and steam production, the ratio of recirculation flow to main feed pump flow reduces, until finally the steam production matches the minimum furnace flow requirement. At this point the BCP flow is close to zero and it is turned off. The boiler is now in once-through mode, and the control system regulates the fuel and feedwater systems to ensure the steam at the separator outlet remains dry, with a margin of superheat. Ch 6 provides details of these controls.

Another significant, although not readily visible, difference in many once-through boilers is the spiral arrangement of the lower section of the water walls. This arrangement ensures even heat distribution to the tubes and avoids localised steam/water separation which can cause uneven water flows and tube failure through overheating.

Figure 2.4 is a view of a furnace's internal walls during construction, with spiral tube panels being clearly visible.

The major remaining plant items forming part of the steam/water cycle will now be briefly described and their operations explained.

2.4 The steam turbine

In plants using a turbine, the energy in the steam generated by the boiler is first converted to kinetic energy, then to mechanical rotation and finally to electrical energy. On leaving the turbine the fluid is fed to a condenser which completes the conversion back to water, which is then passed to further stages of processing before being fed to the feed pumps. In the following paragraphs, we shall examine this process (with the exception of the conversion to electrical energy in the alternator).

Figure 2.4 A once-through boiler's spiral water walls. Photo courtesy of Doosan Babcock

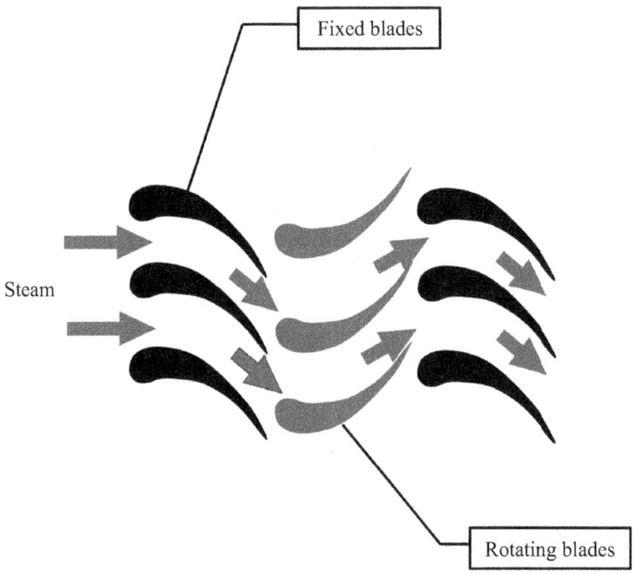

Figure 2.5 Turbine blading

In the turbine, the steam is fed via nozzles onto successive rows of blades, of which alternate rows are fixed to the machine casing with the intermediate rows attached to a shaft (Figure 2.5). In this way the heat energy in the steam is converted first to kinetic energy as it enters the machine through nozzles, and then this kinetic energy is converted to mechanical work as it impinges onto the rotating blades.

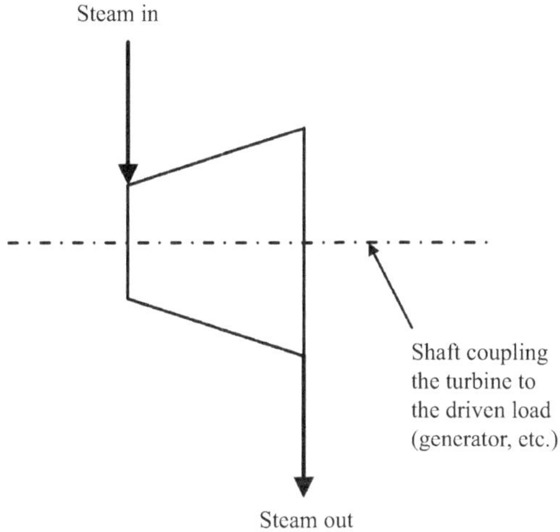

Figure 2.6 Symbolic representation of a turbine

Further work is done by the reaction of the steam leaving these blades when it encounters another set of fixed blades, which in turn redirect it onto yet another set of rotating blades.

As the steam travels through the machine in this way it continually expands, giving up some of its energy at each ring of blades. The moment of rotation applied to the shaft at any one ring of blades is the multiple of the force applied to the blades and mean distance of the force. Since each stage of rings abstracts energy from the steam, the force applied at the subsequent stage is less than it was at the preceding ring and, therefore, to ensure that a constant moment is applied to the shaft at each stage, the length of the blades in all rings after the first is made longer than that of the preceding ring. This gives the turbine its characteristic tapering shape. The steam enters the machine at the set of blades with the smallest diameter and leaves it after the set of blades with the largest diameter. On the control diagrams presented in this book, this is indicated by the usual symbol for a turbine, a rhomboidal shape (Figure 2.6).

Turbines may consist of one or more stages, and in plant which uses reheating the steam exiting the high-pressure or intermediate stage of the machine (the HP or IP stage, respectively) is returned to the boiler for additional heat to be added to it in a bank of tubes called the reheater. The steam leaving this stage of the boiler enters the final stage of the machine, the low pressure (LP) stage. Because the energy available in the steam is now much less than it was at the HP stage, this part of the turbine is characterised by extremely long blades.

By the time it leaves the final stage of the turbine, the steam has exhausted almost all of the usable energy that was added to it in the steam generator, and it is therefore passed to a condenser where it is finally cooled to convert it back to water which can be reused in the cycle. The condenser comprises a heat exchanger through which cold water is circulated. A simplified representation of the complete circuit is shown in Figure 2.7.

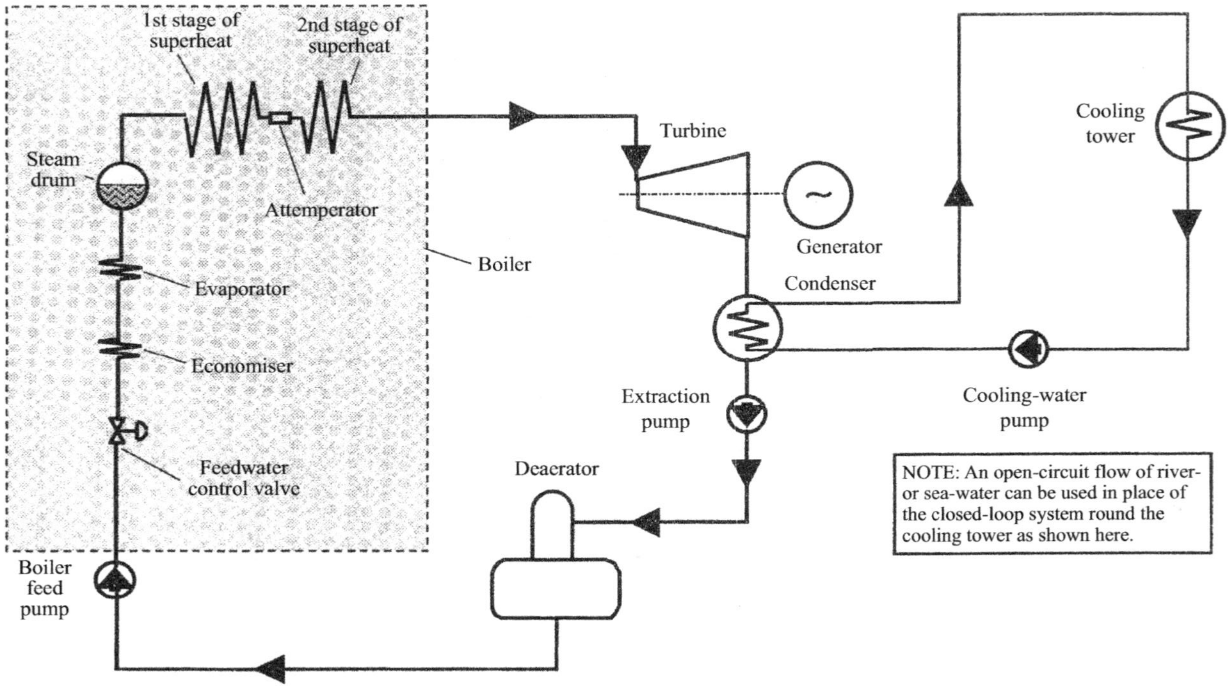

Figure 2.7 The main steam and water circuits of a drum boiler-turbine unit

The cooling water that is pumped through the condenser to abstract heat from the condensate may itself be flowing through a closed circuit. Alternatively, it may be drawn from a river or the sea to which it is then returned. In the latter cases, because of the heat received from the condenser, care must be taken to avoid undesirable heating of the river or sea in the vicinity of the discharge (or outfall).

In a closed circuit, the heat is released to the atmosphere in a cooling tower. Within these, the air that is used for cooling the water may circulate through the tower by natural convection, or it may be fan-assisted. It is usually desirable to minimise the formation of a plume since, as well as being very visible, such plumes can cause disturbance to the nearby environment by falling as a fine rain and possibly freezing on roads.

2.5 The condensate and feedwater system

Inside the plant, the steam and water system forms a closed loop, with the water leaving the condenser being fed back to the feed pumps for reuse in the boiler. However, certain other items of plant now become involved, because the water leaving the condenser is cold and contains entrained air which must be removed.

Air becomes entrained in the water system at start-up (when the various vessels are initially empty), and it will appear during normal operation when it leaks in at those parts of the cycle which operate below atmospheric pressure, such as the condenser, extraction pumps and LP feed heaters. Leakage can occur in these areas at flanges and at the sealing glands of the rotating shafts of pumps. Air entrainment is aided by two facts: one is that cold water can hold greater amounts of oxygen (and other dissolved gases) than can warm water; and the other is that the low pressure parts of the cycle must necessarily correspond with the low-temperature phases.

The presence of residual oxygen in the feedwater supply of a boiler or HRSG is highly undesirable because it will cause corrosion of the boiler pipework (particularly at welds, cold-worked sections and surface discontinuities), greatly reducing the serviceable life span of the plant. For this reason great attention must be paid to its removal.

Removal of dissolved oxygen is performed in several ways, and an important contributor to this process is the deaerator which is shown in Figure 2.7, located between the condenser extraction pump and the boiler feedwater pump.

2.5.1 The deaerator

The deaerator removes dissolved gases by vigorously boiling the water and agitating it, a process referred to as 'stripping'. One type of deaerator is shown in Figure 2.8. In this, the water entering at the top is mixed with steam which is rising upwards. The steam, taken directly from the boiler or from an extraction point on the turbine, heats a stack of metal trays, and as the water cascades down past these it mixes with the steam and becomes agitated, releasing the entrained gases. The steam pressurises the deaerator and its contents so that the dissolved gases are vented to the atmosphere.

Minimising corrosion requires the feedwater oxygen concentration to be maintained below 0.005 ppm or less, and although the deaerator provides an effective

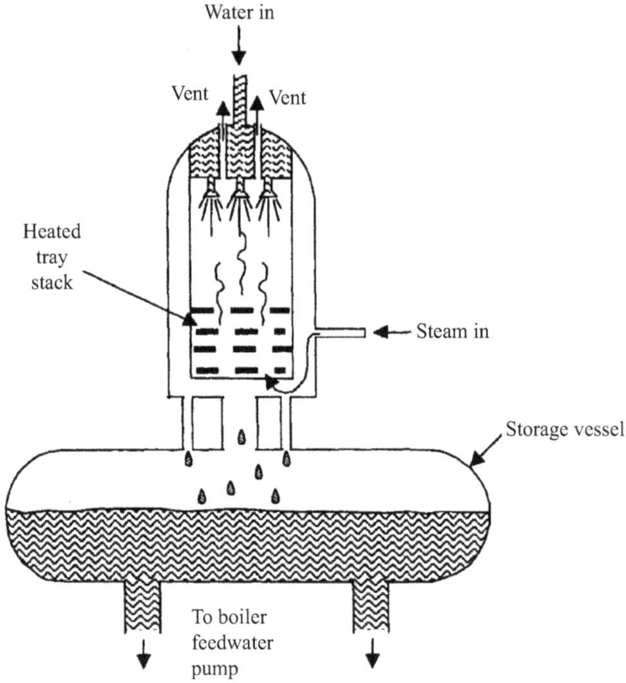

Figure 2.8 Principle of a deaerator

method of removing the bulk of entrained gases, it cannot reduce the concentration below about 0.007 ppm. For this reason, scavenging chemicals are added to remove the last traces of oxygen.

2.5.1.1 Chemical dosing

Volatile oxygen scavengers such as hydrazine (N_2H_4) and sodium sulphite (Na_2SO_3) have been used for oxygen removal (although hydrazine is now being treated as a probable carcinogen). Whatever their form, the chemical scavengers are added in a concentrated form and it is necessary to flush the injection pipes continually or on a periodic basis to prevent plugging.

Similarly, *blowdown*, a process of bleeding water to drains or a special vessel, is used to continually or periodically remove a portion of the water from the boiler, with automatic or manual chemical sampling being used to ensure that the correct concentration is maintained in the boiler water.

From a control and instrumentation viewpoint, the aforementioned chemical dosing operations are highly specialised and are therefore usually performed by equipment that is supplied as part of a water-treatment plant package. The control system [often based on a programmable-logic control system (PLC)] will generate data and alarm signals for connection to the main plant control system [typically referred to as the distributed control system (DCS)].

After the water has been deaerated and treated, it is fed to feed pumps which deliver it back to the boiler via the HP feedwater heaters.

2.6 The feed pumps and valves

The feed pumps deliver water to the boiler at high pressure, and the flow into the system is controlled by one or more feed-regulating valves. The feed pumps are often driven by electric motors, but small steam turbines are also used (although, clearly, these cannot be used at start-up unless a separate source of steam is available for their operation).

The pressure/flow characteristic of pumps and the various configurations that are available are discussed in Chapter 6 but it should be noted here that with any pump the pressure tends to fall as the throughput rises. On the other hand, due to the effect of friction, the resistance offered by the boiler system to the flow of water increases as the flow rate increases. (The system resistance is the minimum pressure that is required to force water into the boiler.) Therefore, the pressure drop across the regulating valve will be highest at low flows.

It is wasteful to operate with a pressure drop that is significantly above that at which effective control can be maintained, both because this entails an energy loss and also because erosion of valve internals increases with high pressure-drops. With fixed-speed pumps there is nothing that can be done about this, but an improvement can be made if variable-speed pumps are used. These are more expensive than their fixed-speed counterparts, but the increase in cost tends to be offset by the operational cost savings that can be achieved (due to more efficient operation and reduced wear on the valve). Such savings are increased if the plant operates for prolonged periods at low throughputs and are most apparent with the larger boilers.

From the control engineer's viewpoint, variable-speed pumps are an attractive option because they enable the control-system dynamics to be linearised over a wide range of flows, leading to improved controllability. However, the decision on their use will generally be made by mechanical and process engineers, and will be based purely on economic grounds.

2.7 The water and steam circuits of HRSG plant

In the combined-cycle plant the task of boiling the feedwater and superheating the steam so produced is achieved by using the considerable heat content of the exhaust from a gas turbine, sometimes with and sometimes without supplementary firing.

The variety of plant arrangements in use is very wide, and although the following description relates to only one configuration, it should enable the general nature of these systems to be understood.

In some plants the gas and steam turbines and the generator are on the same shaft, others have separate generators for the gas and steam turbines. The installation shown in Figure 2.9 is of the latter variety, and the diagram shows just one gas turbine and HRSG from several at this particular plant.

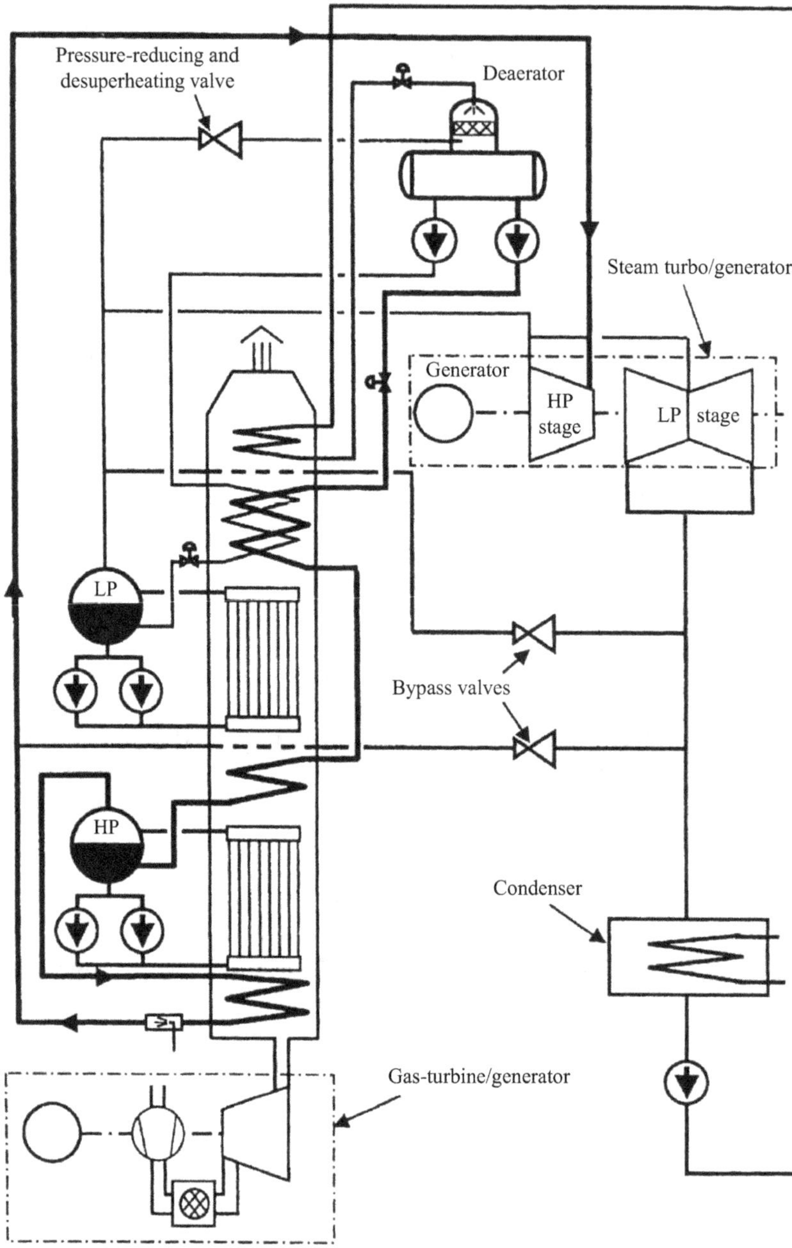

Figure 2.9 Gas and steam turbines and HRSG in a combined-cycle system

Starting at the condenser outlet, the circuit can be traced through the extraction pump and via the economiser to the deaerator. From here two circuits are formed, one feeding the LP section and the other the HP section.

These systems are of the forced-circulation type and are quite similar to each other in layout, but the steam leaving the HP side passes to a superheater bank which is positioned to receive the hottest part of the exhaust from the gas turbine.

The superheated steam goes to the HP stage of the steam turbine and the steam leaving this stage goes to the LP stage. Saturated steam from the LP section of the HRSG also enters the turbine at this point. Bypass valves are employed during start-up and shutdown and enable the plant to operate with only the gas turbine in service, under which condition the steam from the HP and LP stages is bypassed to the condenser.

2.8 Summary

So far, we have studied the nature of steam, and the plant and auxiliaries that are employed in the process of generating and using the fluid. Now we need to understand the mechanisms involved in obtaining the heat that is required to generate the steam. This process involves the fuel, air and flue gas circuits of the plant, and all the major equipment required for clean and efficient operation.

Chapter 3 describes the combustion chamber (or furnace) and the plant and firing arrangements that are employed in burning a variety of fuels. In addition, the chapter outlines how the air required for combustion is obtained, warmed and distributed, and discusses the characteristics and limitations of the plant involved in this process.

Chapter 3

The fuel, air and flue–gas circuits

David Lindsley[1] *and John Grist*[2]

Having looked at the steam and water circuits of boilers and heat-recovery steam generator (HRSGs), we now move on to examine the plant which is involved in the combustion of fuel in boilers.

The heat used for generating the steam is obtained by burning fuel in a furnace, or combustion chamber, but to do this requires the provision of air which is provided by a forced draught (FD) fan (in larger boilers, two such fans are provided). After the fuel has been burned, the hot products of combustion are extracted from the furnace by another fan, the induced draught (ID) fan, and fed to the chimney. Again, two ID fans are provided on larger boilers.

In this chapter, we shall examine not only the burners or other equipment used to burn the fuel but also the fans and air heaters. Finally, we shall briefly examine how gas turbines are used in combined-cycle plant.

3.1 The furnace

In boiler plant the heat used for boiling the water is obtained by burning a fossil fuel (unlike the HRSG, where the heat is delivered by the exhaust of a gas turbine). This process of combustion is carried out in the furnace and comprises a chemical reaction between the combustible material and oxygen. If insufficient oxygen is available some of the combustibles will not burn, which is clearly inefficient and polluting. On the other hand, the provision of too much oxygen leads to inefficient operation and to corrosion and undesirable emissions from the stack due to the combination of the surplus oxygen with other components of the flue gases.

The oxygen for combustion is provided in air, which contains around 21% of the gas. However, air also contains around 77% nitrogen, and the combustion process results in the production of nitrogen dioxide (NO_2) and nitric oxide (NO). These gases (plus nitrous oxide, N_2O) are collectively called nitrogen oxides, or NOx for short, and because they are often blamed for various detrimental effects on the environment a high level of attention must be given to minimising their production.

[1]Retired
[2]Consulting Engineer

Unfortunately, high combustion efficiencies typically correspond with the production of high levels of nitrogen oxides, and therefore NOx reduction involves careful design of the burners so as to yield adequate combustion efficiency with minimal smoke and carbon monoxide generation.

NOx production is affected by flame temperature and the fuel itself. As part of a drive towards greater reductions in NOx emissions, suppliers have provided two-stage combustion systems, that is, the secondary combustion air is provided in two or more stages. The first stage is supplied at the burner, but there is insufficient air for complete combustion and so the temperature is lower. The remaining air is added at the second stage and called over-fire air (OFA). See Section 3.1.4.

High furnace temperatures are needed when firing low-volatile coals, and consequently, NOx levels are generally higher when firing low-volatile coals in downshot boilers compared to normal coal in a conventional wall-fired boiler.

3.1.1 Firing arrangements and burners

The combustion of oil, gas or pulverised coal (PC) is performed in burners. These may be arranged on one wall of the combustion chamber (which is therefore called 'front-fired'), or on facing walls ('opposed fired') or at the corners of it ('corner-fired' or 'tangential') or on the shoulder of a downshot boiler, and the characteristics of combustion will be very different in each case. The burners may be provided with individually controlled fuel and air supplies, or common control may be applied for all the burners, or they may be operated in groups, each group having dedicated and separately controlled supplies of fuel and air.

Combustion of raw coal or other solid fuels such as municipal waste, clinical waste or refuse-derived fuels (RDF) is often carried out in fluidised beds or on stokers consisting of moving grates or platforms.

The methods of controlling these various arrangements are very different. With front-fired or opposed-fired boilers the temperature of the flue gases and the resulting heat transfer to the various banks of superheater tubes are adjusted by bringing burners into service or taking them out of service, and this may be done individually or in banks. This method provides a step-function type of control and fine adjustment of steam temperature is provided by spray-water attemperation.

3.1.2 Wall-fired boilers and wall burners

Most large utility boilers are of the opposed wall-fired type. The majority of coal burners are therefore designed for wall-fired application. A typical burner is shown in Figure 3.1. It comprises a core air tube in the centre which houses the oil burner used to ignite the coal and transports air required to light up the oil burner. This tube is surrounded by a concentric pipe for the primary air (PA) that carries the pulverised fuel (PF). Outside of this is the secondary air supply, usually taken directly from the windbox. Depending upon the specific burner the total secondary air may be split into secondary and tertiary or even quaternary air for NOx control. There may be spinners, straighteners and flame holders. For the control engineer

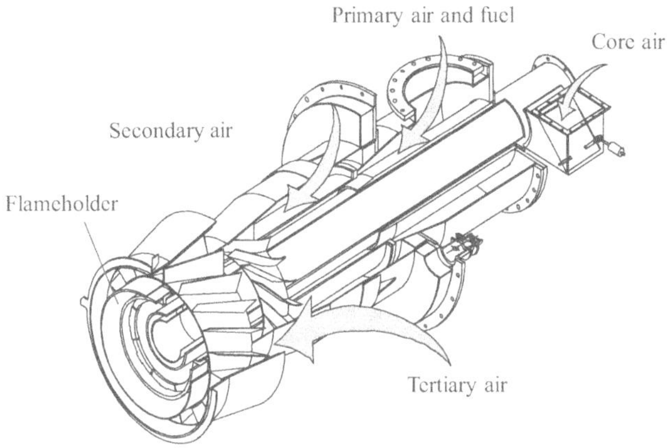

Figure 3.1 Wall burner. © Doosan Babcock Ltd, reproduced with permission

the subdivision of secondary air is preset during commissioning, and no distributed control system (DCS) control is required. The control of the relationship between PA, coal and secondary air flows is explained in Chapter 5. In some cases, metal temperature thermocouples are fitted to the burners to monitor performance. When used on a common windbox an extra damper or register is used to isolate any non-firing burners. In all cases, oil and coal flame monitors are required.

3.1.3 Corner-fired boilers and tilting burners

Corner-fired (tangential) boilers are arranged in such a way that the burning fuel circulates around the furnace, forming a large swirling ball of burning fuel at the centre. With this type of boiler, the manufacturers usually employ tilting mechanisms to direct the fireball to a higher or lower position within the furnace, and this has a significant effect on the temperature of the various banks of superheater tubes, and therefore on steam temperature. The maximum degree of tilt that is available within the basic design is typically $\pm 30^\circ$, although the degree of movement employed in practice is usually restricted during commissioning.

Refer to Figure 3.2 that shows a tilting burner in both horizontal and tilted positions. The oil gun is at the bottom with the coal nozzle in the middle and the secondary air nozzles at the top.

The downside of tilting is that burners – with their fuel and air supplies, igniters, flame monitors, etc., – are complex things, and tilting them requires very careful engineering if it is to be successful. Also, the tilting mechanisms must be rigorously maintained if they are to continue to operate effectively over any length of time.

The control systems that regulate burner tilting mechanisms must ensure that exactly the same degree of tilt is applied to the burners at all four corners of the furnace since any misalignment will cause the fireball to circulate helically rather than as required.

(a) (b)

Figure 3.2 Tilting burner in horizontal and tilted position. © DURAG Group, reproduced with permission

For the boiler's control and instrumentation engineers, the tilting burner presents two unique problems with the flame monitoring system. Firstly, because the burner moves, the oil flame igniter and both the oil and coal burner flame monitor connections must be flexible. The latter requires fibre optic connections. Refer Chapter 5 for more details. Secondly, as the flames form a fire ball in the centre of the furnace, the discrimination of individual established burner flames is very difficult, and so monitoring logic to determine overall 'combustion zone' flame security may be applied by some boiler suppliers.

3.1.4 Downshot (also down-fired or W-fired) boilers and their burners

Downshot boilers are designed to burn low-volatile fuels such as anthracite, which are difficult to fire in a conventional wall-fired furnace. The lower part of the furnace is deeper than on a conventional boiler to accommodate burners mounted on the furnace shoulders. The furnace is generally also wider than for other boilers as all the burners and windboxes are mounted on a single level across the furnace shoulders.

The burners fire downwards, hence the name downshot. The coal is concentrated (i.e. delivered at high fuel to air ratio) by removing most of the PA in high-efficiency cyclones and dropped at low velocity into the furnace via the burner openings, which can be circular or made up of slots. The stripped-off PA is passed back into the furnace. Figure 3.3 shows a side elevation of a downshot boiler. This shows the wider bottom half of the furnace. The size of the cyclones can be judged by comparing them with the operating personnel shown on the firing floor. A more detailed view of the cyclones and oil burners is shown in Figure 3.4.

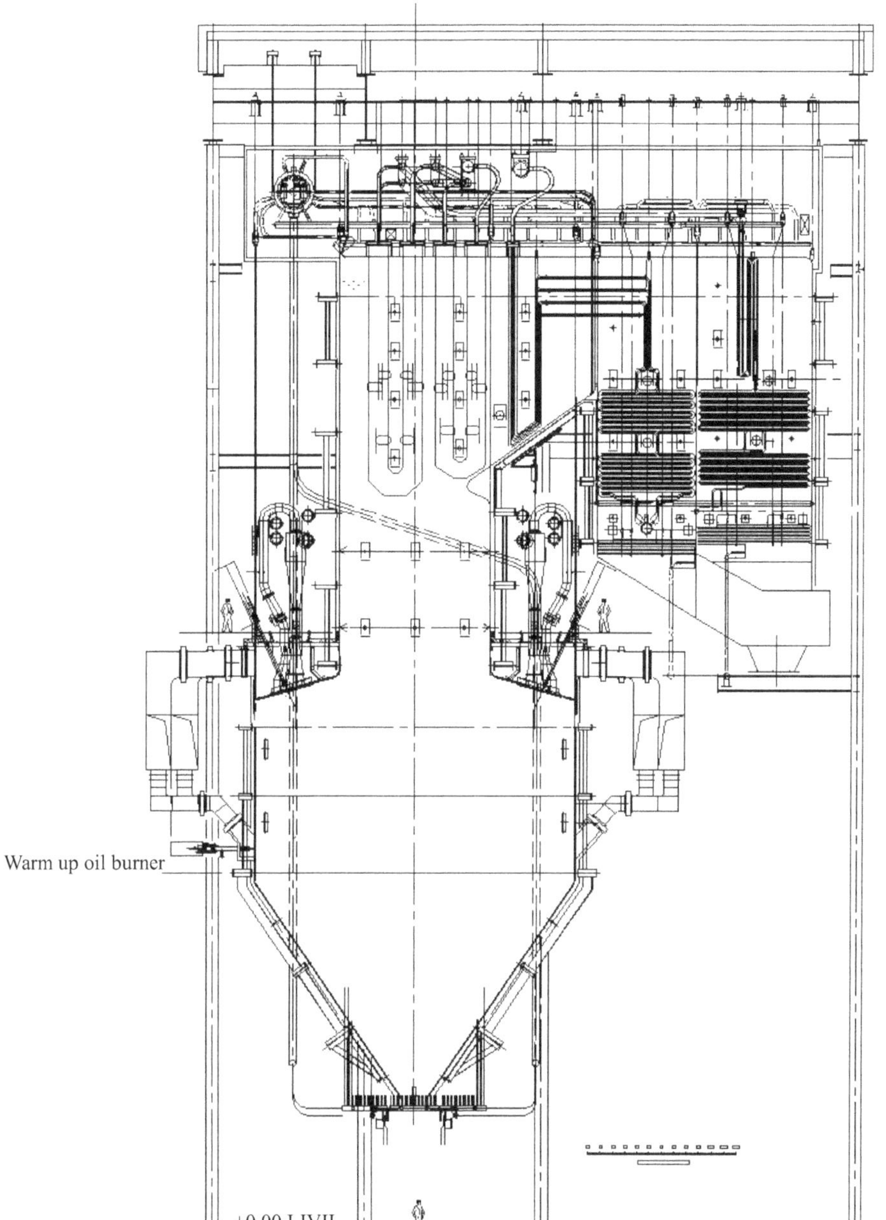

Figure 3.3 Side elevation of downshot boiler. © Doosan Babcock Ltd, reproduced with permission

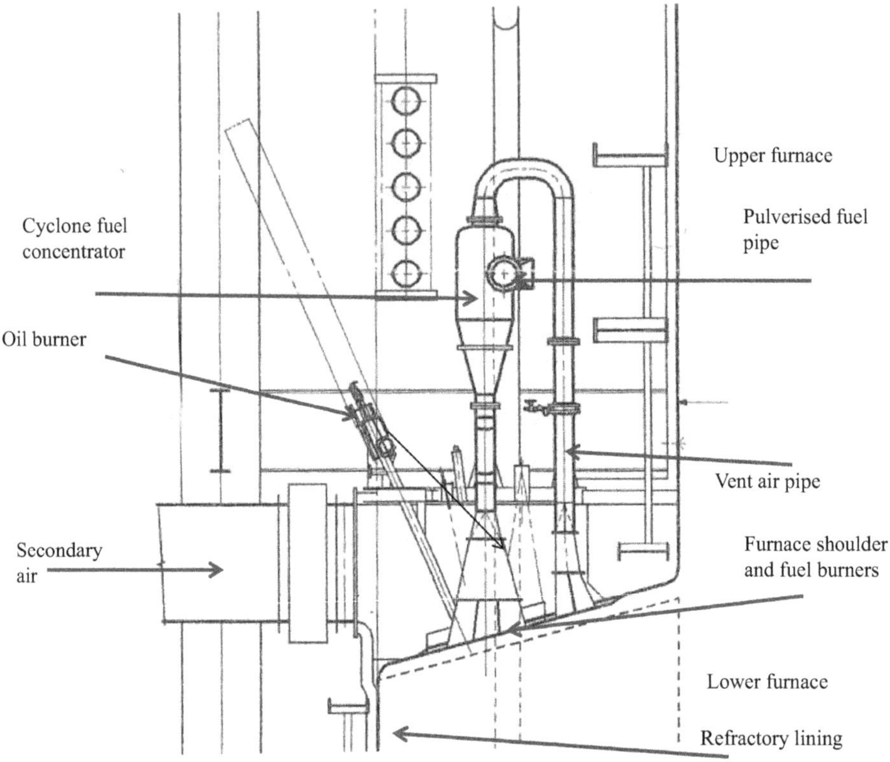

Figure 3.4 Furnace shoulder cyclone and burners. © Doosan Babcock Ltd, reproduced with permission

The low fuel/PA velocity promotes fast heating and ignition of the fuel. The lower furnace walls are partially refractory lined to reduce heat absorption and to increase the furnace gas temperature.

The fuel/PA stream is surrounded by a higher velocity secondary air stream. This produces a longer flame development downwards towards the bottom of the lower furnace before the gases turn upwards forming a typical 'W' pattern. This downward and then upward path increases the residence time in the furnace, allowing the coal to completely burn.

The control design for furnace pressure drum level, steam temperature, etc., is the same in a downshot and wall-fired boiler. There are some changes in the combustion logic.

3.1.5 OFA and boosted OFA

As part of a drive towards greater reductions in NOx emissions, suppliers have provided two-stage combustion systems, i.e. the secondary combustion air is provided in two or more stages. The first stage is supplied at the burner, but there is

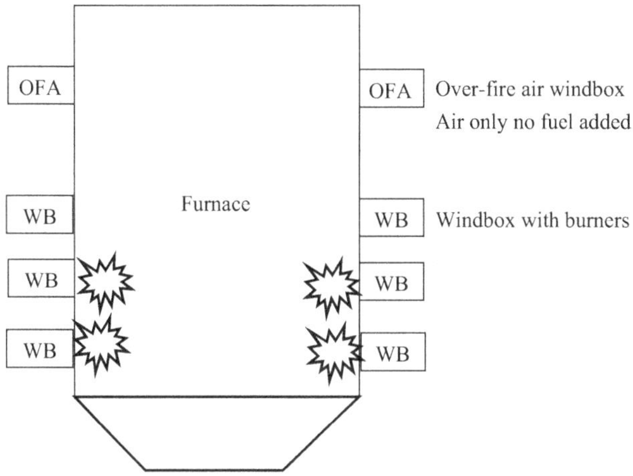

Figure 3.5 Over-fire air concept

insufficient air for complete combustion. The remaining air is added at the second stage. This is described more in Chapter 5. Figure 3.5 shows a typical location of the OFA supply.

For downshot boilers, the furnace arrangement makes the provision of OFA supply challenging but not impossible.

3.2 The air and gas circuits

The combustion process requires the provision of fuel and air in the correct ratio to each other. This is known as the stoichiometric ratio, and under this condition enough air is provided to ensure complete combustion of all the fuel, with no surplus or deficit. However, this is a theoretical ideal, and practical considerations may necessitate operating at a fuel/air ratio that is different from the stoichiometric value. In addition, it must be understood that the efficiency of the combustion process will also be affected by the temperature of the air provided.

In the following sections, we shall see how air is delivered to the furnace at the right conditions of flow and temperature, starting with the auxiliary plant that warms the air and moving on to the types of fan employed in the draught plant. This description is applicable to oil, gas, coal and biomass boilers. Gas-fired boilers may additionally have fans pressuring the penthouse to prevent unburnt gas collecting there and subsequently exploding.

3.2.1 The air heater

In a simple-cycle plant, air is delivered to the boiler by one or more FD fans and the products of combustion are extracted from it by ID fans. Figure 3.6 shows this plant in a simplified form and illustrates how the heat remaining in the exhaust gases

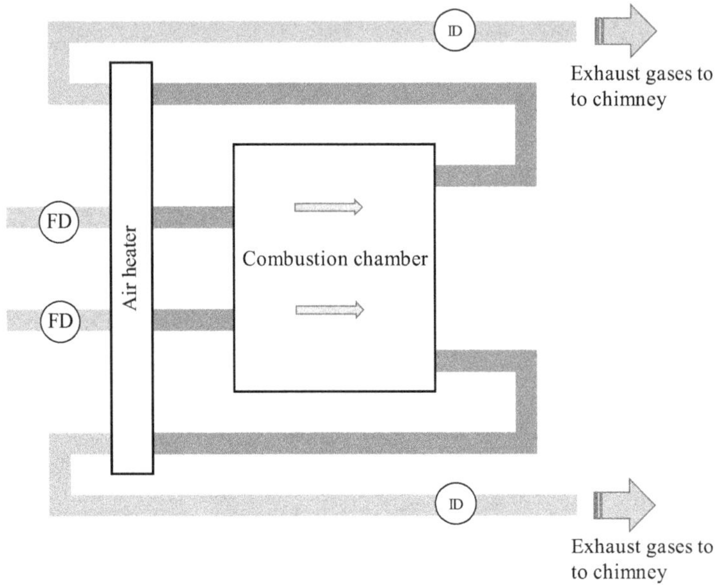

Figure 3.6 Draught plan arrangement

leaving the furnace is used to warm the air being fed to the combustion chamber. This function is achieved in an air heater, which can be either regenerative, where an intermediate medium is used to transfer the heat from the exhaust gases to the incoming air, or recuperative, where a direct heat transfer is used across a dividing partition.

One variety of regenerative air heater is the Ljungstrom type, where metal plates mounted on a rotating frame are passed through the hot gases and then to the incoming air.

From a control engineer's point of view, an important consideration is the efficient combustion of the fuel, and here it is necessary to consider the losses and leakages that occur in an air heater. Figure 3.7 shows a typical trisector air heater. Leakages occur across the circumferential, radial and axial seals, as well as at the hub. The higher-pressure PA duct leaks to the medium pressure secondary air duct and the lowest pressure flue gas duct. The secondary air duct leaks to the flue gas duct. These leakages are minimised when the plant is first constructed, but become greater as wear occurs during prolonged usage. When the sheer physical size of the air heater is considered (Figure 3.7) it will be appreciated that these leakages can become significant. Modern air heaters have increased the number of seals and, if required, offer an automated seal air system.

Alternatively, a quad sector air heater can be used where the PA duct is located between two secondary air ducts, thereby ensuring that the PA does not leak to the flue gas.

Figure 3.7 Tri sector air heater.

3.2.2 Types of fan

In addition to the FD and ID fans mentioned earlier, another application for large fans in a power station boiler is where it is necessary to overcome the resistance presented by plant in the path of the flue gases to the stack.

In some cases, environmental legislation has enforced the fitting of flue gas desulphurisation equipment to an existing boiler. This involves the use of absorbers and/or bag filters, plus the attendant ducting, all of which present additional resistance to the flow of gases. In this case this resistance was not anticipated when the plant was originally designed, so it is necessary to fit additional fans to overcome the draught losses. These are called 'booster fans'.

Whatever their function, as far as the fans themselves are concerned, two types are found in power station draught applications: centrifugal (Figure 3.8) and axial-flow (Figures 3.9 and 3.10). In the former, the blades are set radially on the drive shaft with the air or flue gas directed to the centre and driven outwards by centrifugal force. With axial-flow fans, the air or gas is drawn along the line of the shaft by the screw action of the blades. Whereas the blades of a centrifugal fan are fixed rigidly to the shaft, the pitch of axial-flow fan blades can be adjusted. This provides an efficient means of controlling the fan's throughput, but requires careful design of the associated control system because of a phenomenon known as 'stall', which is described in the next section. The two types of fan have different pressure/flow characteristics, which makes them best suited for certain duties. Centrifugal fans can generate a high pressure rise at low flows, and so are best for supplying PA flow to the pulverisers, since a minimum pressure is required regardless of whether one or all mills are in service. On the other hand, great care is required if they are used as ID fans since they may be capable of imploding the furnace at very low air flows!

Figure 3.8 Centrifugal fan. © Howden Group Limited. All rights reserved. 2017. Howden and the flying H logo are registered trademarks belonging to Howden Group Limited

Figure 3.9 Axial flow fan assembly. © Howden Group Limited. All rights reserved. 2017. Howden and the flying H logo are registered trademarks belonging to Howden Group Limited

Figure 3.10 Large axial flow fan. © Howden Group Limited. All rights reserved. 2017. Howden and the flying H logo are registered trademarks belonging to Howden Group Limited

The pitch of axial fan blades is adjusted by a hydraulic mechanism that slides along the axis of the hub to rotate turning levers attached to each blade. The drive is adjusted by a pilot shaft that is moved by the turning of a modulating-duty actuator located outside the fan casing. The basic arrangement is shown in Figure 3.11.

3.2.2.1 The stall condition

The angular relationship between the air flow impinging on the blade of a fan and the blade itself is known as the 'angle of attack'. In an axial-flow fan, when this angle exceeds a certain limit, the air flow over the blade separates from the surface and centrifugal force then throws the air outwards, towards the rim of the blades. This action causes a build-up of pressure at the blade tip, and this pressure increases until it can be relieved at the clearance between the tip and the casing. Under this condition the operation of the fan becomes unstable, vibration sets in and the flow starts to oscillate. The risk of stall increases if a fan is oversized or if the system resistance increases excessively.

The stall line is shown on the fan curves. The curves typically show the head generated by the fan against the volume passing through the fan for each of the vane positions between 25° and 80° of opening. The fan efficiency curves and the stall line are also included. For any given vane position stall is approached at maximum head and minimum flow. Sometimes the fan supplier may use 'per unit' curves instead and these require converting to engineering units.

For each setting of the blades there is a point on the fan characteristic beyond which stall will occur. If these points are linked, a 'stall line' is generated

Figure 3.11 *Pitch drive assembly for a small fan. © Howden Group Limited. All rights reserved. 2017. Howden and the flying H logo are registered trademarks belonging to Howden Group Limited*

(Figure 3.12), and if this is built into the plant control system (DCS), it can be used to warn the operator that the condition is imminent and then to actively shift operation away from the danger region. The actual stall-line data for a given machine should be provided by the fan manufacturer. In addition to this stall prevention logic, the fan supplier will provide stall probes to actually detect the stall condition, when the fan should be tripped. The fan manufacturer will not permit operation in a stall condition at any time. If the end user operates an axial fan in a stall condition it will be necessary to re-blade the fan.

3.2.2.2 Centrifugal fan surge

The stall condition affects only axial-flow fans. However, centrifugal fans are subject to another form of instability. If they are operated near the peak of their pressure/flow curve a small movement either way can cause the pressure to increase or decrease unpredictably. The point at which this phenomenon occurs is known as the 'surge limit', and it is the minimum flow at which the fan operation is stable.

The system designer needs to be aware of the risk of surge occurring since it may be necessary to adapt the control system design. However, this is generally not a problem if the fan is properly designed in relation to the overall plant. During the initial design of the control system, dialogue with the process engineer or boiler designer will show whether or not surge protection will be required.

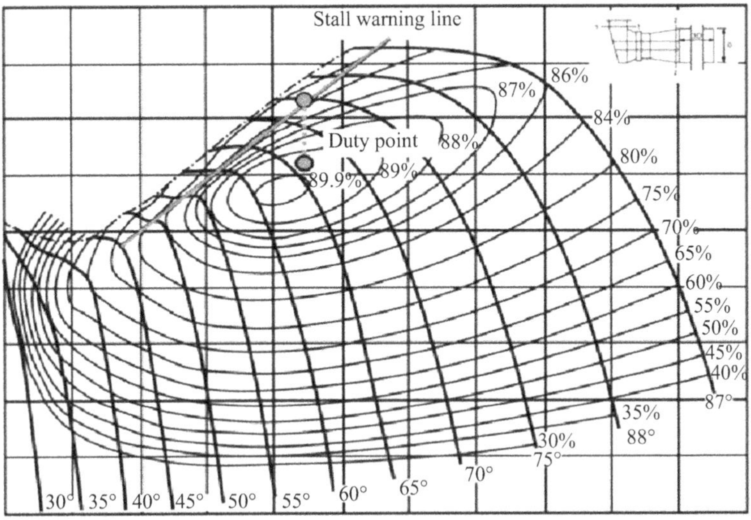

Figure 3.12 The stall line of an axial flow fan. © Howden Group Limited. All rights reserved. 2017. Howden and the flying H logo are registered trademarks belonging to Howden Group Limited

3.2.3 Final elements for draught control

Reference has already been made to the use of pitch control in axial-flow fans to regulate the throughput of the machine. Other means of controlling flow are dampers, vanes or speed adjustment. Each of these devices has its own characteristics, advantages and disadvantages, and the selection of the controlling device which is to be used in each application will be a trade-off between the technical features and the cost.

3.2.3.1 Types of damper

The simplest form of damper consists of a hinged plate that is pivoted at the centre so that it can be opened or closed across the duct. This provides a form of draught control but it is not very linear, and it is most effective only near the closed position. The linearity of a damper control is related to the ratio of the system pressure drop to the damper's pressure drop when fully open. For instance, a ratio of 3 would give even a single-leaf damper almost linear control. But to achieve this a control damper would need to be sized to deliberately constrict the flow, requiring a much larger fan and stronger ducting to handle the higher system pressure. Practical ratios are in the region of 20–50.

Once a typical damper is more than about 40%–60% open it can provide very little additional control. Another form of damper comprises a set of linked blades across the duct (like a Venetian blind). Such multibladed dampers are naturally more expensive and more complex to maintain than single-bladed versions, but they offer better linearity of control over a wider range of operation.

Contra-rotating blade dampers have the odd and even blades moving in opposite directions. These have even more complex linkage arrangements, but provide the best linearisation for practical applications.

The task of designing a control system for optimum performance over the widest dynamic range will be simplified if the relationship between the controller output signal and the resultant flow is linear. Although it is possible to provide the required characterisation within the control system, this will usually only be effective under automatic control. Under manual control a severely nonlinear characteristic can make it difficult for the operator to achieve precise adjustment.

It is possible to linearise the command flow relationship under both manual and automatic control by the design of the mechanical linkage between the actuator and the damper. However, this requires careful design of the mechanical assemblies and these days it is generally considered simpler to build the required characterisation into the DCS. This approach provides a partial answer, but it should not be forgotten that such a solution is only effective under automatic control.

3.2.3.2 Vane control

The second form of control is by the adjustment of vanes at the fan inlet. These are operated via a complex linkage which rotates all the vanes through the same angle in response to the command signal from the DCS.

3.2.3.3 Variable speed drives

Finally, control of fan throughput can be achieved by the use of variable speed motors (or drives). These may involve the use of electronic controllers which alter the speed of the driving motor in response to demand signals from the DCS or they can be hydraulic couplings or variable speed gearboxes, either of which allows a fixed speed motor to drive the fan at the desired speed. Variable speed drives offer significant advantages in that they allow the fan to operate at the optimum speed for the required throughput of air or gas, whereas dampers or vanes control the flow by restricting it, which means that the fan is attempting to deliver more flow than is required. The mathematical formulas describing the relationship between variables describing fan performance are referred to as the fan affinity laws. One of the fan laws states that power is proportional to the cube of the speed, so operating at a lower speed has significant cost saving. Another fan law states that the pressure rise across the fan varies as the square of the speed, so there may not be sufficient pressure to overcome the system resistance when the speed is reduced. It is important that the fan supplier is involved in choosing between variable pitch, damper control, variable speed or a combination of these.

For smaller fans, less than 3 MW, it is possible for end users to buy standard motors and variable speed drives. For larger motors suppliers may prefer to supply the complete 'set' of fan motor and drive to provide optimisation of the overall system. This simplifies the interface design process and commissioning.

3.2.3.4 Additional instrumentation and drives

It may appear on an overview boiler P&ID that the ID fan has only a single motor and vane pitch controls: it is actually far more complex.

The bearings are lubricated via main and standby lube oil pumps. The lube oil is cooled by main and standby circulating pumps that push the oil through a main and standby heat exchanger each with its own cooling fan. The vane pitch is moved hydraulically powered from main and standby hydraulic pumps. This circuit also has main and standby cooling fans. The main shaft has main and standby seal air fans, sometimes at each end of the shaft. A total of 12 or 14 auxiliary motors.

In addition to the large number of supporting fans and pumps, each drive has its own control logic within a hierarchy of interlocking and sequence controls, so, for example, a standby cooler fan might be started if the duty fan trips. In addition, the main fan motor and vane position are interlocked with the other main fans.

3.3 Fuel systems

Fossil fuels that are burned in boilers can be used in solid, liquid or gaseous form, or a mixture of these. Naturally, the handling systems for these types of fuels differ widely. Moreover, the variety of fuels being burned is enormous. Solid fuels encompass a wide spectrum of coals as well as wood, the waste products of industrial processes, municipal and clinical waste and RDF. (The last are produced by shredding or grinding domestic, commercial and industrial waste material.) Liquid fuels can be heavy or light oil, or the products of industrial plant. Gas can be natural or manufactured, or the by-product of refineries.

Each of these fuels requires specialised handling and treatment, and the control and instrumentation has to be appropriate to the fuel and the plant that processes it.

3.3.1 Coal firing (PF)

Although coal can be burned in solid form on grates, it is more usual to break it up before feeding it to the combustion chamber. The treatment depends on the nature of the coal. Some coals lend themselves to being ground down to a very fine powder called pulverised fuel (PF) or pulverised coal (PC) which is then carried to the burners by a stream of air. Other coals are fed to impact mills which use flails or hammers to break up the material before it is propelled to the burners by an air stream. The type of mill to be used on a particular plant will be determined by the process engineers, and it is the task of the control engineer to provide a system which is appropriate. To do this it is necessary to have some understanding of how the relevant type of mill operates. Indeed, understanding pulveriser operation, control design, output dynamics and fault conditions is crucial for anyone who is responsible for coal-fired boiler controls commissioning, optimisation or maintenance.

All PF mills and their associated feeders perform the same basic set of functions: metering the fuel input as required by the fuel demand system, grinding, heating (and thereby drying), classifying to the desired fineness and transporting the PF out of the mill to the burners. To achieve this, most mills are supplied with heated PA as a drying and transport medium.

Various types of PF mill will be encountered, but two are most commonly used: the pressurised vertical spindle mill and the horizontal tube mill. Vertical spindle mills fall into two basic categories, defined by their crushing mechanisms. These are described in the following sections.

3.3.1.1 Vertical spindle ball-and-race mills

Figure 3.13 shows the operating principle of a typical ball-and-race mill. In this device, the coal that is discharged from the storage hoppers is fed down a central chute onto a table where it is crushed by rotating steel balls. Air is blown into the crushed coal and carries it, via adjustable classifier blades, to the PF pipes that transport it to the burners.

The air that carries the fine particles of coal to the burners is supplied from a fan called a 'primary-air fan'. This delivers air to the mill, which therefore operates under a pressure which is slightly positive with respect to the atmosphere outside. Because of this and because of its other constructional features, this type of mill is properly called a 'vertical spindle, pressurised ball-and-race mill'. The air supply system for this type of mill is discussed in more detail in Section 3.3.1.3.

3.3.1.2 Vertical roller mills

Vertical spindle roller mills are manufactured by a large number of companies worldwide, each with particular features. In general the coal enters and leaves the mill in the same manner as the ball-and-race mill. But instead of the relatively large balls, there are typically three, even larger rollers. See Figure 3.14 for a typical roller retracted for maintenance. The roller is usually either a tyre shape, or conical, the latter lying in a deeper 'bowl' grinding table arrangement.

It is held in position above a rotating table. Its height is adjustable by spring and/or hydraulic loading.

An overview of a Loesche vertical spindle mill is shown in Figure 3.15. This shows the PA entering underneath the rotating table, passing the rollers and entraining the PF that is carried to the dynamic (rotating) classifier. PF of the correct fineness passes to the burners while oversized particles are returned to the grinding table inside the central cone. The coal bunker sits vertically above the coal feeder which discharges a measured amount of coal directly onto the rotating table.

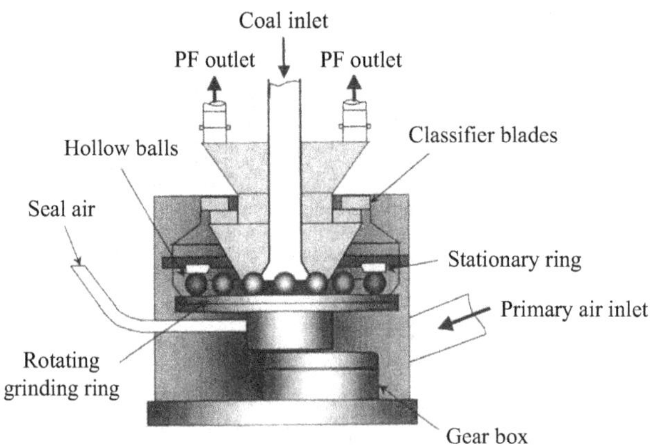

Figure 3.13 Pressurised ball-and-race mill

Figure 3.14 Roller retracted for maintenance. © Loesche, reproduced with permission

In most designs the rollers cannot actually touch the table, even when empty so avoiding metal to metal sparks and potential coal-dust ignition. It has a further advantage in that once the rollers are lifted the motor is able to restart the mill with a full inventory. (Some boiler designers do not allow the mill to restart when full. It must be emptied via the rejects system.) The National Fire Prevention Association (NFPA) codes relating to boiler design and operation sets out strict guidelines on mill operation and requires the mill to be 'inerted' during start-up and shutdown and following a mill fire. Inerting is a process to reduce the risk of spontaneous ignition within the mill. Steam injection is the most common medium employed. The mills can also be fitted with an ammonium phosphate explosion suppression system. These are standard for biomass operation but can also be used on a coal mill.

Modern roller mills designed for wide load operation and fast load ramping have several methods to control the grind and classification process: adjustable pressure hydraulic rollers and classifiers that are either rotating with variable speed, or static classifiers with variable vanes.

3.3.1.3 General aspects of vertical spindle mills

All vertical spindle mills operate with the lifting and returning classification process described earlier. This airborne 'recirculating load' varies with coal throughput. It is possible for a mill to reach a 'choking' point where the recirculating load is too high for the PA flow and the coal output to the burners is less than the inflow from the feeder. This is a dangerous operating condition and is detected by a high air inlet to PF outlet differential pressure.

As with the fans and air heaters, the mill supplier provides control interlock requirements for all the equipment supplied with the mill, for example, the minimum seal air to PA differential pressure setting and the maximum differential pressure across the grinding table and classifier. However, it is up to the boiler supplier's C&I engineer to interface and take overall responsibility for the integration of the mill with the coal feeder, the pyrites (mineral stone components of the stone) rejection system, the seal air system, the PA supply and mill inerting – this is one of the most challenging aspects of boiler controls for the designer.

3.3.1.4 Horizontal tube mills

In a tube mill (Figure 3.16) sometimes referred to as a ball-tube mill the coal is fed into a cage that rotates about a horizontal axis, at a speed of 18–35 rpm. This cage

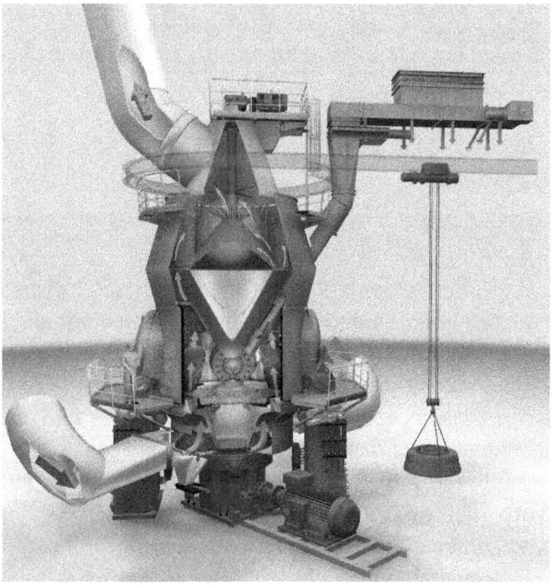

Figure 3.15 Overview of Loesche mill. © Loesche, reproduced with permission

contains a charge of forged-steel or cast alloy balls (each of which is between 25 and 100 mm in diameter) which are carried up the sides of the cage by the rotation, until they eventually cascade down to the bottom, only to be carried up again. The coal is pulverised by a combination of the impact with these balls, attrition of adjacent particles and crushing between the balls and the cage and between one ball and another.

Figure 3.16 shows the tube mill, its motor, the trunnion gear that surrounds the mill and the two classifiers with their reject chutes. An end view is shown in Figure 3.17. This shows more details of the PA supply and includes the coal feeder. You see that the hot PA and cold coal enter through one half of the end bearing while the PA mixed with coal leaves via the other half. This is repeated at each end.

Some tube mills do not have separate classifiers above the rotating tube. Instead, they are integral to the mill body and located at either end.

Figure 3.18 provides a cutaway view of the mill that identifies the inching drive and internal linings. Sometimes the motor is sin series with the mill body instead of in parallel as shown. The mill itself is often covered in an acoustic hood to reduce the noise

There can be significant additional challenges with operating tube mills. Most of these relate to the process rather than the control design and so are described in this section.

The first is that the tube mill is designed for continuous operation. While vertical spindle mills usually achieve boiler maximum continuous rating (MCR) with, say, five mills operational there is a spare mill and spare row of burners that are not operational. (Referred to as having $n+1$ mills.) Boilers with tube mills are sometimes, but not always, installed with no spare mill and no spare rows of burners. They are made more reliable by having static classifiers and dual feeders, so that if one feeder fails the other feeder can supply sufficient coal flow to supply output from both classifiers. These mills can also operate in a single-ended mode, supplying only half of the mill's burners.

Double-ended operation refers to two feeders, two classifiers and the PA going to both ends. It is easy to understand if all function correctly but difficult to understand when all the failure possibilities are considered. In addition to failure modes single-ended operation may be required during start-up if the two mill ends feed opposite walls of the furnace.

Table 3.1 shows the effect of any single failure of the major double-ended components. The operating modes are summarised next.

Mode 1. All components fully functional 100% mill output achievable.

Mode 2. Feeder A failed. Feeder B ramps up to provide total mill input. 100% mill output achievable but only for a short time as the hot PA passing over the A side bearing will cause it to overheat. However, if the hot PA to the A side is isolated then the operational time can be extended.

Mode 3. Feeder B failed similar to mode 2.

Figure 3.16 Horizontal tube mill. © Metso, reproduced with permission of Metso Minerals Industries, Inc

Mode 4. Outlet A failed. No motorised failure of the mill is possible but a damper may have failed shut or the classifier reject system failed. Now because fuel is not going to one set of burners the output must be reduced to 50%.

Mode 5. Outlet B failed. Similar to mode 4.

Mode 6. PA to A side failed, typically a damper has failed closed. The total PA enters via the B side and allows continued operation at 100% load.

Mode 7. PA to B side failed. Similar to mode 6.

The actual periods of operation in the various modes also depend on the fuel and PA temperatures so these must be defined by the mill supplier not the controls engineer. However, the control engineer has to develop the logic to cope with these different operating scenarios and the transfer between single- and double-ended operation.

In discussing vertical spindle mills we saw that some could start with a mill full of coal. However, the tube mill is much heavier and with some designs the motor does not have sufficient torque to start it. This is overcome by using a clutch so the motor can be started with no load and the load added as the clutch is released.

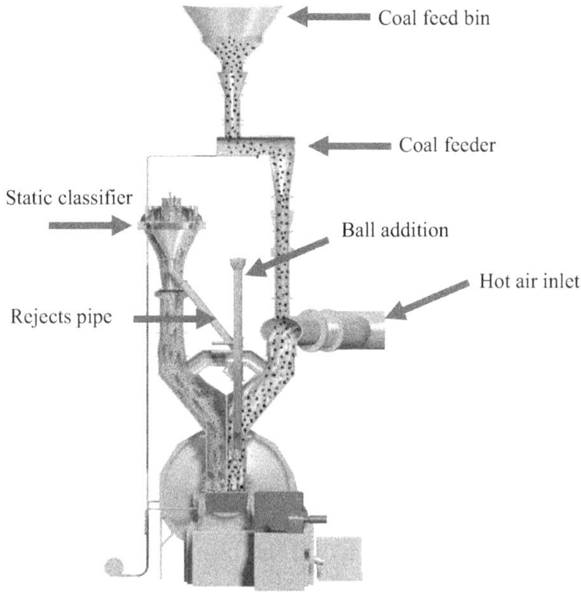

Figure 3.17 End view of pressurised tube mill. © Metso, reproduced with permission of Metso Minerals Industries, Inc

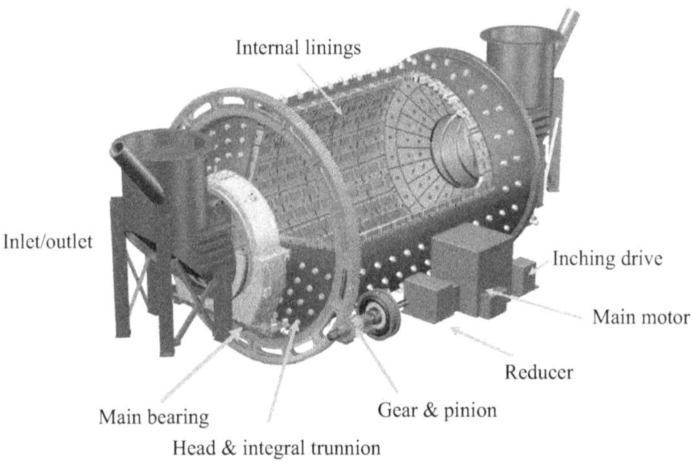

Figure 3.18 Mill main components. iMetso, reproduced with permission of Metso Minerals Industries, Inc

The instrument engineer needs to ensure that his air pressure is high enough to operate the clutch.

Secondly, tube mills exhibit a different loading response when overloaded with coal. As the coal level rises, the impacting performance of the falling balls reduces,

Table 3.1 Double-ended operating combinations

	Feeder A	Feeder B	Outlet A	Outlet B	PA to A side	PA to B side	Mill capacity (%)
1	OK	OK	OK	OK	OK	OK	100
2	Failed	OK	OK	OK	OK	OK	100
3	OK	Failed	OK	OK	OK	OK	100
4	OK	OK	Failed	OK	OK	OK	50
5	OK	OK	OK	Failed	OK	OK	50
6	OK	OK	OK	OK	Failed	OK	100
7	OK	OK	OK	OK	OK	Failed	100

and the motor current will reach a peak, and then begin to fall as more coal is simply rotated, rather than being crushed.

Thirdly, while the sequence logic for starting and stopping the lube oil system is similar for all mills, the tube mill uses high-pressure oil to lift the shaft, and if this is lost, the load is taken directly on the bearing and can damage it. This is overcome by fitting hand pumps to the mills.

The large gear wheel needs to be greased regularly, and this is achieved via a grease pump controlled from the DCS. The mill supplier will provide the logic for this.

During maintenance if, say, the internal linings are being replaced, it will be necessary to inch the mill body. This is not possible with the air clutch, and so either an inching motor is added to each mill or one per boiler is available for maintenance.

Where the classifier is not integral with the mill body rejects are returned via a dedicated pipe.

The balls in the tube mills wear, as do the rollers in a vertical spindle mill. However, the tube mill balls can be replaced online. There is a chute provided at one end of the mill down through which replacement balls can be added. There are two valves in series, and they are interlocked so only one can be opened at a time.

It is possible to increase the turndown on a tube mill by adding a bypass system that allows some of the PA to bypass the mill and so not pick up any coal but rejoin before the classifier to ensure that there is adequate transport velocity to stop coal falling out in the pipework. Figure 3.19 shows the bypass, and Figure 3.20 plots the different air flows while the bypass is in operation.

For the control engineer, switching between mill ends and understanding how the PA flow and bypass flows are measured or calculated can be a real challenge.

3.3.1.5 Air supply systems for mills

Sometimes in this type of mill the crushed mixture is drawn out of the cage by a fan, which is called an exhauster. Because of this configuration, the tube mill runs under a negative pressure, which prevents the fine coal dust from escaping (as it tends to do with a pressurised mill). However, the exhausters have to handle the dirty and abrasive mixture of coal and air that comes through the mill and they

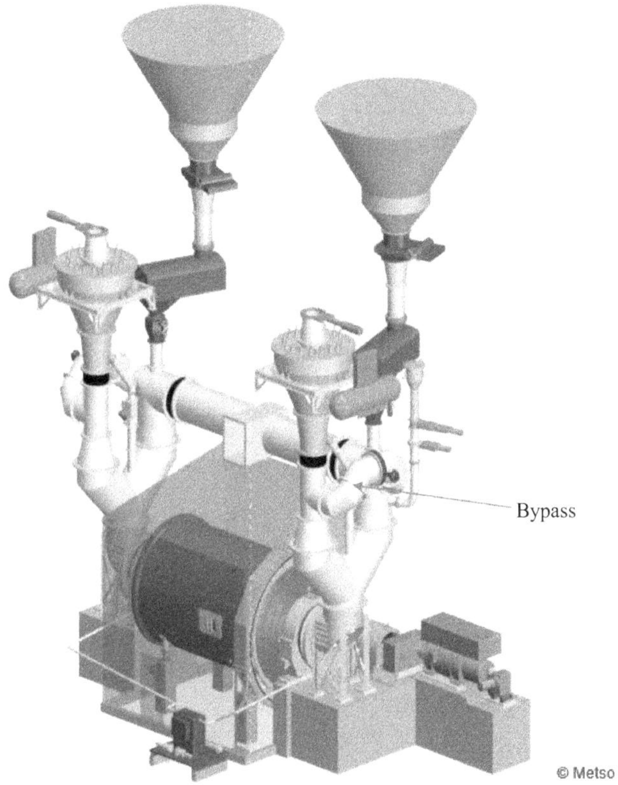

Figure 3.19 Ball tube mill with bypass. © Metso Minerals Industries Inc., reproduced with permission

therefore require more frequent maintenance than the fans of a pressurised ball mill, whose function is merely to transport air from the atmosphere to the mill. As the sealing technology has improved most of these mills now have conventional PA fans and are pressurised.

As stated earlier, the crushed coal in a pressurised ball-race mill is propelled to the burners by a stream of warm air. Cool air and heated air are mixed to achieve the desired temperature. This temperature has to be high enough to partially dry the coal, but it must not be so high that the coal could overheat (with the risk of the coal/air mixture igniting inside the mill or even exploding while it is being crushed). The warm air is then fed to the mill (or a group of mills) by means of yet another fan, called a 'primary air fan'. It should be noted that the cooler of the two air streams is commonly referred to as 'tempering air' because its function is to attemperate the mixture, although because it is obtained from the FD fan discharge it may be slightly warm.

Figure 3.21 shows the system that is used with one type of tube mill. Here, hot air and cold air are again mixed to obtain the correct temperature for the air stream but because the mill in this case operates under suction conditions a PA fan is not

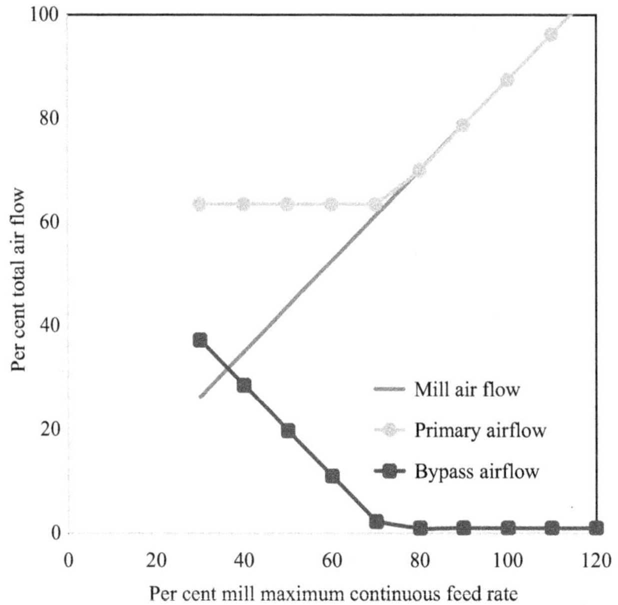

Figure 3.20 Per cent airflow versus per cent feed rate air swept ball mill system. © Metso, reproduced with permission of Metso Minerals Industries, Inc

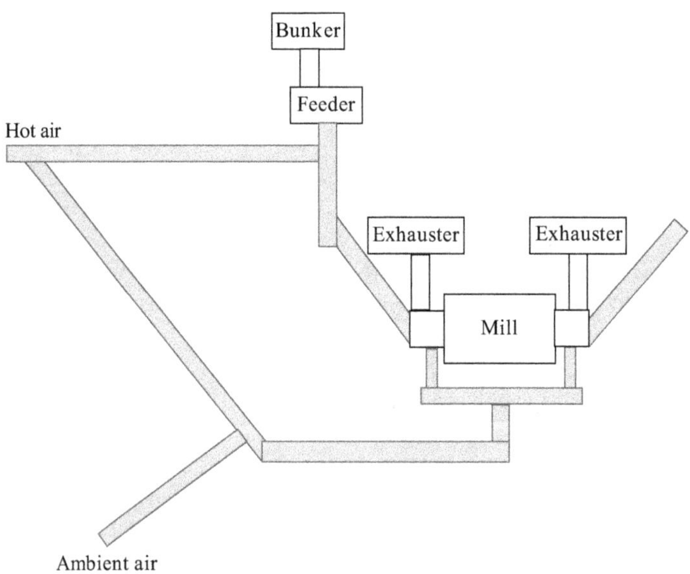

Figure 3.21 Suction mill air supply

needed and the cold air is obtained directly from the atmosphere. The warmed air mixture is again fed to the mill as 'primary air' but in addition a stream of hot air is fed to the feeder for transportation and drying purposes.

3.3.1.6 Oil-firing systems

In comparison with coal, oil involves the use of much less capital plant. On the face of it, it would appear that all that is required is to extract the oil from its storage tank and pump it to the burners. But in reality life is more complicated than that!

Proper ignition of oil depends on the fuel being broken into small droplets (atomised) and mixed with air. The atomisation may be achieved by expelling the oil through a small nozzle (a 'pressure jet') or it may be achieved by the use of compressed air or steam.

The fuel oil itself may be light (such as diesel oil or gas oil), or it may be extremely viscous and tar-like (heavy fuel oil, commonly 'Bunker C'). The handling system must therefore be designed to be appropriate to the nature of the liquid. With the heavier grades of oil, prewarming is necessary (shown schematically in Figure 3.22).

Depending upon the load that the oil can carry there may be two or three pumps and heaters. The fuel oil pumps are positive displacement pumps so a spill back valve(s) (1 and 2) is required to control the discharge pressure. This system may be configured on a boiler basis or may be common to more than one boiler. There is a control valve (3) per boiler. Heavy fuel oil is heated to 90 °C or above. This is achieved by steam heaters at the tank discharge and after the main pressure pumps. The fuel lines are also warmed by steam or electric trace heating to overcome any heat loss. Prior to start-up the oil is retuned via valve (4) in the recirculation line to prevent it cooling and thickening the fuel is continually circulated to the burners via a recirculation system. The latter process is sometimes referred to as 'spill-back'. When a burner is not firing, the oil circulates through the pipework right up to the shut-off valve, which is mounted as close to the oil gun as possible.

From the point of the C& I engineer, the control systems involved with oil firing may include any or all of the following: controlling the temperature of the fuel, the pressure of the atomising medium and the equalisation of the fuel pressures at various levels on the burner front.

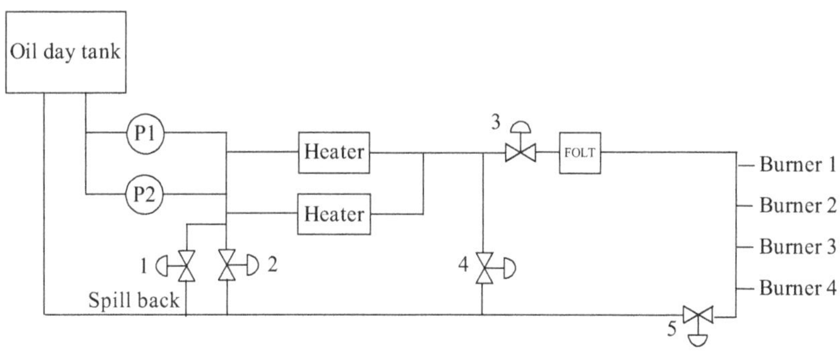

Figure 3.22 Simplified oil pumping, heating system

3.3.1.7 Gas-firing systems

Although inherently simpler than either oil- or coal-fired systems, gas-fired boilers have their own complexities. Any escape of gas, particularly into confined areas, presents considerable hazards, and great care must therefore be taken to guard against leakage, for example, from flanges and through valves. But natural gas is colourless, and any escape will therefore be invisible. Also, it is not safe to rely on odour to detect leakages. By the time an odour has been detected sufficient gas may have already escaped to present a hazard. It is therefore necessary for gas leak detectors to be fitted along the inlet pipework wherever leakage could occur and to connect these to a comprehensive, central alarm system.

It is also necessary to prevent gas from seeping into the combustion chamber through leaking valves. If gas does enter undetected into the furnace during a shutdown period, it could collect in sufficient quantities to be ignited either by an accidental spark or when a burner is ignited. The resulting explosion would almost certainly cause major damage and could endanger lives. (It should be noted that this risk is present with propane igniters such as those used with fuels other than oil.)

Protection against leakage into the furnace through the fuel supply valves is achieved by the use of 'double-block-and-bleed' valve assemblies which provide a secure seal between the gas inlet and the furnace. The operation of this system (see Figure 3.23) is that before a burner is ignited both block valves are closed and the vent is open. In this condition any gas which may occupy the volume between the two block valves is vented to a safe place and it can therefore never develop enough pressure to leak past the second block valve. When start-up of the burner is required, a sequence of operations opens the block valves in such a way that gas is admitted to the burner and ignited safely.

Many boilers have undergone a conversion from coal to gas firing, generally for economic reasons. In the USA, for example, there has been an unprecedented growth in the supply of natural gas that has become available from hydraulic fracturing, or fracking, of shale deposits. Natural gas produces lower CO_2 emissions than coal due to its high hydrogen component that combines mostly with oxygen to form water. Environmental challenges exist including avoiding groundwater contamination and leakage (natural gas has many times the effect of CO_2 as a greenhouse gas).

In addition to the substantial burner, pipework, isolation and leak detection systems that must be installed, there are strict requirements for instrumentation and protection logic described in codes and standards, which in some jurisdictions are mandated and in others considered best practice. These include the relevant gas sections of NFPA 85 and the safety instrumented systems standard IEC 61511 and its US equivalent ANSI/ISA-84.00.01-2004 Parts 1-3 (IEC 61511 Mod).

Conversion to natural gas will also result in changes to the flame shape and temperature as well as air/fuel ratio. This will modify the heat distribution in the furnace and gas pass zones, affecting start-up procedures, steam pressure and temperature control, excess O_2 control and operation of selective catalytic reactors used for NOx control.

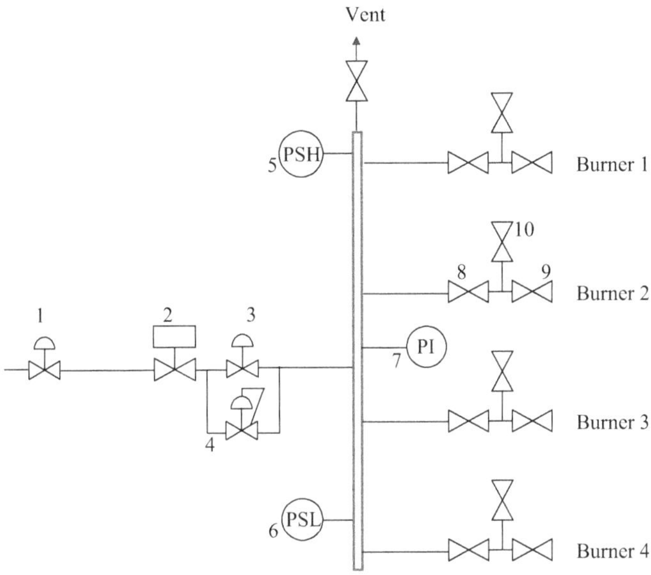

Figure 3.23 *Simplified schematic of gas-firing system. Note that the gas system includes a vent valve on the main header that is opened to during initial start-up of the first burner*

3.3.2 Waste-to-energy plants

3.3.2.1 Overview

There has been a steady development of plants that incinerate waste material of various types and use the heat thus produced to generate electricity. Early units suffered from the unpredictable nature of the waste material and the severe corrosion resulting from the release of acidic compounds during the combustion process. But the problems have been largely overcome through the application of improved combustion systems and by better knowledge of the materials used in the construction of the plant.

Waste material may be obtained from any of several sources, including the following:

- municipal
- industrial
- clinical
- agricultural

The material may be burned after very basic treatment (shredding, etc.) or it may be processed in some way, in which case the end result is termed RDF.

Several types of waste-to-energy plant are in existence, and we shall look at two of them, so that its nature and characteristics can be appreciated. Other plants will differ in their construction or technology.

3.3.2.2 The bubbling fluidised-bed boiler

Early waste-to-energy plant were based on the use of a bubbling fluidised-bed boiler. First, the incoming waste is sorted to remove oversized, bulky or dangerous material. The remainder is then carried by a system of conveyors to a hammer mill where it is broken down until only manageable fragments remain. After a separator has removed incombustible magnetic items, the waste is held in a storage building, from where it is removed as required by a screw conveyor and transferred via another conveyor to the boiler. Immediately before entering the boiler, nonferrous metals are removed by a separator.

The boiler itself comprises a volume of sand which is kept in a fluidised state by jets of air. A portion of dolomite is added to the sand to assist in the reduction of corrosion and to reduce any tendency of the sand and fuel to coalesce (a process known as 'slagging'). After the sand/dolomite mixture has been heated by a system of start-up burners, combustible waste material added to it ignites. The heat released is used to generate steam in a way that is similar to conventional boilers such as those described in Chapter 2.

3.3.2.3 Circulating fluidised boilers

Larger and more modern waste to energy plants are based on circulating fluidised beds where the bed of sand and refuse is circulated around the furnace. The refuse is broken down as described above. This type of boiler is also used to burn anthracite, biomass or waste coals (see Figure 3.24). Limestone and coal are used to form the circulating bed. Limestone helps to reduce the sulphur content of the flue gas. As with conventional two-pass wall-fired boilers, fuel oil is used during the start-up phase. Primary air enters the base of the furnace and carries the limestone and coal mixture through the furnace where secondary air is added to complete the combustion. The circulating fluidised bed leaves the furnace and enters the cyclone, where the bed is separated and returned to the furnace while the flue gas enters the back pass and heats the steam as in a conventional boiler. The fluidised bed exiting the cyclone is shared between the furnace and the fluidised bed heat exchanger (FBHE). This is achieved using a Spiess valve. The volume passing via the fluidised bed heat exchanger is adjusted by the steam temperature control system.

With the exception of using the FBHE, the steam and water circuits are similar to a conventional boiler. NFPA 85 includes a chapter on fluidised beds but does not go into the detail of how the bed is mixed, nor the use of the FBHE. These are considered proprietary information and are outside the scope of this book. While the flow diagram shows a single cyclone, larger boilers may have four cyclones in parallel. This technology is not limited to drum boilers and has also been applied to once through boilers.

3.3.2.4 Biomass systems

Early attempts at biomass firing added low levels of biomass, up to 5%, with the coal before the mill and the biomass was crushed with the coal. Alternatively, for

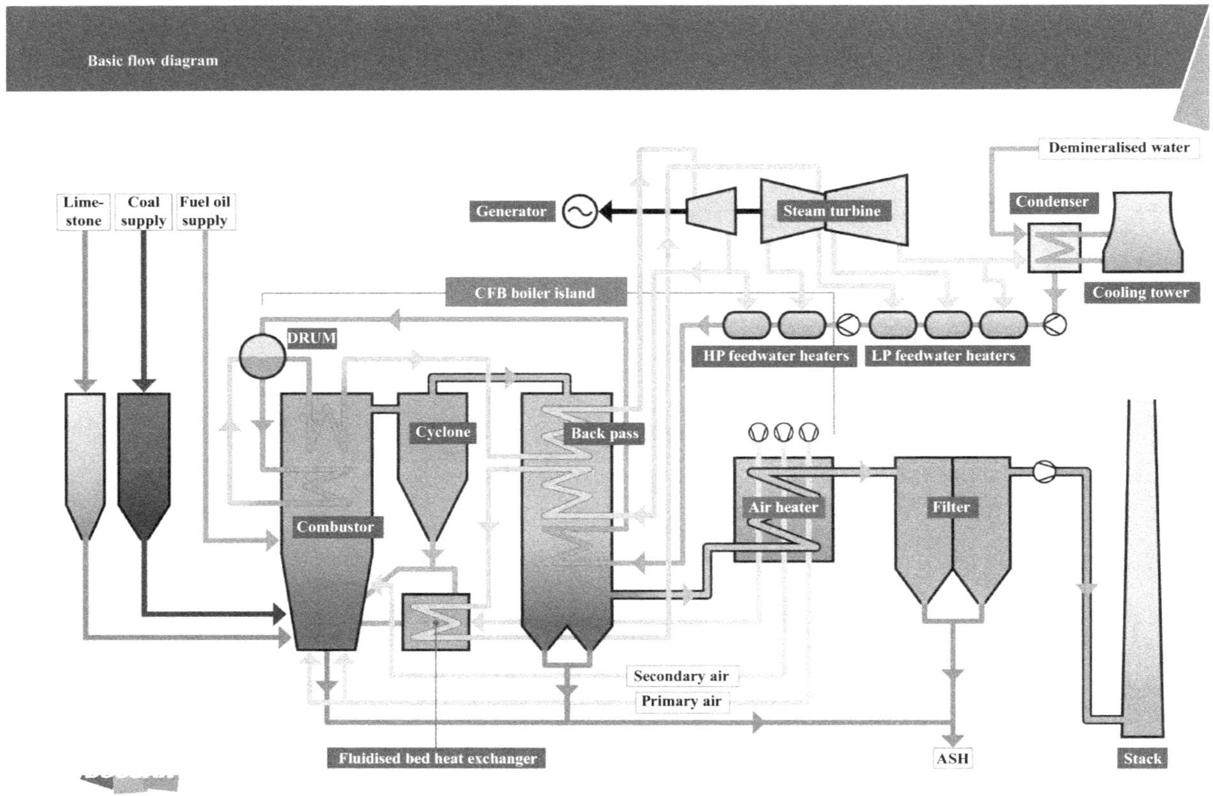

Figure 3.24 Basic flow diagram of CFB. © Doosan Lentjes, reproduced with permission

higher levels of co-firing, up to 15%, the biomass could be milled separately and then added into the PF lines between the mill and the burner. They had a true economic benefit of being able to burn unwanted vegetable waste. More recently environmentalists have agreed that burning wood is better than burning fossil fuels as it is considered carbon neutral, and some governments have offered financial incentives for operators to retrofit to 100% biomass. To reach such a level on a utility boiler, wood must be used and for reasons of transport and to be able to use existing mills, the wood dust is pelletised. Most of the plant and control logic described throughout this book is applicable to both coal and biomass fired boilers. In particular fuel air ratio, oxygen trim, OFA, furnace pressure, drum level, spray water control, and master pressure control are the same for coal and biomass. At the implementation stage, while the logic drawings may look the same, actual air to fuel ratios, etc., will change. The principal differences in approach between coal and wood pellets are summarised next.

As wood is lighter than coal, a retrofit may be limited by the actual volume of fuel that can be brought on to the site. As a kilogramme of wood has a lower energy for combustion, or calorific value (CV) than a kilogramme of coal, then even if the same mass throughput can be achieved, the overall MWs that can be produced is usually reduced or spare mills (available for use in the case of breakdowns when firing coal) must be pressed into service for full load on biomass.

Transporting is usually by enclosed lorries or railway trucks to avoid being wetted by rain. They may be emptied by tipping or pneumatic conveying systems. When the pellets are moved during the conveying process very fine wood dust is produced that ignites very easily. Because of this explosion risk much of the transfer system will be classified as a hazardous area. The system may be divided into sections by rotating valves and screw conveyors. These reduce the risk of an explosion or fire propagating through the system. Tramp metal is removed by magnetic separators along the conveyor route and this is essential to avoid ignition sources from sparks.

Wood pellets are supplied at about 5% moisture; if they get wet they lose their strength and easily turn to dust or worse, porridge, so wood pellets are stored under cover. Steam inerting is inappropriate for biomass mills and to avoid the danger of dust explosions on start-up, when the mill operates in the explosible region, explosion suppression systems are used. As it is easier to start a wood fire than a coal fire, there is greater risk of the stockpile being destroyed, additional monitoring and firefighting is provided. Carbon monoxide and temperature monitoring is used on the silos, as with coal silos. Additionally, explosion vents and inert gas injection may be fitted.

Wood pellets can be milled in a conventional vertical spindle mill with some modifications to increase local velocities, eliminate fuel hold ups where it could start to smoulder. Changes to the classifier are needed to accommodate the larger biomass particles compared to coal. PA is used in a similar manner to when using coal, but with a higher flow rate to lift the larger particles and a lower temperature to avoid fuel ignition.

Alternatively, hammer mills can be used to mill biomass. There are two distinct types of hammer mills used in the power industry. Where boilers fire brown coal the coal is also broken up using a hammer mill, but this type of mill is more like a centrifugal fan that draws the coal through it and chops it up with its blades.

These blades wear quickly and brown coal-fired boilers will usually have two spare mills and a team continuously replacing the worn blades. By contrast the hammer mills used for biomass have a say a hundred small blades that are spun on a shaft and hit the wood as it passes by. The blades are just metal rectangles that impact the biomass. The edges wear quickly and from time to time the direction of rotation is changed so the opposite edge wears. In some designs it is necessary to replace all of the hammers every few days, although more robust designs are available at a price. When a hammer mill is used the biomass is fed by a rotating valve or screw conveyor into the PA system, rather than the PA passing through the mill.

Wood dust has a lower ignition temperature and so is more likely to explode than coal dust and this results in all fuel transfer stations being classified as hazardous areas. Once again magnetic separators are used in the mill to remove any tramp metal. All systems where wood dust may be present must either be designed for a dust explosion over pressure of ~13 barg or incorporate safe venting via rupture discs or explosion suppression systems which rapidly detect the very incipient pressure rise at the start of an explosion and inject an inert powder to cool the wood dust below its ignition temperature. This can react within 30 ms. These are specialist systems and not designed by the boiler C&I engineer. If rupture discs are used operators are not allowed in the area in the path of the explosion vent.

Normally there is no change to the spray water control logic, although biomass may result in a lower steam temperature. Many conversions to biomass have been able to accommodate this. If it were a problem on a particular unit then it could be countered by adding gas recycling, to the furnace design, to increase the steam temperature. See Chapter 7 for gas recycling control logic.

Some budget-end biomass conversions have merely modified burners to reduce the fuel inlet velocity and these have suffered with biomass flame stability, particularly at low loads. Sometimes the flame may require oil support to achieve the required turndown. The instability also makes flame monitoring more difficult. Even with bespoke burners providing a stable flame, replacement flame monitors are required because of the different flame signatures between wood and coal. Most owners will only use flame monitors that have been proven to work reliably when detecting wood flames.

Wood pellets have lower inherent moisture than coal and ignite at a lower temperature. Because of this the PA entering the mill or used to transport the biomass needs to be at a much lower temperature than for coal. On a new boiler PA ducts and the air heater could be designed to allow for this but on a retrofit it is necessary to cool the PA before it is used in the mill. The PA is cooled either by cold tempering air as with a coal-fired boiler or by inserting cooling coils into the duct. Feedwater is passed through these coils instead of one of the LP feed heaters. This saves steam bled from the turbine and reduces the boiler flue gas exit temperature, leading to a greater station efficiency than by using tempering air.

The control logic for the cooler varies between suppliers and depends on the physical system configuration. In all systems there are two principal loops, one controls the PA temperature and the other the water temperature. Precautions are needed to prevent the water in the cooler steaming.

3.4 Igniter systems

Whatever the main fuel of the boiler may be, it is necessary to provide some means of igniting it. A variety of igniters are used, but most modern systems comprise a means of generating a high-energy electric spark which lights a gas or light-oil supply which in turn lights the main fuel.

In addition to igniting the fuel, the igniter may sometimes be used to ensure that the fuel remains alight under conditions where it may otherwise be extinguished. This is referred to as providing 'support' for the main burner.

Like many aspects of power station burner operations, the requirements for igniters are defined in standards such as those developed by the National Fire Protection Association (e.g. NFPA 85 [1]). In these standards, igniters are divided into three categories, each of which is defined in detail. In essence, the three classes have the following characteristics.

A class 1 ignitor is guaranteed to always support combustion of the main burner. An oil burner with a thermal input greater than 10% of the coal burner usually satisfies this requirement. For the control engineer the 'guaranteed to support' is particularly important as it means that if, when you start the first mill, the coal flames go out, you can restart the mill without tripping the boiler. If you did not have a class 1 igniter (you have a class 2 igniter) you would need to trip the boiler and re-purge the furnace.

Sometimes there is confusion between the titles of burner and igniter. In a normal burner start-up sequence, a class 3 high-energy spark igniter may be used to light the oil burner. An analogy is using the spark ignition on a gas cooker. This source is intermittent and less than 4% of the energy of flame it is trying to light (the oil burner). Once the oil burner is lit, it can have two duties: to warm the furnace, ignite and stabilise the coal flame, or to warm the furnace, ignite and stabilise the coal flame *and* actually carry load. If the latter, the total oil thermal capacity may be up to 100% of the coal capacity.

Once lit this oil burner is used to light the coal burner. Now it is treated as an igniter and depending on its thermal capacity relative to the coal it is class 2 if between 4% and 10% and class 1 if above that rating. The igniter classification also affects the need for a 'dark check' in the burner management system (BMS). The dark check requirement needs careful reading, and you must refer to the current NFPA 85 code for clarification. Gas igniters may be used instead of oil.

The type of igniter in use will define the methods of operation of the burner and the sequences that are to be employed in the associated BMS.

3.5 Burner management systems

Safe operation of the burner and its associated igniter must be ensured, and in most cases, this requires the use of a sophisticated BMS. In outline, these systems include a means of monitoring the presence of the flame and a reliable method and procedure for operating the associated fuel valves in a sequence that provides safe ignition at start-up and safe shutdown, either in the event of a fault or in response to an operator command.

This section introduces the main codes and function of the BMS. It is not intended to be a detailed review, and BMS should only be designed by competent engineers.

The BMS supervises the operation of the burners; its exact requirements are determined by the applicable international standard. The most popular is NFPA 85 and is applied in the USA, South America, China, India, South Korea, Australia and Africa. In Europe, the governing standards are EN12952. Part 8; Requirements for firing systems for liquids and gaseous fuels for the boiler, and Part 9 Requirements for firing systems for pulverized solid fuels for the boiler.

Many European plant owners ask for general compliance with NFPA 85 and EN12952 but accept some variations where they conflict. Applicable on all continents is IEC61511: Functional safety – safety instrumented systems for the process industry. This does not directly refer to BMSs but deals specifically with safety instrumented design and the maintenance of its integrity over the life cycle of the system being protected. See Chapter 10 for some of the procedures associated with IEC61511.

NFPA 85 is prescriptive and very descriptive. Some boiler makers also use it in as their in-house standard as it has more detail than EN12952. Generally following NFPA 85 will also comply with EN12952. However, there are some significant differences. The main ones are:

1. NFPA demands one local oil burner safety shut-off valve per burner plus a master shut-off valve for the boiler. EN 12952 demands duplicate shut-off valves for each oil burner. Both codes require double isolation for gas burners.
2. NFPA requires a formal leak test before firing the boiler. EN 12952 required no automatic leak test, but you must be able to test the valves when required.

   One interpretation is that NFPA 85 relies on single isolation, having proved that it works. Also, if it fails or there is a master fuel trip (MFT) the master shut of valve is closed providing double isolation.
3. NFPA 85 demands a MFT relay panel. A group of hardwired relays connected directly to burner shut-off valves, feeder, PA fans, etc., that are initiated either by the BMS or the operator to ensure that the fuel is properly isolated. EN12952 does not ask for this but expects SIL-rated system as determined by following EN61511.
4. NFPA 85 specifically allows some measurements to be shared between the BMS and the DCS, including air flow measurements and furnace pressure. EN12952 does not specifically allow this, so the necessary duplication is required to satisfy the SIL rating as determined by following EN61511.
5. NFPA 85 prescribes that the purge to be carried out with a boiler air flow between 25% and 40% of Boiler Maximum Continuous Rating (BMCR). EN12952 allows the boiler designer to set these limits but specifies a minimum airflow of 20% BMCR if you want to extend the ignition period of 1st burner to

30 minutes following purge. It does also specify a 50% minimum position for the FGR damper.

6. The latest version of EN12952 expressly prohibits the use of tip recirculation burners.

The following description is based on NFPA 85 and assumes a coal-fired boiler with oil burners to ignite and stabilise the coal burners. The oil is lit by a high-energy spark igniter. (Defined as class 3 special by NFPA 85.)

3.5.1 Furnace purge and fuel oil leak tests

The procedure for lighting the first burner depends on checking that it is safe to light it at all. This means that, if no other burner is firing, confirmation has been received that any flammable mixtures have been exhausted from the furnace by means of a purge. Such a purge involves the operation of FD and ID fans for a minimum of five minutes, so that at least five volumes of air has passed through the furnace. On 800 MWe boilers this time is likely to be greater than 10 minutes.

3.5.1.1 Fuel oil leak test part 1

While the purge is in progress a bypass valve around the main safety shut-off valve is opened and used to pressurise the fuel header. The flow through this small valve is limited by a restriction orifice. If the header is not pressurised in a brief time a leak is assumed and the test fails. Once pressurised the leak test valve is closed and the pressure monitored. If it falls the leak test fails. Note some oil headers are fitted with accumulators to help prevent the pressure fall when starting a burner. These are effective but can also stop the pressure falling if there is a leak. Therefore, automatic isolation valves must be fitted to the accumulators so they can be isolated during the fuel oil leak tests (Figure 3.25).

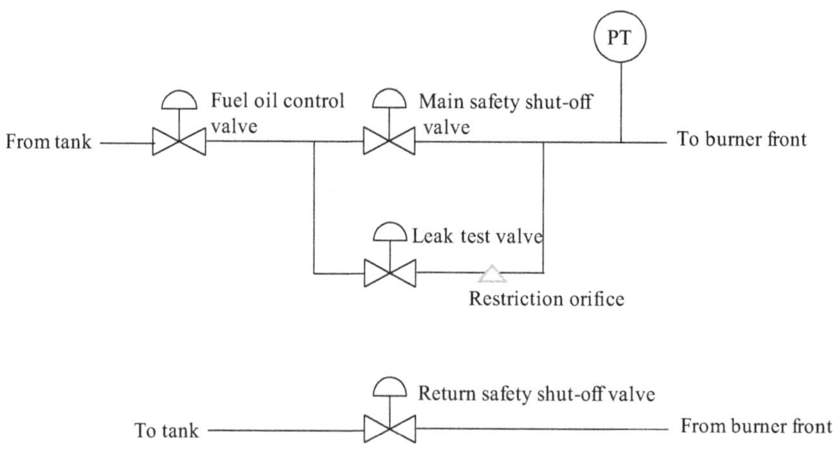

Figure 3.25 Fuel oil leak test arrangement

3.5.1.2 Fuel oil leak test part 2

On completion of the purge the second stage of the leak test is initiated. This is to check that the main safety shut-off valve itself is not leaking as this could mask any leak on the system. The return safety shut-off valve is part opened and closed again. This drops the header pressure. If the pressure subsequently rises the main safety shut-off valve or the leak test valve is assumed to be leaking. This test follows the purge as NFPA 85 does not allow the shut-off valves to be opened during the purge.

3.5.2 Starting the oil burners

Once confirmation has been received that the furnace purge is complete and MFT is reset, the oil system is put into service. Ignition of the first oil burner will depend on the successful operation of the spark igniter and, once the oil burner has been successfully lit, its operation must be continuously monitored because an extinguished flame may mean that unburned fuel is being injected into the combustion chamber. If such fuel is subsequently ignited it may explode.

During this start-up the burner goes through several stages as defined here.

		Oil burner running		
Stage	Oil burner ready	Oil burner starting	Oil burner firing	Oil burner purging
Definition	Burner start permit satisfied	Start initiated	End of trial for ignition and flame on	Burner purge initiated

Having lit the first oil burner additional oil burners are lit to warm through the boiler and raise steam pressure.

Once a burner has ignited, the BMS must ensure that safe operation continues, and if any hazard arises, the system must shut off the burner, and if necessary, trip the entire boiler.

If at least one other burner is firing then on shutdown of a burner steps must be taken to ensure that any unburned fuel is cleared from the burner pipework. This procedure is known as burner purging, and in an oil burner it may involve blowing compressed air or steam through the pipework and burner passages.

The individual oil burners may be started by the operator or started sequentially if the logic is provided. To start the first mill all of its oil burners must be firing. It is normal to include logic to group start all the oil burners associated with any mill. Note that for tube mills with single-ended capability only the burners for the end to be used are lit.

Safe operation of the boiler depends on proper design of the BMS, including the flame scanner and on careful siting of the scanner so that it provides reliable and unambiguous detection of the relevant flame under all operational conditions. After installation, the system can be expected to perform safely and reliably only if constant and meticulous attention is paid to maintenance. This important matter is all too often ignored, and the inevitable result is that the system malfunctions, leading to failure to ignite the fuel, which may in turn delay start-up of the boiler. In the extreme, malfunctions could even endanger the safety of the plant if they result in fuel being admitted to the combustion chamber without being properly ignited. A properly designed BMS will not allow this to happen, but if repeated malfunctions occur it is not unknown for operators to ignore the warning signs and even to override safety systems. In such cases it is usual to blame the BMS and/or the flame monitors, which could be fully functional if they were not misused or badly maintained. This important subject is discussed in greater depth in Chapter 5.

Each component of the BMS is vital to the safety of the plant and to the reliability of its operation, but the most onerous responsibility rests with the flame detector: an electronic device which is required to operate in close proximity to high-energy spark ignition systems, and in conditions of extreme heat and dirt. Moreover, it must provide a reliable indication of the presence or absence of a particular flame in the presence of many others and it must discriminate between the energy of the flame and high levels of radiant energy from hot refractory materials and pipes. The sighting of the flame may also be affected by changes in flame pattern over a wide range of operating conditions, and it may also be obscured by swirling smoke, steam or dust.

3.5.3 Starting and stopping a mill

3.5.3.1 BMS operation during mill start

Typically, oil burners that are used as a support fuel to PF supply about 30% BMCR. Occasionally they will be able to provide 70% or 100% BMCR if the end user is committed to providing power and sometimes limited to only 15% BMCR. Before their limit is reached a mill must be started. The mill sequence logic was traditionally included with the BMS logic, but is now more likely to be in a separate processor. The start-up sequence includes starting lube and hydraulic oil pumps, establishing a minimum PA flow, warming the mill, starting the mill motor and classifier and finally starting the coal feeder.

As soon as the classifier outlet valves are opened to permit the flow of PA there is a risk of coal being carried into the furnace and subsequently exploding. The BMS prevents this by ensuring that all the oil burners for that mill are firing before the classifier valves are opened. The oil burners are kept in service until the feeder is started and the coal flame is established as proven by the coal flame monitors plus a short stabilisation time.

Once the oil burners are shutdown the BMS continues to monitor the coal flame monitors, and if any flame fails, the BMS initiates a trip of that mill.

3.5.3.2 BMS operation during mill shutdown

The safest way to shut down a mill is to introduce oil burners, reduce the mill load and temperature to a minimum and then trip the feeder. Leave the mill motor and classifier running until the mill is empty and then close the classifier valves and stop these motors.

A mill trip, which requires all PF flow to the furnace to be stopped, closes the classifier outlet valves and trips the motors leaving the mill full of coal. The mill is inerted to prevent a fire and subsequently emptied either into the furnace or via the pyrities system.

The mill logic described is the safest, but may cause unnecessary trips. NFPA 85 allows some variation in tripping scenarios where agreed between the boiler maker and the mill supplier. For example, if firing at high loads a mill maybe allowed to continue with one flame monitor showing a flame failure. This is because the total heat in the furnace will ensure that all the PF is burnt and therefore this is treated as an instrument failure and not a flame failure.

3.5.4 Fuel oil trip

If after the last mill is started and then when the oil burners are shutdown, a burner shut-off valve fails to close, this will initiate a trip of all oil burners and the main oil safety shut-off valves. This is fuel oil trip. It does not trip the running mills. However, if only oil is being fired it results in a MFT.

3.5.5 Master fuel trip

A MFT can be initiated by about 20 different conditions. The major trips relating to potential furnace explosion are loss of all flames or sufficient to cause instability, high furnace pressure or low combustion air flow. EN12952 also asks for a trip on low air to fuel ratio. While high or low drum level or low feed flow on a once through boiler will lead to tube damage. Although not an NFPA requirement some operators also ask for a MFT on high superheater or high reheater steam temperature.

Other trips include high or low fuel pressure, and maloperation of fuel valves. Some operators ask for a MFT on high superheater or reheater temperature.

If any of these occur the MFT logic and MFT relays, where provided, trip all fuel supplies and isolate them from the furnace. Following the trip many more potential causes may be displayed to the operator. For example, the flames will have failed. NFPA 85 specifies a first up alarm system to identify the cause of the MFT.

3.5.6 Boiler trip

The MFT does *not* trip the FD and ID fans. However, if after the MFT the furnace pressure continues to increase or decrease then both FD fans and one ID fan are tripped. This is known as a boiler trip.

Note that while this is the NFPA definition it is common practice for others to use boiler trip to mean an MFT. Such ambiguities need to be clarified in the BMS specification.

3.5.7 Special considerations for external plant

In addition to the boiler proper the control engineer must also consider external plant such as the flue gas desulphurisation (FGD) unit. While this has no interaction with boiler combustion, temperature control and drum level, there is a physical interaction in the flue gas flow path and so it can interact with the furnace pressure control. For example, if the temperature is too high for the rubber-lined vessels then the FGD process is tripped, stopping the flow of exhaust gas. Either a bypass must be opened or the boiler tripped.

3.5.8 Special considerations for downshot or W-fired boilers

During start-up of a downshot boiler the heat from the shoulder-mounted oil burners might potentially cause high steam temperatures in the pendant superheater sections. This is prevented by fitting start-up oil burners at a lower level in the wall of the furnace, as was seen in Figure 3.3. The BMS logic for these extra burners requires great care as the secondary air is directed either to these or to the shoulder-mounted burners but not both.

NFPA 85 applies to all boiler types but NFPA 85 Chapter 9 (9.1.1), which refers back to other chapters, suggests strictly the application of the code is only applicable to fuels with >8% volatiles on a dry basis, which would exclude many anthracitic fuels used in downshots. In practice, with the absence of specific alternative specifications most owners apply the NFPA code.

Ideally downshot flames form the classic W shape; however, if one flame is longer, the pattern changes and the combustion 'hot spot' moves in the flame. This makes flame detection very difficult and individual flame monitors may show a flame-out, while its adjacent monitor continues to show a stable flame. For this reason, the BMS flame-fail logic may be configured to look at a pair of burners from each mill end.

3.6 Ancillary systems

There are several subsystems to a boiler that often come complete with their own Motor Control Centre (MCC) and Programmable Logic Controller (PLC); from a C&I perspective these are sometimes called 'black boxes' or just 'stand–alone' systems. In designing the overall control system architecture, the EPC contractor needs to know how many inputs and outputs are required to connect these systems to the main DCS and what electrical load each requires.

In many cases, it is possible to completely control these systems from the DCS; however, this would involve extra cabling, more expensive I/O, special programming and present difficulties in Factory Acceptance Test and commissioning. The downside is that the power station staff now need to look after the DCS and potentially a range of PLCs. The end user will often limit the range of PLCs to, say, Siemens and Allen Bradley. Often the logic is sequential and repetitive so best suited to a PLC. It is expected that the trend to total integration in the DCS will continue.

Use a PLC if	Because
Predominately digital I/O	Cheaper than DCS.
Sequential control	May be easier in a PLC depending on sequence capability of the DCS. Will be faster in the PLC.
Needs to be factory tested as a package	All in one location.
Contains proprietary algorithms	The PLC allows the supplier to embed and protect his IP.
Specialised scope	Using SAMA to control the movement of a water cannon is a challenge!

- Bottom ash systems
  A coal-fired boiler produces two types of ash, some falls to the bottom of the furnace called bottom ash and some is carried away from the furnace and collected in the precipitator or bag house called fly ash.
  The bottom ash system comprises a hopper at the bottom of the furnace that can collect say eight (or 24) hours of ash, this sits above a conveyor of some sort that carries the ash away, usually then up an incline and into a crusher. The crushed ash is the taken away by lorry or conveyed to an ash lagoon. There are two types of bottom ash system: the traditional wet system and a more modern dry ash system.
  With the wet system, large chunks of ash drop into the water where they are quenched. The heat transferred to the water is lost as the water is recycled. The wet system operates independently of the boiler. That is, it can be turned off for a shift, and then run to remove the ash. It is usually supplied with a local MCC, PLC, and operator panel.
  The dry bottom ash system is considered to have many advantages. The most obvious is that it does not use water – removing the need for a complex water supply and clean-up system. As the ash is dry it can continue to burn, thereby reducing its carbon content and returning heat to the furnace. The dry bottom ash system was invented by Magaldi in 1985, and there are now over 200 installations. This Magaldi dry bottom ash system improves the boiler efficiency between 0.1% and 0.6% or higher depending on the coal properties.
  The dry ash may be resold while the wet ash is often directed to a dedicated settling pond as part of a reclamation process.
- Sootblowers
  Each sootblower as is supplied as a self-contained unit with its own motor and limit switches. A large coal-fired boiler may have a hundred sootblowers. These are sequenced from a PLC, the logic for each sootblower is similar, for example, insert the sootblower until maximum limit reached, then retract sootblower. The steam is automatically admitted by a poppet valve as the sootblower moves and no poppet valve control logic is required. A simple pressure control loop is required to reduce the steam pressure to that required. If a sootblower optimiser is used then then the pressure set point might vary for

different types of sootblowers. During start-up, the lines are warmed through and soot blowing only allowed when they are warm enough and the boiler load is stable enough to avoid the sootblowers disturbing the combustion. The sootblowers are grouped in sections and the steam to each section controlled by an on–off section valve. Each section also has a trap and/or an operated drain valve. The section valves allow one section at a time to be heated saving steam and to speed up the initial warm up. The sootblowers are automatically withdrawn following an MFT.

The logic can be reproduced in a DCS if required, but is easily achieved in a PLC.

- Water cannon
  Water cannon may be used as an alternative to sootblowers to clean the furnace walls. As the name suggests each cannon fires a jet of water across the furnace when it hits the slag on the opposite wall the slag contracts and breaks off. Each cannon fires water at several areas. The heat flux is measured in each area and the area is only cleaned after slagging has been detected. There is some simple logic associated with the pump that pressurises the water. The system supplier typically provides a stand-alone digital control system with proprietary control application software.
- Electrostatic precipitators (ESP)
  There are usually two per boiler, one on each side. They consist of a number of plates suspended from the roof. The plates are connected to a high voltage and attract the dust. From time to time rappers strike the plates and the dust falls to hoppers. When the hoppers fill up they are emptied via a fly ash system. The C&I scope includes a PLC to control the rapper, some basic interlocks to prevent man entry unless the fields are off and earthed and bespoke controls to vary the field voltage depending on the dust burden.
- Bag filters
  Used in place of, or in addition to, ESP. It comprises many filters. Simple cleaning logic blows back each filter, in turn allowing the dust to drop into hoppers.
- Selective catalytic reactors
  Selective catalytic reactors (SCRs) are placed in the flue gas stream to reduce NOx emissions. The catalyst is activated by the injection of a mist of ammonia (NH_3). The flow of ammonia is determined by measurement of NOx before the SCR, the current steam flow and the desired NOx after the SCR. The unit DCS typically provides the ammonia injection control for this system.
- HP & LP bypass systems
  When used to assist start-up or to survive a load rejection these are often controlled directly from the DCS. Because of the high forces required the actuators are often hydraulic and the system is supplied with its own hydraulic pack. This contains duty and standby pumps, filters and hydraulic accumulators. However, if the bypass systems are used instead of safety valves then a separate safety instrumented system (SIS) is provided. Some suppliers also offer a complete package including a customised PLC. This enables them to embed their own control algorithms and protect their intellectual property (IP).

3.7 Gas turbines in combined-cycle applications

In the combined-cycle plant, the heat used for boiling the water and superheating the steam is obtained from the exhaust of a gas turbine, as described in Chapter 2. In such plant, unless supplementary firing is used, the combustion process occurs entirely in the gas turbine. Where supplementary firing is used the relevant control systems take on many of the characteristics of the oil-or-gas-firing systems discussed earlier.

3.8 Summary

So far, we have looked at the operation of the boiler and studied in outline the boiler's steam, water and gas circuits, and all the major items of plant required for their operation. With this understanding, we can now look at the control and instrumentation systems associated with the plant. This survey will be structured in much the same way as the preceding chapters, starting with an overview of an important fundamental: the method by which the demand for steam, heat or electrical power is obtained. Afterwards, we shall see how this demand is transmitted to all the relevant sections of the plant so that the requirements are properly and safely addressed.

Reference

[1] NFPA 85: Standard for the prevention of furnace explosions/ implosions in multiple burner boilers. National Fire Protection Association, Batterymarch Park, Quincy, MA, USA, 2015.

Chapter 4

Setting the demand for the steam generator

David Lindsley[1] *and Don Parker*[2]

4.1 Nature of the demand

The steam generated by the boiler may be used to drive a turbine in a thermal power plant or it may be delivered to an industrial process or a district-heating scheme (or it may be provided for a mixture of these uses). Alternatively, the primary purpose of the plant may be to incinerate industrial, domestic or clinical waste, with steam being generated as a valuable by-product, to drive a turbo-generator or to meet a heating demand. In each case, the factor that primarily determines the operation of the plant is the amount of steam that is required. Everything else is subsidiary to this, although it may be closely linked to it.

The determinant that controls all the boiler's operations is called the 'master demand'. In thermal power plant the steam is generated by burning fuel, and the master demand sets the burners firing at a rate that is commensurate with the steam production. This in turn requires the forced draught (FD) fans to deliver adequate air for the combustion of the fuel. The air input requires the products of combustion to be expelled from the combustion chamber by the induced draft (ID) fans, whose throughput is also related to the fuel and air flows. At the same time, water must be fed into the boiler to match the production of steam.

As stated previously, a boiler is a complex, multivariable, interactive process. Each of these parameters affects and is affected by all of the others.

The way in which the master demand operates is determined both by the general nature of the plant (is it a power station, an incinerator or a provider of process steam?) and also by the way in which the boiler is configured within the context of the overall plant (is there only one boiler meeting the demand, or are several combined?). The nature of the master demand system depends on the type of plant within which the boiler operates, and it is therefore necessary to examine it separately for each type of application. In the following

[1]Retired
[2]Provecta Process Automation LLC, USA

sections we shall deal with the master demand as used in the following classes of plant:

- power stations
- combined heat and power (CHP) plants
- waste-to-energy (WTE) plants

We shall see that although all of these require the boiler to be operated to generate steam each has its own requirements and constraints.

4.2 Setting the demand in power station applications

A boiler producing steam for an operating turbo-generator has to ensure that the machine continually delivers the required electrical energy to the load. With a combined-cycle gas turbine plant it is frequently the case that the power generated by the gas turbines is adjusted to meet the demand, with the steam turbine making use of all of the waste heat from the turbines.

With all types of power-generating plant, however, the requirement for generation will be set, directly or indirectly, by the grid control centre (or the 'central dispatcher'), and the amount of power that is generated will be related to the local or national demand and other available generation at that time.

In national networks, power stations are linked together to generate electrical power in concert with one another. Together they must meet a demand that is made up of the combined needs of all the users that are connected to the system (domestic, commercial, agricultural, industrial, etc.). The overall demand will vary from minute to minute and day to day in a way that is systematic or random, dictated by economic, operational and environmental factors. This pattern of use relates to the entire network, and the fact that a large number of power generators and users are linked via the network has little bearing on the overall demand, although the extreme peaks and troughs may well be smoothed out. The interlinking does, however, have operational implications. For example, a sudden failure of one generating plant will instantly throw an extra demand on the others.

In a cold or temperate climate the demand will be based predominately on the need for light, heat and motive power. In warmer climates and developed areas it will also be determined by the use of air-conditioning and, possibly, desalination plant (for drinking water production). In climates with both summer and winter extremes two demand peak periods will be experienced annually.

Figure 4.1 shows how the total electrical demand on the United Kingdom's grid system typically varies from hour to hour through the day, and from a warm summer day to a cold winter day. Clearly, in addition to being affected by normal working patterns, the demand is determined by the level of daylight and the ambient temperature, both of which follow basic systematic patterns but which may also fluctuate in a very sudden and unpredictable manner. Even television broadcasts can dramatically affect the demand for power; for example, during breaks in international sports matches the switching on of kettles and microwaves can impose a sudden, brief rise in demand.

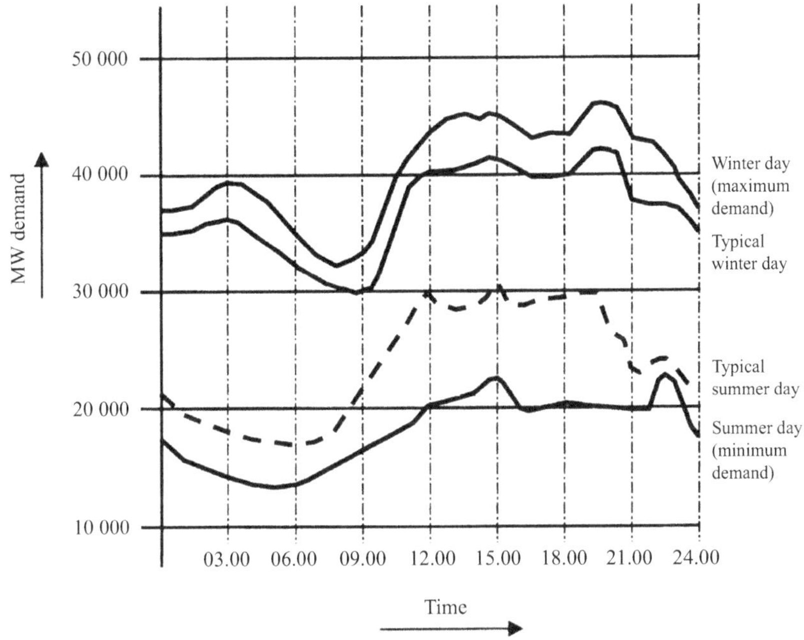

Figure 4.1 Typical electrical demand in the United Kingdom

Similar demand profiles can be developed for each country and will be determined by climate as well as the country's industrial and commercial infrastructure.

Today, an individual power plant's expected output demand relates far less closely to the total grid load demand than in the past. While in earlier years most demand fluctuation came from changes in consumer load, the output demanded from conventional, synchronous generating plant is also now being affected by the penetration of intermittent renewable generation such as wind and solar power, which can fluctuate dramatically and with limited predictability. Power plants are now required to vary over wider ranges and ramp more quickly in response to widespread changes to wind and cloud patterns. These changes are creating challenges for the design and performance of power plant controls, particularly for older 'base load' plant that were often designed for slow-ramping high-range operation.

4.2.1 Operation of the UK electricity trading system

The electrical grid system has to be managed so that the demand for electricity is met within statutory limits at all times and under all conditions, and the available generating plant has to be used in the most economic manner.

Since the privatisation of the electricity supply industry in the United Kingdom, the generation of electricity is based on the demands of a power trading system. Up until 2001, England and Wales participated in a trading system known as the Electricity Pool. The Pool, managed by the National Grid, took bids from generators who set the price of wholesale electricity via the system marginal price. The demand to each generator was set according to price ranking with appropriate security margins and issued by the National Grid Control Centre on a minute-by-minute basis.

The demand trading system has since been replaced by the British Electricity Trading and Transmission Arrangements (BETTA) which now includes Scotland and involves both generators and the transmission network suppliers trading in the wholesale electricity market.

4.2.2 Frequency response services

The grid's system frequency is a natural indication of the balance between electricity supply and demand. Large excursions in frequency can occur when a generator trips offline (frequency falls) or when many consumers are disconnected at once (frequency rises). The relationship between load balance and frequency is a result of the operation of synchronous rotating generators which always remain in lock-step with the system's cycles.

If the load demand exceeds supply, all of the synchronous machines will begin to slow, and the system frequency will fall. Some rotating kinetic energy will be released in the process which helps to provide the shortfall, but this will be quickly depleted unless additional energy support is provided. For steam turbines, the addition comes from the steam and water energy storage in the boiler, which can provide about 5% extra steam for up to around 40 seconds before the reduced steam pressure becomes significant.

By coordinating the opening of the turbine's valves with adjustments to the fuel, air and water supply to the boiler, additional power of around 3% in 6 seconds and 10% in 60 seconds can be supplied to the grid in times of large frequency events.

Generally, both steam and combustion turbine generators are required to provide a minimum level of frequency response, typically around 5% for a brief period. With the growth of non-synchronous generation (such as photovoltaic (PV) inverters and most wind generators), increased emphasis is being placed on additional frequency services which are provided on a commercial basis. These may include:

1. Frequency support
   - Fast raise and lower services: set output responses in proportion to a frequency change in 6–10 seconds, either to be sustained during the excursion or may be allowed to return after a period.
   - Slow raise and lower services: set output responses achieved over a longer period such as one minute.
2. Frequency control and restoration
   - Automatic generation control (AGC) frequency ancillary service: provision for the generator load to be raised above and below its bid output within an agreed band on a fast-update basis (such as each four seconds).

It is important to realise that during the transition to energy sources with reduced CO_2 emissions conventional fossil-fired generation is providing a vital role in maintaining grid security, at least until other major storage technologies such as hydro pumped storage, molten salt thermal solar and advanced large-scale battery systems are established. The capacity of boilers and heat-recovery steam generator (HRSG) systems to manage faster load ramps and responses to frequency events without sustaining life-reducing stresses or equipment damage is often influenced by the design and performance of the control system.

4.3 The master demand in a power station application

The output response of a boiler/turbine unit in a power station is determined by the dynamic characteristics of the two major items of plant. These differ quite significantly from each other. The turbine, in very general terms, is capable of responding more quickly than the boiler to changes in demand. The steam generation response of the boiler is determined by the thermal inertia of its steam and water circuits and by the characteristics of the fuel system. For example, a coal burning boiler, with its complex fuel-handling plant, will be much slower to respond to changes in demand than a gas-fired one.

Also, the turndown of the plant (the range of steam flows over which it will be capable of operating under automatic control) will depend on the type of fuel being burned, with gas-fired units being inherently capable of operating over a wider dynamic range than their coal-fired equivalents.

The common factor in all these systems, however, is the master demand which, in addition to setting the firing rate, regulates the quantity of combustion air to match the fuel input and the quantity of feed water to match the steam production. In this chapter we shall examine the master system in overview. Chapters 3, 5–7 look at how the commands from the master system are acted upon by the fuel, draught, feedwater and steam temperature systems. Chapter 11 then describes in detail how the demand to each system is configured to account for process dynamics and provide the coordination of the boiler and turbine processes for load ramping, frequency response and unloading during contingency events.

The design of the master system is determined by the role which the plant is expected to play, and here three options are available. The demand signal can be fed primarily to the turbine (boiler-following control); or to the boiler (turbine-following control); or it can be directed to both (coordinated unit control). Each of these results in a different performance of the unit in a manner that will now be analysed.

4.3.1 Boiler-following operation

With boiler-following control, the power demand signal modulates the turbine throttle valves (or are manually adjusted) to meet the load, while the boiler systems are modulated to keep the steam pressure constant. The principles of this system are illustrated in simplified form in Figure 4.2.

In such a system, the plant operates with the turbine throttle valves partly closed. Sudden load demand changes are met by opening the valves to release some of the stored energy within the boiler. When the demand falls, closing the valves

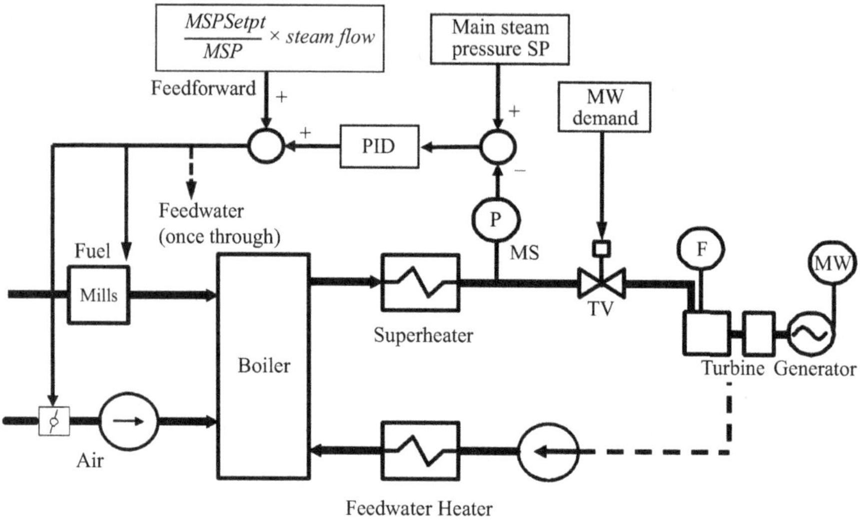

Figure 4.2 'Boiler-following' system

increases the stored energy in the boiler. The turbine is thus the first to respond, providing the required steam flow change. The boiler control system then reacts by increasing or reducing the firing to restore the steam pressure to the set value.

Since the boiler firing rate acts in response to the turbine's effect on steam pressure, and flow, there is no anticipatory signal available to the boiler to account for the inherent delay between fuel flow and steam production. The mode is therefore not suitable for fast ramping and is often only used in emergency situations such as during turbine protective unloading.

4.3.2 Turbine-following operation

In the turbine-following system (Figure 4.3), the demand is fed directly to the boiler and the turbine throttle valves are left to maintain a constant steam pressure. Particularly in the case of coal-fired plant, this method of operation offers slower response because the turbine output is adjusted only after the boiler has reacted to the changed demand. The mode is typically used for testing where constant fuel input is required or during fast boiler unloading conditions such as after loss of a major fan or feed pump.

However, by carefully calibrating additional fuel input during ramps to compensate for the slow boiler reaction, MW demand can be controlled with the boiler to a reasonable accuracy, leaving the turbine to either control pressure or simply remain wide open to achieve more efficient unit operation. This more advanced mode is often labelled 'Coordinated Turbine Follow'. The mode is worth considering for large base load power plant if close MW-following capability is not required, or with gas-fired plant where the fuel response is comparatively rapid. The design is discussed in more detail in Chapter 11.

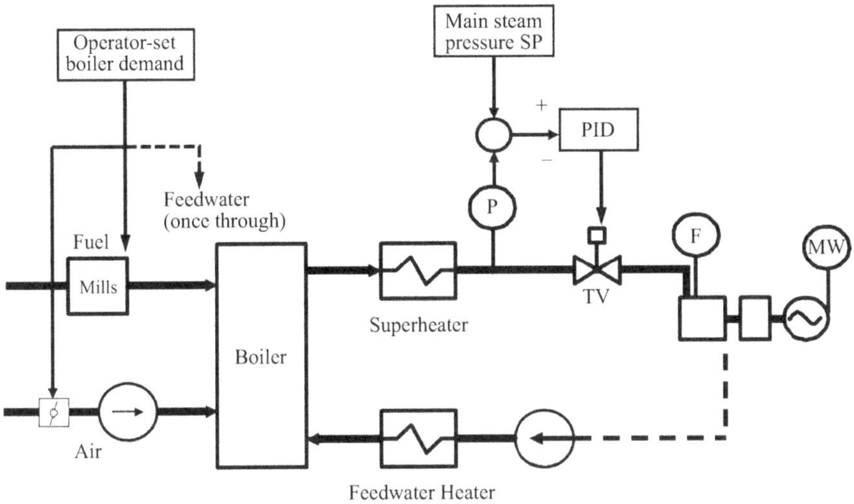

Figure 4.3 Basic 'turbine-following' system

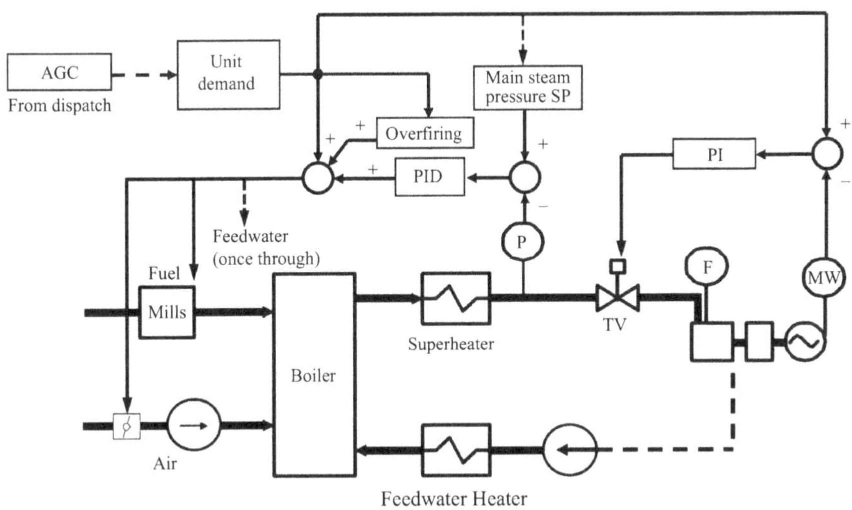

Figure 4.4 'Coordinated unit control' system

4.3.3 Coordinated unit control

With coordinated unit load control (Figure 4.4), the power demand is fed to the boiler and turbine in parallel, with various constraints built into each channel to recognise and allow for any dynamic limitations of the relevant plant. This is a sophisticated technique which, with the development of modern digital control systems, has become the 'standard' for normal load ramping operation. It combines the best features of both the boiler-following and the turbine-following systems.

However, its design demands considerable knowledge of the characteristics and limitations of the major plant items. Also, commissioning of this type of system demands great skill and care if the full extent of the benefits is to be obtained. Further details of design variations and commissioning and tuning principles are described in Chapter 8.

4.3.4 Relative performance

Of the three options described earlier, the coordinated unit load control system provides the best possible response to changes in demand, within the constraints of the plant itself. Transient components ('kickers') are added to the demand for fuel and air which are both carefully calibrated at several load points, while steam pressure setpoint may be also varied to make use of energy storage during ramps. The mode also provides the best response to system frequency deviations as a frequency-based dynamic bias can be applied to both the fast MW controller and the boiler demand signal. A simulation trend of unit ramping in coordinated mode is provided in Figure 4.5, where MWD refers to the MW demand signal.

While this mode provides the best ramping capability, its performance is heavily dependent on the accuracy of the many pieces of information on which its operation is based. Deterioration in performance of items of plant such as fans, pumps and coal mills can affect the accuracy of anticipatory signals and reduce the ramping capability over time.

Providing robust designs to enable coordinated control systems to accommodate such deterioration has been a challenge many control vendors have been addressing through model-based and supervisory control schemes (often referred to as optimisers).

4.3.4.1 Response of the boiler-following system

When a change occurs in its steam-flow demand, a boiler must overcome its own thermal inertia before it can match the change. Therefore, by using the turbine's ability to respond more rapidly, the boiler-following system provides a better response to load changes than the turbine-following system. After the turbine has responded to the change in demand, the boiler is commanded to follow on, correcting the steam pressure error as quickly as it can.

Load changes invariably generate pressure variations as the boiler is continually responding to match steam requirements, and so the mode is not commonly used for fast ramping or AGC operation. Boiler-following system can provide a reasonable frequency response as the boiler will restore any pressure loss from fast turbine responses to system frequency deviations. A simulation trend of unit responding to fast turbine valve changes in boiler-following mode is provided in Figure 4.5.

4.3.4.2 Response of the turbine-following system

In the simplest version of the turbine-following system the boiler-firing rate and the rate of air and feedwater admission, etc., are all fixed (or, at least, held at a set value, which may be adjusted from time to time by the boiler operator), and the

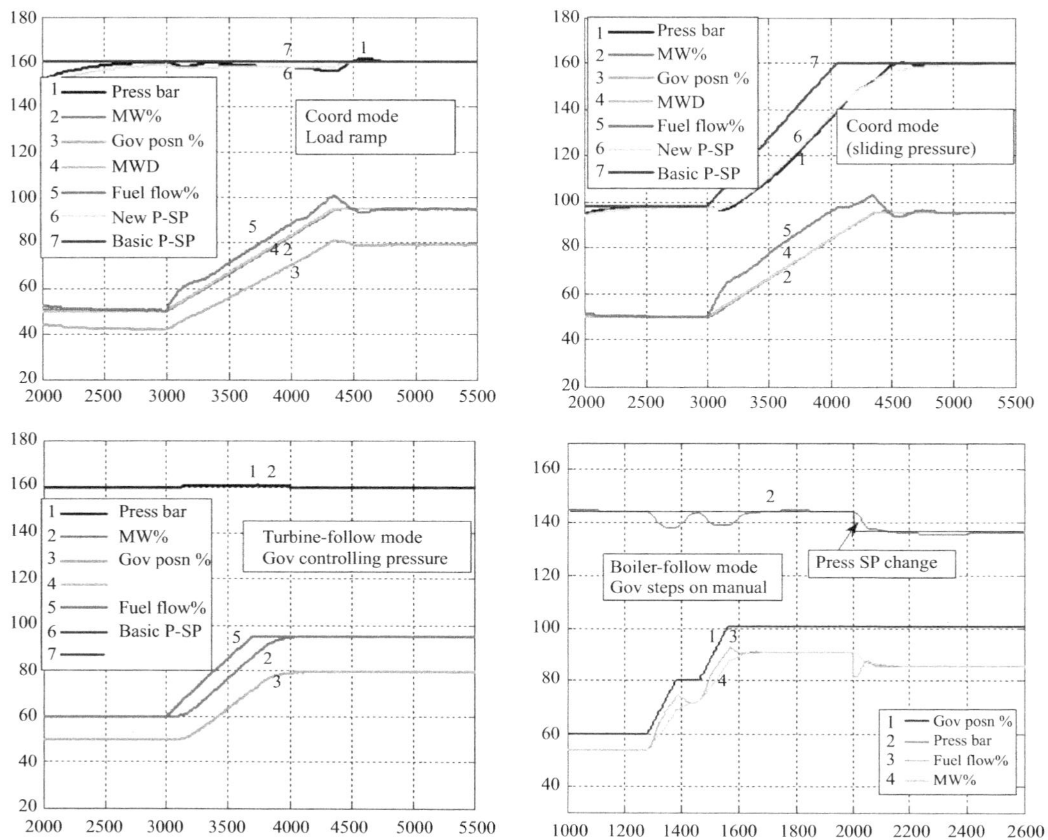

Figure 4.5 Simulations of load ramps in each control mode

turbine throttle valves are modulated to keep the steam pressure constant. However, when boiler demand is held at a constant value the amount of steam that is generated will not, in general, remain constant, mainly because of the inevitable variations that will occur in parameters such as the calorific value of the fuel, the temperature of the feed water, etc. In the simple turbine-following system, the resulting steam pressure variations are corrected by modulation of the turbine throttle valve, but this results in variations in the power generated by the turbine.

Since the load demand to the unit in this mode is only directed to the boiler, changes in load demand adjust the boiler's firing rate, relying on the resulting changes in pressure to produce a correcting adjustment by the throttle valve and so change the amount of power that is generated. As might be expected, because of the slow reaction time of the boiler, this results in a slower response to load changes than that of the boiler-following system. A simulation trend of a unit responding to a fuel demand ramp in turbine-following mode is provided in Figure 4.5 (bottom-left graph).

Because the steam generation rate of its boiler is not automatically adjusted to meet an external demand, the power generated from plant operating in turbine-following mode will not relate to the short-term needs of the grid system. Such a plant is therefore incapable of operating in a frequency-support mode, although this mode of operation may be used where it is not easy, or desirable, to adjust the fuel input, for instance, in industrial waste incineration plants. Indeed, if the turbine valves respond to a frequency event through the speed governing system, the loss of steam pressure will not only direct the turbine demand to reduce, but to fall *below* the initial output in order to restore steam pressure. The mode can therefore have a detrimental effect on system recovery from frequency events.

4.4 Load demand in CHP plants

Reference has already been made to the use of gas turbines in combined-cycle installations. This is a particular example of a 'co-generation' scheme: a term applied to dual-purpose plants where heat which would otherwise be wasted from one process is used in another. In the case of combined-cycle gas turbine CCGT plant, the heat exhausted from a gas turbine is used to generate steam. In CHP plant, heat from a power station is used in another process. The heat may be taken from the power plant as steam extracted from the turbine or it may be the heat abstracted from the condensate. While the latter may be considered to only have 'low-grade' heating value, being typically around 40 °C, it can prove very useful for space heating support in cold climates.

Co-generation plants are either 'topping' or 'bottoming' systems. With the former, the first priority is to generate electricity, and as much use as possible is made of the heat that would otherwise be wasted in the process. With the latter, waste heat from some industrial process is used to generate electricity via a steam generator and turbine.

A steam generator employed in a CHP plant has to serve two masters: the need for heat, and the demand for electricity generation. In most cases the former predominates because the entire raison d'être of the plant was probably the need to serve a community or an industrial plant, and the plant's ability to generate electricity is of secondary importance (even though, as a spin-off, it is extremely valuable).

For this reason, the development of a truly effective master demand signal for a CHP plant is much more complex than it is with a plant whose only function is to generate electricity. The needs of all the users have to be taken into consideration, as must the cost of the steam, heat and electricity that is produced. Furthermore, it is possible that the way in which the master demand is configured may need to be modified at some time over the life of the plant because of changes in fuel prices or alterations in the requirements of the industrial, commercial or domestic complexes which benefit from the process.

The wide range of possibilities of interconnecting the various systems in CHP plant gives rise to very diverse methods of organising the master demand. Configuring a master demand signal that takes all the requirements into account ought not to be a significant problem, bearing in mind the power and flexibility that is offered by the modern distributed control system, but the difficulty is to obtain enough data on these requirements, and then to ensure that the information is correct. Quite often it seems that, even if the options might have been considered at some time, reconciling the various requirements has proved to be intractable and so a cheap and simple compromise has been employed. This may be reasonably effective, and the plant that is so developed continues to generate heat and power for days on end, the response to changes in demand seems adequate, and the operational staff are unprepared to alter anything for fear of rocking the boat.

None of this alters the fact that the expenditure of a few more days (or even weeks) of effort in front-end definition could have yielded, over the operational life of the plant, efficiency and performance improvements that would have amply recovered the cost of development.

4.5 WTE plants

The design of master demand signals of WTE plants, as described in Chapter 3, requires very careful attention. The requirements in this area are somewhat similar to those relating to CHP plants: the reconciliation of differing operational requirements.

The design of the system may favour consumption of the waste material, with electricity generation treated as a useful and revenue-earning by-product or it may try to maximise the power generation capability of the plant. In both cases, however, it is important to recognise the special characteristics of the plant, in particular, the boiler response.

Because of the nature of its complex fuel-handling system (see Figure 3.24), a WTE plant cannot be expected to be very responsive to demand changes. Therefore it is largely impractical to consider the application of advanced control logic to the master demand system for this type of plant (although various attempts have been made to do so). The most cost-effective solution is to apply a simple boiler-following system as described in Section 4.3.1, designed for only slow load-changing capability. The reduction in efficiency is negligible (and even somewhat academic since not only is the fuel in this type of plant easily obtained, but also the user is paid for consuming it!). Also, the difficulties of tuning the system (due to the interaction between the steam generator and the steam user) are less of a problem in this type of installation because of the very different dynamic responses of the turbine and boiler. The difference between the slow response of the boiler and the quick response of the turbine also simplifies decoupling one from the other in the optimisation process.

These factors ease the selection of a master system in WTE plants. A basic boiler-following design provides speed and simplicity of commissioning and usually performs adequately.

4.6 Heat recovery steam generators

HRSGs, as described in Chapter 2, deliver steam to a turbine which typically operates in turbine-following mode to maintain steam pressure at a fixed or varying ('sliding') setpoint. Frequently the steam turbine is set to operate with throttle valves in the fully open position, only modulating closed if steam pressure falls quickly due to a loss of heat input to the HRSG.

The steam pressure and turbine output thus follow the steam flow produced by the heat from the combustion gas turbine (CGT). Power production from the steam turbine will therefore be delayed behind that being generated by the associated combustion turbine.

In many cases the steam turbine generator (STG) receives steam from two or even three HRSGs. For example, with two CGTs, two HRSGs and one steam turbine (designated 2:2:1 configuration), each generator will supply approximately one-third of the total power output. The dispatch system will typically set the power demand as a single combined output of the three generators. The MW controller will therefore generate a master demand and adjust the output of the two fast-acting CGTs to equal the total load demand, minus the current output load of the steam turbine's generator.

In order to follow MW demand closely, the CGTs will need to ramp more quickly than their own proportion of required generation to make up for the delay in power output from the steam turbine. This 'overfiring' effect can cause large variations in gas temperatures which can produce thermal stresses in the HRSG if not carefully controlled. Figure 4.6 shows trends from a simulation of a combined cycle plant ramping to follow a total unit demand signal. Note the delay in the STG's generated load compared with the CGTs.

With the growing need for faster ramping from these combined cycle plants has come an increased emphasis on good steam temperature and feedwater control to ensure margins for safe long-term operation are not exceeded.

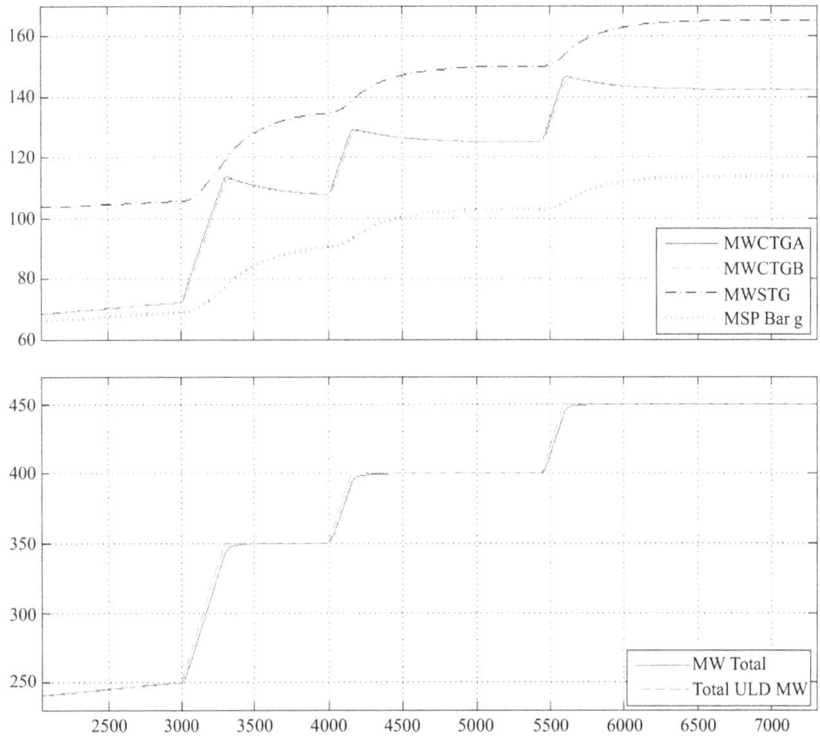

Figure 4.6 Simulation trends of a 2:2:1 combined cycle plant ramping to follow total MW demand

4.7 Summary

In this chapter we have seen how a 'master demand' signal is generated with respect to the nature of the duties that the plant is designed to undertake. This signal is responsible for ensuring that the boiler reacts to changes in demand, and it must also coordinate the operation of each of the subsidiary systems. The main areas involved in this process are the combustion and draught systems in conventional boilers, the feedwater and the steam temperature control systems in both conventional boilers and HRSGs.

In the next chapter we shall see how the combustion and draught systems of a boiler react to the demands of the master signal to produce the required firing rate and how the supply of air keeps in step with the changes to produce the correct conditions for the combustion of the fuel. In addition, we shall consider how the exhaust fans are regulated to maintain the correct pressure in the furnace while all this is occurring.

Chapter 5
Combustion and draught control

David Lindsley[1] *and John Grist*[2]

Part 1: Control

When considering fired boilers and heat-recovery steam generators it is clear that in the areas of their steam and water circuits there are many similarities between them (although the heat-recovery steam generator (HRSG) may have two or more pressure systems). But when the systems for controlling the heat input are examined, the two types of plant take on altogether different characteristics. The reason for this is fundamental: within the HRSG, no actual combustion process is involved since all the heat input is derived from the gas–turbine exhaust (except where supplementary firing is introduced between the gas–turbine and the HRSG). The subject of combustion control, which we shall be examining in this chapter, is therefore only relevant to fired plant.

Naturally, in a fired boiler the control of combustion is extremely critical. To maximise operational efficiency combustion must be *accurate,* so that the fuel is consumed at a rate that exactly matches the demand for steam, and it must be executed *safely,* so that the energy is released without risk to plant, personnel or environment. (The amount of energy involved in a power plant is considerable: in each second of its operation a large 800–1000 MWe boiler requires a heat release in excess of 2 GJ, and in a process of this scale the results of an error can be catastrophic.)

In this chapter, we shall see how the combustion process is controlled to meet the two objectives defined in the previous paragraph. We shall also examine the subsidiary systems that maintain the correct operational conditions in the fuel-handling plant of coal-fired boilers.

5.1 The principles of combustion control

In Chapter 3 we saw that the theoretically perfect combustion of a fuel requires the provision of exactly the right amount of air needed for complete combustion of the fuel. For the boiler, as a whole this means that the total amount of air being

[1]Retired
[2]Consulting Engineer

delivered to the combustion chamber at any instant matches the total amount of fuel entering that chamber at that time. For an individual burner, it means that the fuel and air being delivered to the burner are always in step with one another.

On the surface, therefore, it appears that the matter of combustion control merely involves keeping the fuel and air inputs in step with each other, according to the demands of the master, and if this were true this role would be adequately addressed by a straightforward flow ratio controller. Unfortunately, when the realities of practical plant are involved, the situation once again becomes far more complex than this simple analysis would suggest.

When the relationship between the fuel and air flowing at any instant into the furnace is chemically ideal for combustion, the relationship between the two flows is known as the stoichiometric fuel/air ratio. However, as stated earlier, it is usually necessary to operate at a fuel/air ratio that is different from this theoretically optimal value, generally with a certain amount of excess air. All the same, even though more than the theoretical amount of air has to be provided, any overprovision of air reduces the efficiency of the boiler as the air has to be heated and there is extra fan work and must therefore be limited.

The reduction in efficiency is due to losses which are composed of the heat wasted in the exhaust gases and the heat which is theoretically available in the fuel, but which is not burned. As the excess air level increases, the heat lost in the exhaust gases increases, while the losses in unburned fuel reduce (the shortage of oxygen at the lower levels increasing the degree of incomplete combustion that occurs). The sum of these two losses, plus the heat lost by radiation from hot surfaces in the boiler and its pipework, is identified as the total loss.

Figure 5.1 shows that operation of the plant at the point identified at ‘A’ will correspond with minimum losses, and from this it may be assumed that this is the

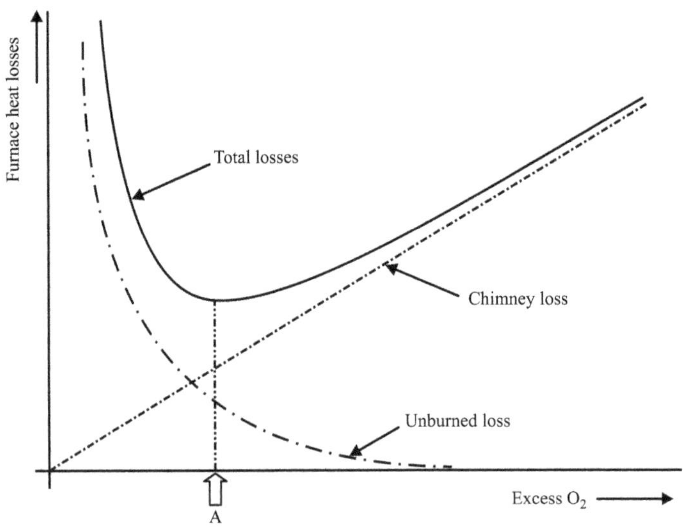

Figure 5.1 Heat losses in a furnace

point to which the operation of the combustion-control system should be targeted. However, in practice air is not evenly distributed within the furnace. For example, operational considerations require that a supply of cooling air is provided for idle burners and flame monitors, to prevent them being damaged by heat from nearby active burners and by general radiation from the furnace. Air also enters the combustion chamber through leaks, observation ports, sootblower entry points, etc. The sum of all this is referred to as 'tramp air' or 'setting leakage'. If this is included in the total being supplied to the furnace, and if that total is apportioned to the total amount of fuel being fired, the implication is that some burners (at least) will be deprived of the air they need for the combustion of their fuel. In other words, the correct amount of air is being provided in total, but it is going to places where it is not available for the combustion process.

Operation of the firing system must take these factors into account, and from then on, the system can apportion the fuel and air flows. If these are maintained in a fixed relationship with each other over the full range of flows, the amount of excess air will be fixed over the entire range.

5.1.1 Cross-limited control

Small oil- or gas-fired shell boilers are supplied with a mechanical linkage between the fuel valve and the air supply. However, for larger boilers where the National Fire Prevention Association (NFPA) 85 is applied the code calls for a metred system [1]. The so-called cross-limited combustion control system addresses these factors in a very comprehensive way, as described in the following section.

Figure 5.2 shows the principles of the cross-limited combustion control system. Individual controllers (7, 8) are provided for the fuel and air systems, respectively. Ignoring for the moment the selector units (5, 6) and the fuel/air ratio adjustment block (4), it will be seen that the master demand signal is fed to each of these controllers as the setpoint, so that the delivery of fuel and air to the furnace continually matches the load. Because fuel flow and air flow are each measured as part of a closed loop, the system compensates for any changes in either of these flows that may be caused by external factors. For this reason, it is sometimes referred to as a 'fully metered' system. The effect of the fuel/air ratio adjustment blocks (4) is to modify the air flow signal in accordance with the required fuel/air relationship.

The maximum and minimum selectors allow for differing response-rates of the fuel and air supply systems, consider what happens when the master demand signal suddenly requests an increase in firing. Assume that, prior to that instant, the fuel and air controllers have been keeping their respective controlled variable in step with the demand, so that the fuel flow and modified air flow signals are each equal to the demand signal. When the master demand signal suddenly increases, it now becomes larger than the fuel flow signal and it is therefore ignored by the minimum selector block which instead latches onto the modified air flow signal (from item 4). The fuel controller now assumes the role of fuel/air ratio controller, maintaining the boiler's fuel input at a value that is consistent with the air being delivered to the furnace. The air flow is meanwhile being increased to meet the new demand, since the maximum selector block (6) has latched onto the rising master signal.

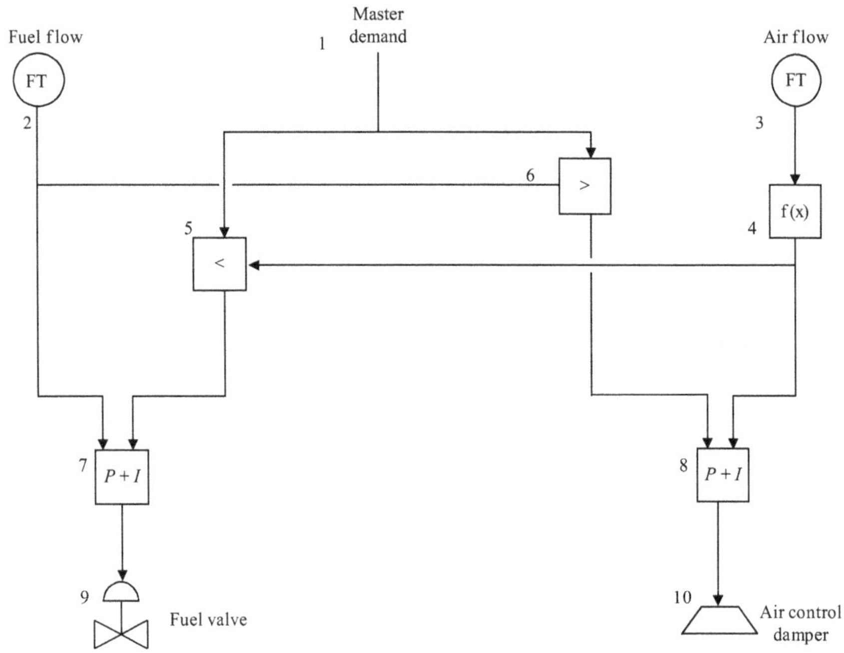

Figure 5.2 Basic cross-limiting

On a decrease in load, the system operates in the reverse manner. The minimum selector block locks onto the collapsing master and quickly reduces the fuel flow, while the maximum selector block chooses the fuel-flow signal as the demand for the air flow controller (8), which therefore starts to operate as the fuel/air ratio controller, keeping the air flow in step with the fuel flow.

Analysis of the system will show that it is much better able to deal with plant or control and instrumentation (C&I) equipment failures. For example, if the fuel valve fails open, the air controller will maintain adequate combustion air to meet the quantity of fuel being supplied to the combustion chamber. This may result in over firing but it cannot cause fuel-rich conditions to be created in the furnace.

The system cannot compensate for all possible failures, but when coupled with self-checking diagnostics and proper fault-detection techniques it provides a high degree of safety.

Figure 5.2 shows the exact hardware blocks used in pneumatic or electronic hardware systems for combustion control. It has two drawbacks:

- The controller (8) shows a corrected air flow rather than the actual air flow.
- The signal selectors only work if their inputs are in the same units. In this case % Boiler Maximum Continuous Rating (BMCR) flow.

Early computer systems converted the 4–20 mA inputs to 0%–100% or 0–1 enabling the same logic configuration to be used. Now most distributed control

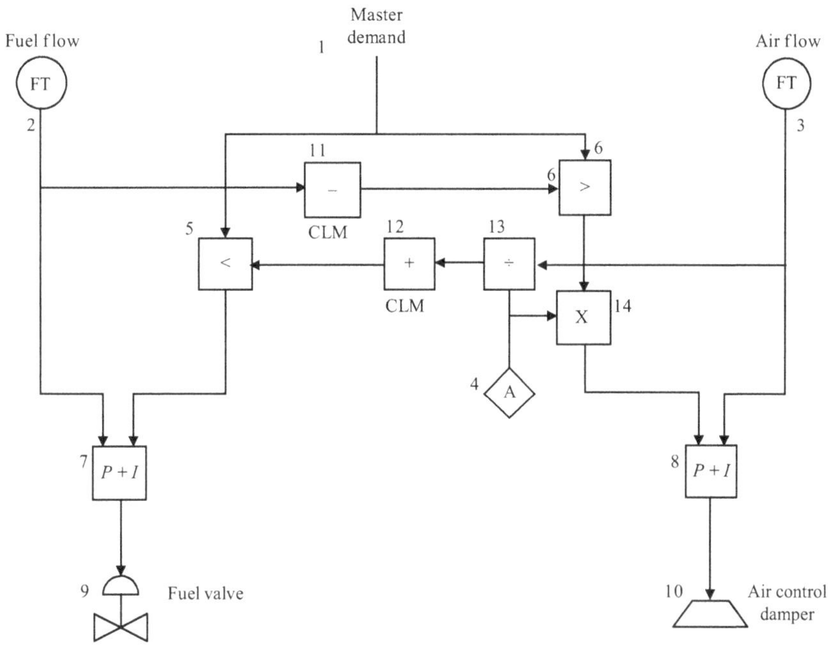

Figure 5.3 Improved cross-limit logic

system (DCS) allow the user to use 0%–100% or to configure in engineering units. This is the approach used by Doosan Babcock, as the calculated values can be directly related to the design values.

To allow this the previous logic is modified as shown in Figure 5.3. Block 4 has been moved and affects blocks 13 and 14. In Figure 5.2 if block 4 was designed to give 20% excess air then in Figure 5.3 it is set to 1.2. Now at 100% master demand the fuel demand is 100 and the air demand is 120. The corresponding air flow measurement (120) is divided by 1.2 in block 13 so the minimum selector (5) see 100 and can instantly respond if the air flow falls.

Blocks 11 and 12 are cross-limit margin blocks. They introduce a small margin and are adjusted during commissioning.

The function of the margins is to ensure that normal control of fuel and air flows are taken from the respective demand signals without unwanted interference from the cross-limits, either due to noise or slight calibration offsets in the air and fuel signals. In both steady-state and during load ramps, the margins should not be reached and cross-limits should not activate. However, if a fault occurs in either fuel control, causing an unexpected fuel increase, or air flow is below demand, the margins are small enough to ensure that air demand or fuel demand, (depending on which cross-limit is activated) will respond well before a fuel-rich condition occurs.

The cross-limit logic is applicable to gas, oil coal and biomass-fired boilers. However, the stoichiometric ratio will change for each of these fuels.

If the DCS calculates in 0%–100% than 60% of fuel oil requires 60% of maximum air. If, however, it uses engineering units then 60 kg/s of fuel oil requires much more than 60 kg/s of air. Typically fuel oil requires 16 times as much air for complete combustion. If we ignore excess air for a moment block 4 could be set to 16 to give the correct stoichiometric ratio, or 19.2 (16×1.2) to give the correct stoichiometric ratio and the excess air. Block 4 could also be used for oxygen trim or setting the working burner lambda (WBλ). In practice the O_2 trim, WBλ, and basic excess air requirements all change with load and blocks 13 and 14 are repeated for each, but block 4 becomes a function of load.

EN12952-9 7.2.4 asks for the measured air to fuel ratio to be measured and alarmed if it decreases too much. If it continues to decrease the standard asks for a master fuel trip (MFT), but not necessarily automatic. The logic to implement this is straight forward but there are some complications.

- Choosing the trip level is a challenge. Picking a conservative value is easy, but will lead to spurious alarms and trips. Making the MFT an operator choice allows the spurious alarms to be ignored during commissioning.
- If the HAZOP determines that this alarm/trip needs a safety instrumented system, then the fuel and air flow transmitters have to be safety integrity level (SIL) rated and may need to be duplicated to satisfy the layer of protection analysis.

5.1.1.1 Using gas analysis to vary the fuel/ air ratio

In the systems shown in Figure 5.3 the relationship between the fuel and air quantities is preset. With such systems, if the adjustment factor is set wrongly or if changes outside the system dictate that the fuel/air ratio should be altered, no provision exists for automatic correction, and the right combustion conditions can only be restored by adjusting block 4. To improve performance and safety, some form of automatic recognition and correction of these factors is preferable.

If the fuel/air ratio is incorrect, combustion of the fuel will be affected and the results will be observable in the flue gases. This indicates that an effective way of optimising the combustion process is to change the fuel/air ratio automatically in response to measurements of the flue gas content.

For all fossil-fuelled boilers, the oxygen content of the flue gases increases as the excess air quantity is increased, while the carbon dioxide and water content decreases. The carbon monoxide content of the boiler's flue gases is a direct indication of the completeness of the combustion process and systems based on the measurement of this parameter have long been recognised as an effective mechanism for improving combustion performance in coal, biomass and oil-fired boiler plant [2]. However, experience indicates that the use of this gas as a controlling parameter is less advantageous in boilers fired on natural gas [3].

Measurement of the flue gas oxygen content often provides a good indication of combustion performance, but it must be appreciated that the presence of 'tramp air' due to leakages into the combustion chamber can lead to anomalous readings.

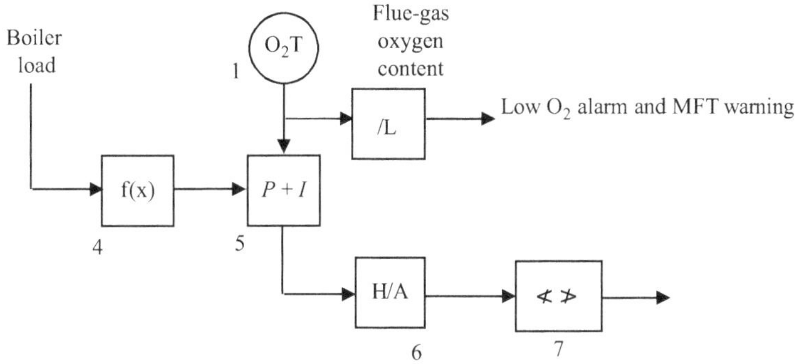

Figure 5.4 Oxygen trimming of fuel/air ratio

In the presence of significant leakage, reducing the air/fuel ratio to minimise the flue gas oxygen content can result in the burners being starved of air. This is an area where systems based on carbon monoxide measurements provide better results since the carbon monoxide content of the gases is a direct indication of combustion performance and is unaffected by the presence of tramp air.

A system which adjusts the fuel/air ratio in relation to the flue gas oxygen content is shown in Figure 5.4. The oxygen measurement is fed to a controller (5) whose output adjusts the fuel/air ratio in place of block 4 in Figure 5.3.

The flue gases leave the combustion chamber through ducts of considerable cross-sectional area and it is inevitable that a significant degree of stratification will occur in the gases as they flow to the chimney. Air entering the furnace through the registers of idle burners will tend to produce a higher oxygen content in the gases flowing along one area of the duct than will be present in another area, where fewer burners may be idle. It is therefore necessary to take considerable care that any gas analysis provides a truly representative sample of the average oxygen content, and this demands that great care should be exercised over the selection of the location of the analyser. With larger ducts, it may be necessary to provide several analysers. The signals from these can be combined, or the operator can be given the facility to select one or more of them for use.

A better option is now available, but not normally used. The power and flexibility of modern computer-based control systems allows for truly intelligent sampling to be applied, where the system recognises the dynamic status of the plant, such as which burners are being fired, and automatically selects the analyser signal to be used, or intelligently mixes the analyser signals to optimise performance. The installation of such a system requires careful observation of the plant performance over an extended period and the development and subsequent application of a suitable system based on those observations.

Although such techniques are possible. Despite the considerable advances that have been made in gas analyser technology over the past few years, fuel/air ratio trimming on the basis of gas analysis is still treated with some reservation. It is generally accepted that the measurements may occasionally fail or be misleading and for this reason it is usual to allow manual intervention in the absence of reliable oxygen control. In Figure 5.4 this facility is provided by the auto/manual station (6). In addition, a maximum/minimum limiter block (7) restricts the amount of adjustment that is permitted, to constrain the effects of anomalous or invalid measurements or incorrect control actions.

This system also characterises the set-value signal for the oxygen controller over the boiler's load range by means of a function block (4), providing for higher excess air operation at low loads. The indication of boiler load may be obtained from steam flow, fuel demand or air flow, and the exact shape and parameters of the oxygen versus load characteristic will be defined by the boiler designer or process engineer.

In practice, facilities may also be incorporated to allow the operator to adjust the system by biasing the load signal upwards or downwards at any given point to yield better combustion with reduced stack emissions.

Because the oxygen content of air is 21% by volume (or roughly 23% by weight), a given change in oxygen content represents approximately five times that change in terms of excess air. Since it is indeed *airflow* that is being controlled, the oxygen loop must be set with the appropriate scaling, relating a change in O_2 to the respective influence to air flow to restore the O_2 deviation.

The integral setting depends upon the controller algorithm, but will be based on the response time constant which is typically 60–90 seconds.

5.1.1.2 Combining oxygen measurement with other parameters

The use of an oxygen-trim signal on its own can be misleading, for the reasons noted earlier, and better performance can be obtained by combining oxygen trim with the opacity of the flue gases since reducing the air flow eventually results in the production of visible smoke. However, it is usually undesirable to operate a boiler in the region where smoke is being produced, and an improvement is to adjust the air flow on the basis of another parameter, such as carbon monoxide.

Figure 5.5 shows how the carbon monoxide and oxygen measurements can be combined to trim the fuel/air ratio. Basically, the system comprises two gas analysis controllers (6 and 10) whose set-value signals are determined in relation to the boiler load (via function generators 5 and 7). However, the set value for the oxygen controller is also trimmed by the output of the carbon monoxide controller [the two signals being combined in summator (9)]. Auto/manual facilities enable the system to operate with both analysers in control, or with only oxygen trim in service (the CO controller being on manual at auto/manual station (8), or with fully manual fuel/air ratio adjustment [auto/manual station (12) being on manual, to isolate both gas analysis controllers]. CO trim is more appropriate to oil- and gas-fired boilers with low excess air levels. Coal and biomass fire with higher excess air levels and in a simple boiler are unlikely to generate CO. However, if over-fire air (OFA) is used the main combustion zone is sub stoichiometric with a real risk of generating CO at the furnace outlet.

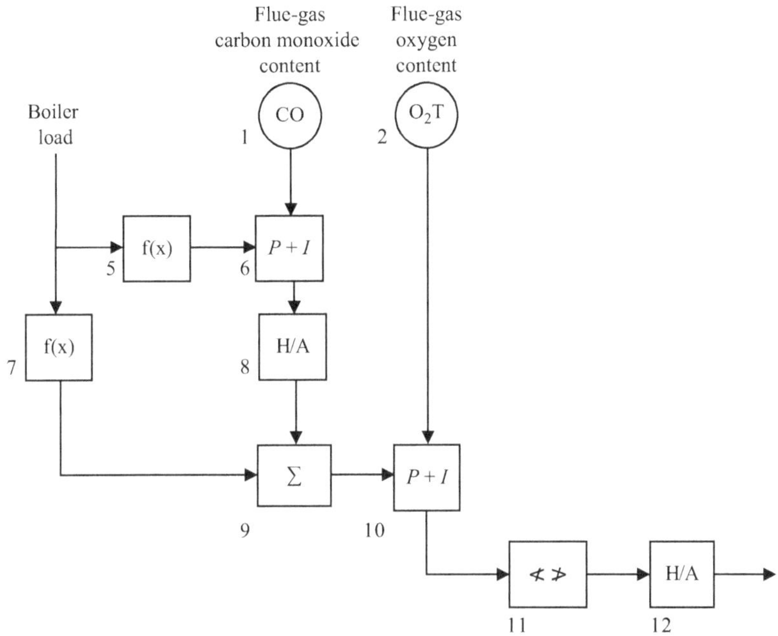

Figure 5.5 Combined CO and O_2 trimming of fuel/air ratio

In another variant of this system, either of the two flue gas analysis controllers can be selected for operation, either by manual intervention or automatically by means of a maximum selection function.

Newer technologies allow for the measurement of combustibles COe as a simple addition to a zirconium oxygen analyser. This can be used more reliably than CO when being used for combustion efficiency.

5.1.1.3 Using carbon-in-ash (CIA) measurements

In boilers burning solid fuels, the carbon content of the ash has traditionally been used to provide an indication of the completeness of combustion, since any carbon remaining in the ash indicates that incomplete combustion has occurred. Until comparatively recently, accurate online carbon-in-ash sampling was not possible and measurement of this parameter required manual sampling and analysis. With the emergence of online analysers, the picture has changed, and tests have indicated that online measurement can play a useful part in optimising the combustion process [4]. In addition, analysis of unburned carbon can indicate whether the coal mills (pulverisers) require adjustment. However, the long transfer time constants of the combustion process coupled with the comparatively slow response of the instruments and problems of stratification [5] suggest that this technique is only useful for long-term correction of firing, where relatively stable load conditions can be maintained for extended periods.

5.1.2 Multiple fuel sources

This section covers single fuel boilers with multiple fuel valves, and boilers firing different fuels. The systems that have been described so far are based on the adjustment of the total quantity of fuel and air that is admitted to the combustion chamber. This approach may suffice with smaller boilers, where adjustment of a single fuel valve and air damper is reasonable, or for larger boilers where a single control valve modulates the flow to all the burners, but larger units will have a multiplicity of, fuel valves and coal feeders, dampers and combustion-air supplies. In such cases, proper consideration is given to the distribution of air and fuel to each burner or, if this is not practical, to small groups of burners.

The arrangement is usually determined between the boiler supplier and end user. The most common are:

- Common windbox where the forced draught (FD) fans control total air to the furnace as supplied by Foster Wheeler.
- Multiple windboxes, usually one per mill, where the FD fans control the common supply pressure to all windboxes and the actual flow per windbox is by secondary air dampers on each side of the windbox, as supplied by Doosan Babcock.
- Individually modulating air registers to each burner as common place in Germany.

These have increasing complexity and increasing cost but potentially a better opportunity to optimise performance.

Figure 5.6 shows how the principles of a simple cross-limited system are applied to a multi-burner, multi-windbox oil-fired boiler. The plant in this case comprises '*N*' rows of burners one row per windbox, and the flow of fuel oil to each row is controlled by means of a control valve per row. The combustion air is supplied to each windbox, and the flow to the firing burners is controlled by a set of secondary air dampers at each windbox.

The system operates in the same way as the improved configuration of Figure 5.3, and it is repeated for each row of burners, so that the ratio of total fuel oil flow to total air flow entering the boiler is maintained at the desired value. The master demand and the oxygen-trim signals are fed to all the rows to keep the firing rate in step with the load demand and the flue gas oxygen content at the correct level.

This basic configuration is not restricted to oil-fired boilers. It can also be used with gas-fired plant and it can be applied to systems burning a mixture of fuels, with suitable modifications as will now be described.

5.2 Working with multiple fuels

The control systems of boilers burning several different types of fuel have to recognise the heat input contribution being made at any time by each of the fuels, and the arrangements become more complicated for every additional fuel that is to be considered.

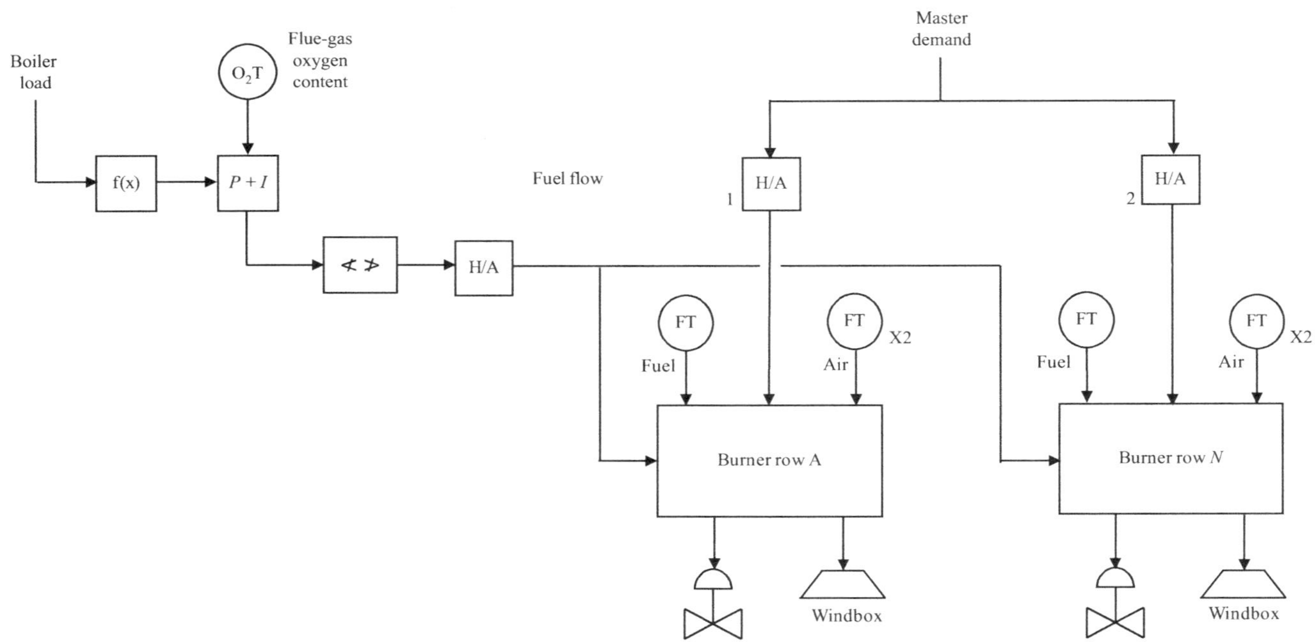

Figure 5.6 Multi-windbox multiple fuel valves but same fuel

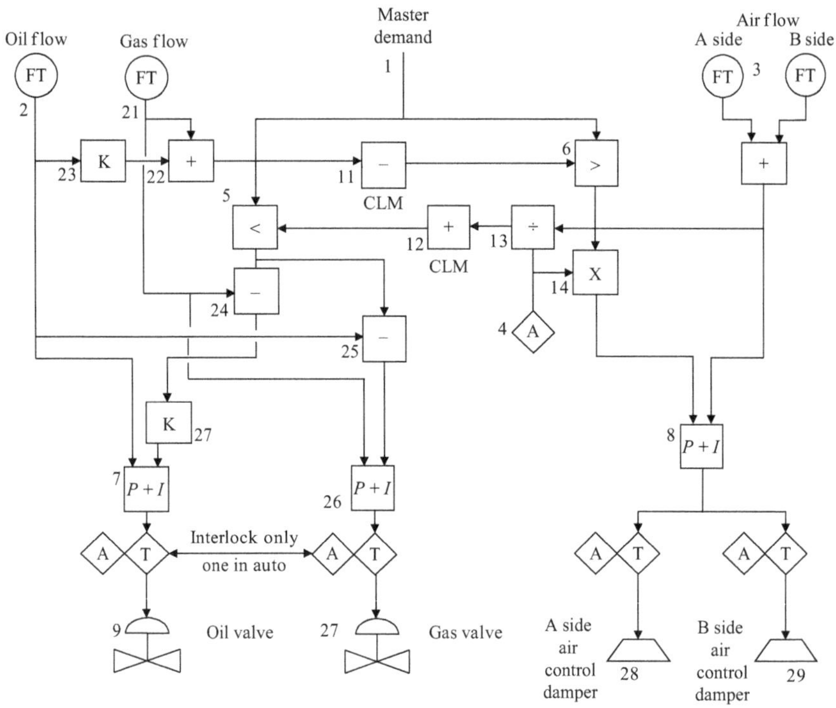

Figure 5.7 Oil- and gas-fired boiler

Figure 5.7 shows a system for a boiler burning oil and gas. The similarities to the improved cross-limited system are very apparent. The cross-limiting function is performed at the minimum selector block (5) which continuously compares the master demand with the quantity of combustion air flowing to the common wind-box of the burner group. The air to fuel ratio is set up by block 4, which acts via blocks 13 and 14. The selected signal (the load demand or the available air) ultimately forms the desired value of both the gas and oil closed-loop controllers. But, before it reaches the relevant controller a value is subtracted from it, which represents the heat contributed by the other fuel (converted to the same heat/kg value as the fuel being controlled). In this example gas is the normal fuel and the conversion of oil flow to equivalent gas flow is performed in a scaler (23), while scaler (27) converts the gas demand to an equivalent oil demand. Each of the two subtract units (24 and 25) algebraically subtracts the 'other-fuel' signal from the demand. This logic assumes that only one fuel is being automatically controlled while the other is controlled by the operator.

Dual fuel is most common on coal-fired power plant where the contribution of the support oil may be significant. The aforementioned logic is also used and within the cross-limited system the fuel demand to the mill is reduced by the contribution from the oil. To give a resulting coal demand. Similarly, both the oil and coal flows are considered when calculating the air requirements.

These diagrams are highly simplified, and in practice it is necessary to incorporate additional features and interlocks. For example, the oil and gas are each shown as measured by a simple flow transmitter. If a volumetric flow transmitter is used then it may be necessary to pressure and temperature compensate the signal.

In some instances, the oil or gas pressure signal is used instead of flow. This is because there is a nominally fixed relationship between pressure and the flow through the burner tips. As before temperature compensation may be necessary since the pressure/flow relationship of the gas is temperature dependent. If pressure is used the pressure flow relationship is fixed for a single burner. The pressure signal represents the flow *per burner,* and so it is multiplied by the number of burners in service in order to represent the total fuel flow. Note that if the burner tips are changed during commissioning the curve will need to be changed. Sometimes different tips are used on different rows.

Similarly, the furnace to windbox DP can be used to represent air flow through a burner. This combination often being used on ships where space was limited.

5.3 The control of mills or pulverisers

So far, we have looked at boilers where the input of fuel can be measured and where its flow can be regulated by means of one or more valves. With boilers burning coal, or biomass the mill (or pulveriser) system must be taken into consideration. The mills have already been described in Chapter 3, now we shall look at how they are controlled. But first it has to be understood that, because the mill has to meet defined performance guarantees, the control strategy to be applied in a given installation must be developed in association with the manufacturer of the mill. Once that strategy has been agreed it must be applied to each of the mills that feed the boiler. The demand is fed in parallel to all the mill sub-systems, with facilities for biasing the signal to any one of them with respect to the others. Conventional mills can be adapted to fire some forms of biomass. Usually a screw conveyer and/or rotary valve is used instead of a conventional feeder to ensure isolation between the bunker and the mill and minimise the propagation of a fire or explosion.

5.3.1 The 'load line'

The drop in pressure experienced by air flowing through a mill will be determined by the geometry of the mill, the amount of coal in it and the volume of air flowing through it. A high pressure drop across the mill may be the result of a high coal load in the mill or a high air flow through it, or a combination of both. The air flow rate will bear a square–law relationship to the differential pressure across the mill, and the differential pressure across a restriction such as a flow nozzle or an orifice plate will also have a square–law relationship with the air flow. From this, it can be appreciated that the characteristic curve relating the mill differential pressure and the primary air differential pressure for a vertical spindle mill will be a straight line. This is called the 'load line' and is specific to a given design of mill operating

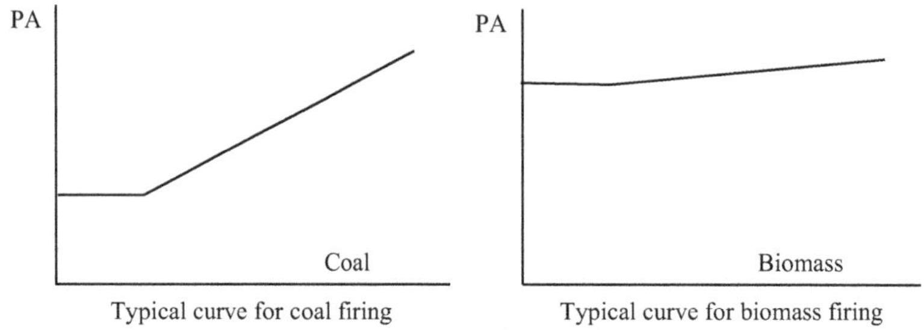

Figure 5.8 Mill load line

under defined conditions. Note that for biomass firing the primary air (PA) velocity is always higher because a high velocity is needed in the furnace to provide a stable flame (Figure 5.8).

5.3.2 Mill control strategies

The following assumes the configuration shown in Figure 5.9. Two FD fans, two ID fans, two PA fans, two trisector air heaters, four mills, four gravimetric feeders, four windboxes and OFA.

To satisfy NFPA 85 all mill control systems will be fully metered with measurements of feeder coal flow, PA flow and secondary air to each end of each windbox [1].

The speed of the feeder multiplied by the weight of coal on the belt is used to calculate coal flow. Alternatively, the screw conveyer speed can be used to infer biomass flow.

Figure 5.10 shows the improved cross-limit logic that adjusts the feeder speed *in parallel with* the PA flow. The PA flow is set as a function (24) of the mill load

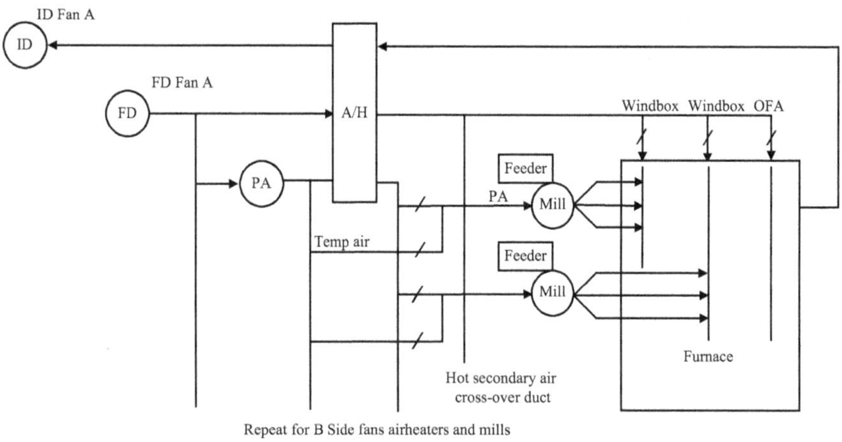

Figure 5.9 Typical boiler arrangement

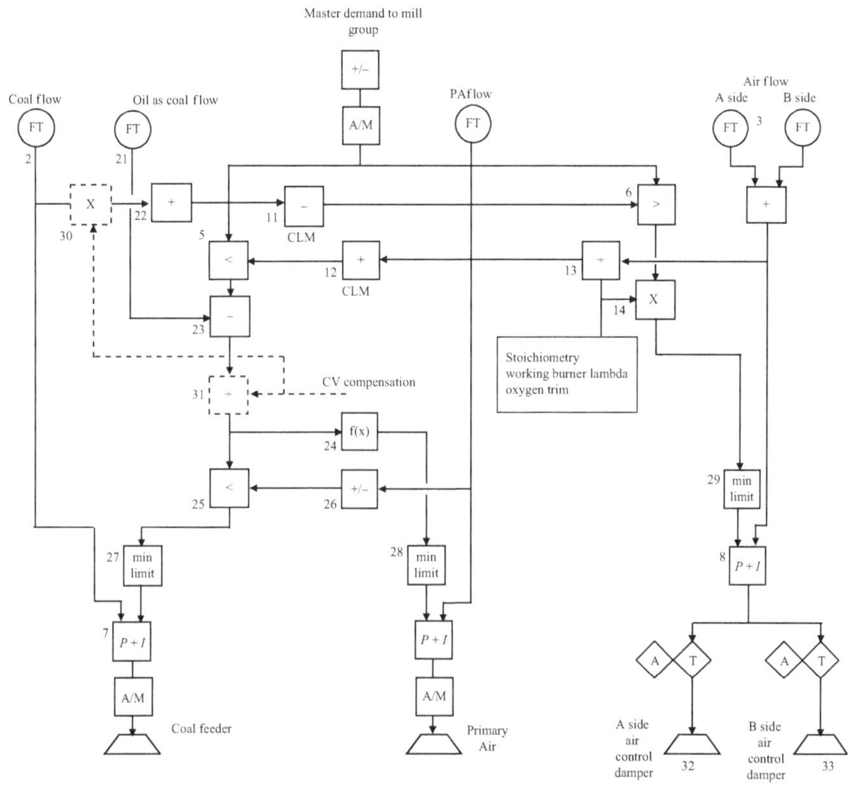

Figure 5.10 Cross-limited system applied to a mill group

line. This also shows some refinements: minimum limit blocks 27, 28 and 29 prevent the coal flow, PA flow and air to the windbox from being reduced below predetermined limits; these are NFPA 85 requirements [1].

A minimum selector block (25) which prevents the coal or biomass demand being increased above the availability of PA. The bias unit (26) sets the margin of air over coal. A correction is also included to allow for changes in the calorific value (CV) of the coal blocks 30 and 31.

As in the multi-fuel example (Figure 5.7) the heat contribution from any oil burners firing in that mill group is taken into consideration (23). By accounting for the oil firing, the opening of the primary air damper and the feeder speed are immediately adjusted, if an oil burner trips, or if one is brought into service, to compensate for the change, without waiting for the heat input effects to be detected via the master pressure controller.

5.3.3 Fuel quality factor

Boilers are designed around a performance coal. Sometimes it is an assumed coal, somewhere between the extremes likely to be encountered. The actual coal that is fired may differ in many ways, hardness, CV [or specific energy (SE)], slagging index. Typically, the CV may differ by ±10% and consequently the air requirement will also change by around 10%. The fuel quality factor logic tries to make this correction. It does not have to be completely accurate as the O_2 logic can make the final trim. A target accuracy of 2%–3% is suggested. The concept is not new, early boilers had steam flow air flow controllers, where the air was kept in a fixed ratio to the steam produced. Higher CV coal 'rocket fuel' produced more steam and hence the air flow was increased.

Modern designs are sometimes called BTU correction logic. The Doosan Babcock approach was called coal quality factor (CQF) but changed to fuel quality factor to allow for biomass firing. In some instances, the CV change with different biomasses may be very small and not require this correction.

Automatic fuel quality correction or BTU correction rescales fuel flow to maintain the relationship between fuel and air flows, and between fuel and feedwater flows. Adjusting BTU automatically provides several control advantages:

- Fuel/air and fuel/feedwater cross-limits can be implemented with narrower margins, giving better protection.
- Offsets in trim controllers to flue gas O_2, boiler pressure and roof temperature (or enthalpy) due to BTU changes are eliminated or minimised. This improves ramping performance as the controllers no longer 'walk' between offset values over the load range.

The detailed CQF logic is IP protected but calculates the heat gain from feedwater to superheated steam, and from cold reheat to hot reheat. This is divided by the quantity of coal being fired to calculate the CV of the coal. This is divided by the CV of the design coal to give the coal quality. For those who want to develop their own approach precautions must be taken to ensure that this is a slow acting loop and to block changes during upset conditions.

Fuel quality factor compensation (also known as SE, CV or BTU compensation) is not always required when firing natural gas, if the supply source is stable. However, where 'off gas' (waste gas) is fired from a refinery, very large changes in CV are possible. Two solutions are often used together. One is to measure the CV using a Wobbe meter and then add natural gas as required to return the CV to the expected range. The other is to install a large storage vessel in the off gas supply line to provide a residence period for mixing, thus acting as a damper on the changes in CV.

Wobbe = CV/sqrt specific gravity

5.3.4 Mill temperature control

It is very important that the temperature of the air in the mill should be maintained within close limits. For many reasons, including inadequate drying of the coal, combustion efficiency will be reduced if the temperature is too low, while too high a temperature can result in fires or explosions occurring in the mill.

The first fleets of 500 MW plus power stations used three control dampers for each mill. Damper B modulated the hot PA, damper A the cold tempering air and after these two air streams were mixed damper C controlled the total flow of the mixed PA. Effectively the first two set the correct ratio of hot and cold PA to achieve the correct mill exit temperature while the third controlled the flow. This had some advantages.

In some instances, the hot air damper is fully open and the system struggles to make temperature. The loop is designed to limit maximum temperature. In this instance the hot damper has minimum impact on the PA and there is no cold PA. A homogeneous flow of the main control damper that disturbs the flow the most is downstream of the flow measurement (Figure 5.11).

The was relatively easy to control as the flow and temperature loops were almost independent. That is, if the proportion of hot and cold air was set correctly the flow damper and flow could be changed without changing the temperature control dampers. This had another advantage in that the flow profile past the flow meter was also unchanged allowing a more consistent flow measurement. The hot PA and tempering air dampers worked in opposition as shown in Figure 5.12.

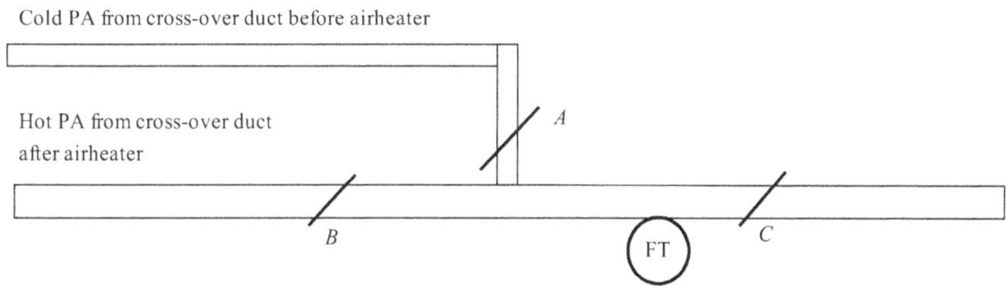

Figure 5.11 Three-damper duct arrangements for mill temperature control

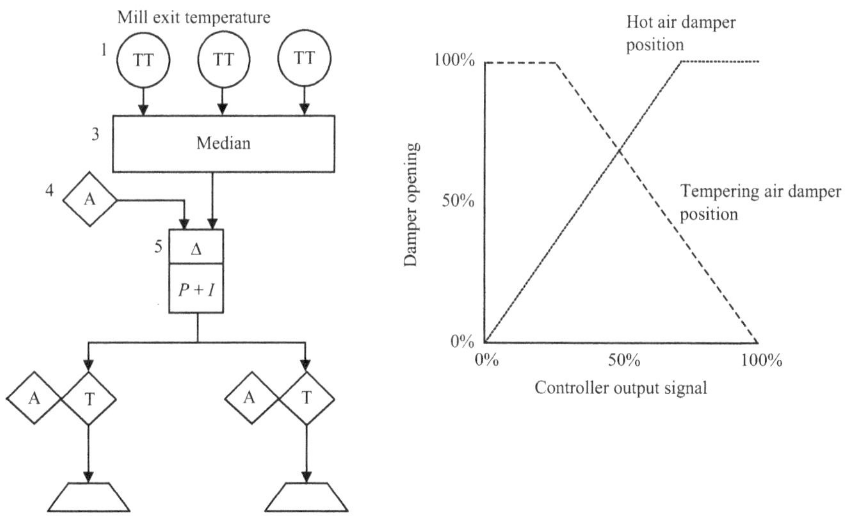

Figure 5.12 Simple mill exit temperature control for three-damper system

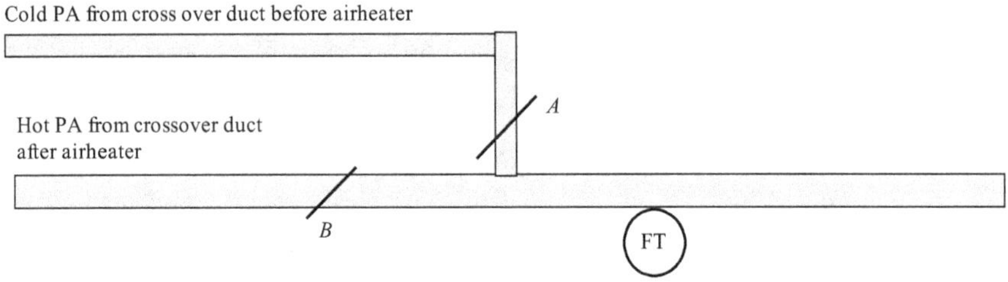

Figure 5.13 Two-damper duct arrangements for mill temperature control

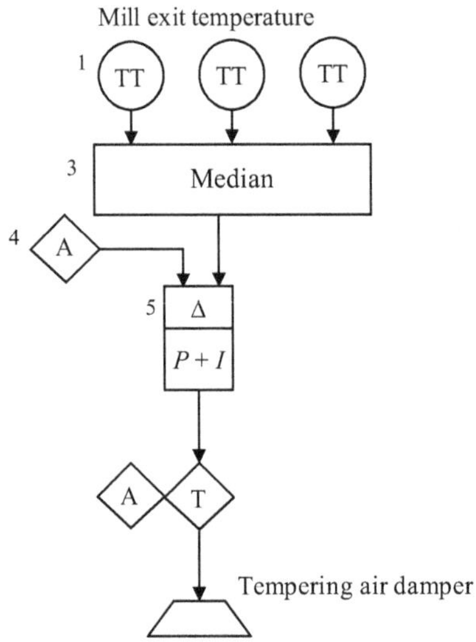

Figure 5.14 Simple mill exit temperature control for two-damper system

To reduce costs most mills are now supplied with only the first two control dampers. This makes control more difficult as moving either of them changes both the flow and the temperature. One approach is to use the hot PA damper to control flow and the tempering air damper to control temperature (see Figure 5.13).

The basic control loop acts only on the tempering air and works under steady load conditions (Figure 5.14).

While the loop as described works well in normal operation changing to a cascade controller has two benefits. It is easier to set a maximum limit on the demand for the slave mill inlet temperature controller. This is particularly useful during start-up, as without this limit the inlet will temperature might rise to over 300 °C shocking the mill. Some suppliers have algorithms that vary the mill inlet temperature dependent upon the fuel being fired (Figure 5.15).

To minimise the interaction between the flow and temperature loops, the movement of either damper acts as a feedforward on the other. Hence if the flow demand increases we must also increase the tempering air to maintain the same

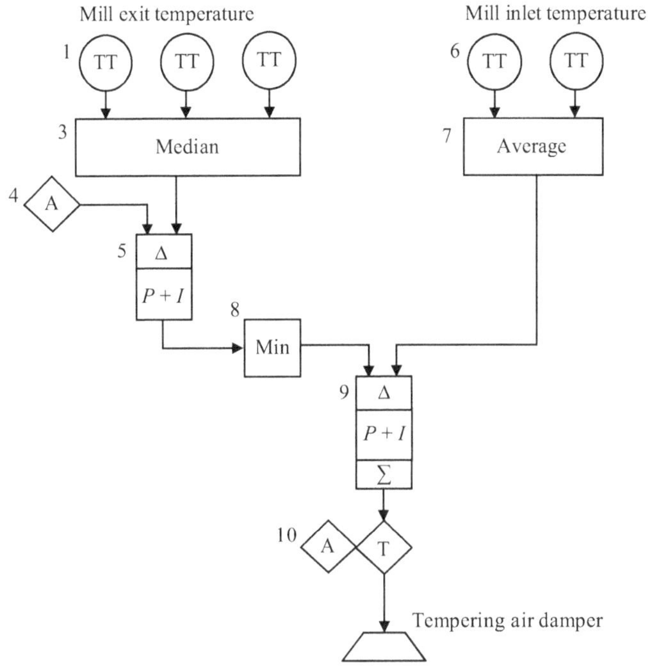

Figure 5.15 Cascade temperature control for two-damper system

ratio of hot to cold air at the mill inlet. However, if the mill exit temperature is too high any increase in tempering air flow must have a corresponding decrease in the hot air to maintain the same flow. The tempering air duct in much smaller than the hot air duct. Hence a 5% movement in tempering air might only require a 2% movement in the hot air.

A possible solution is shown in Figure 5.16 while this loop has only a single temperature controller in may also be used with the cascade version.

Note that the temperature measurements are triplicated for two reasons:

1. The thermowells are abraded by the PF and will eventually wear away.
2. The temperature measurement is also used to indicate a potential fire.

Some boiler makers prefer to use a temperature measurement on each PF line, rather than on the classifier.

5.3.5 Controlling multiple mills and multiple fuels

Large coal or biomass-fired boilers are provided with several mills, each of which has its own control subsystem as described earlier, and in addition they invariably burn other fuels as well as pulverised fuel. Usually oil that is used for early pressure raising, ignition of the coal/biomass and flame stabilisation.

Figure 5.17 shows the mill-control system of such a plant in simplified form. It is presented here to illustrate a requirement that is an integral part of boiler control

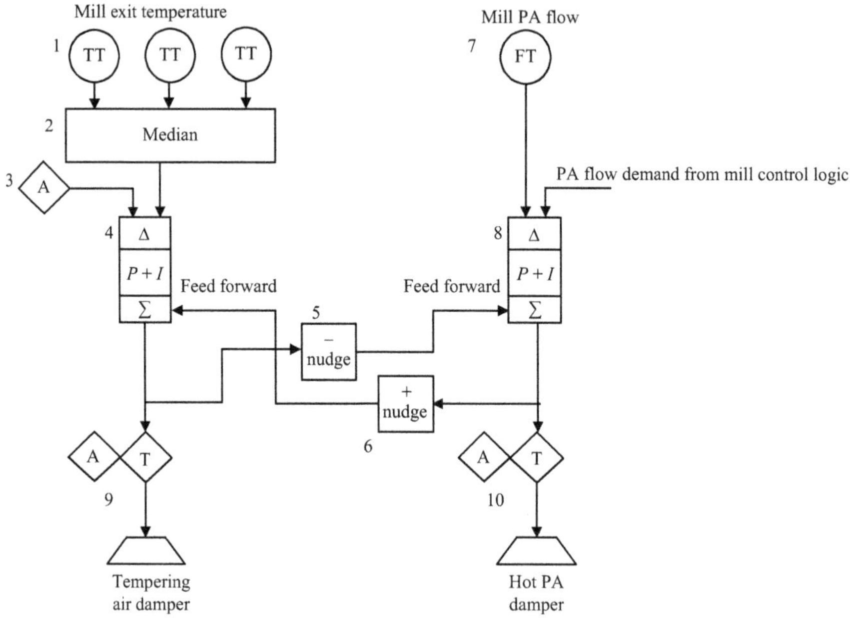

Figure 5.16 Advanced mill temperature control

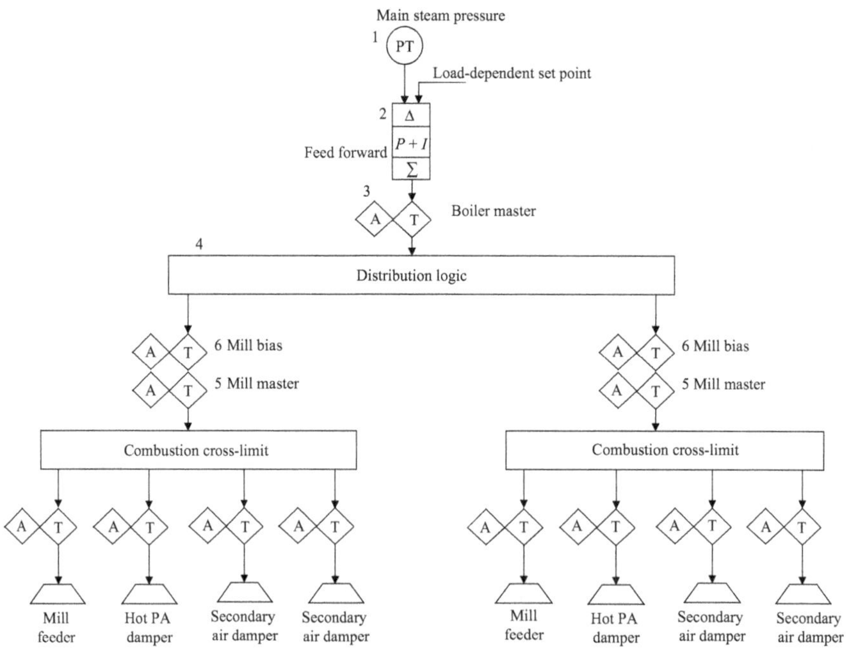

Figure 5.17 Multiple mill control systems

Master demand	Mill A		Mill B		Mill C		Man load	Total auto load	Auto demand per mill in auto
A	B1	C1	B2	C2	B3	C3	D	E	F
Boiler load in T/h	MS $M = 1$ $A = 0$	Manual demand	MS $M = 1$ $A = 0$	Manual demand	MS $M = 1$ $A = 0$	Manual demand	C1+C2+ C3	$A - D$	E/qty mills in auto
40	1	10	1	30	1	0	40	0	0
40	1	10	0	0	1	0	10	30	30
40	0	0	0	0	1	0	0	40	20
40	0	0	0	0	0	0	0	40	13.33

MS = manual status

Figure 5.18 Load sharing logic

systems: the need to handle applications where a single controller sends commands to several subloops in parallel, and where any of the subloops may be isolated at will from the main controller.

Here, each mill feeds a group of burners (say six), and each of these groups may also fire fuel oil. Since, at any given time, any mill group may be out of service, operating at a fixed throughput, or otherwise requiring independence from the other groups, the overall loop gain will change, and this is addressed by the distribution logic (4). The demand signal from this block is fed to each group via individual hand/auto stations, one for each mill group (5). These are known as the mill master and associated with each is a mill bias (6). Typically, this allows the operator to bias the load on an individual mill(s) by ±10%.

The output of each these stations (5) eventually becomes the desired value for the relevant feeder, primary air and secondary air controllers (not shown).

The distribution logic performs the following for the two mills shown, but the principle can be extended for any number of mills (Figure 5.18).

The demand from the master controller is shared between the mills in auto and no additional gain change is required.

5.3.5.1 The challenge of auto/manual changeover

The heat input from a large coal mill can be as much as 160 MWe, but the mechanical design of the mill and its auxiliaries is such that it can vary the throughput by only a comparatively small amount, certainly no more than 50%. Therefore, the introduction of one mill to the heat input of such a boiler amounts to a step change of as much as 80 MWe, and the change in throughput that can be smoothly modulated is also 80 MWe. Such large step-changes require efficient modulation of any other fuels that are being fired at the same time.

These factors make it impractical to consider starting up more than one mill at a time and require the facility of allowing any mill to be operated under manual or automatic control, independent of the others. This brings about a severe challenge to the DCS software.

The master demand is fed in parallel to several subloops, one for each mill group. On start-up of the plant all of these will be under manual control. When the mill has reached a throughput of roughly 50% of its capacity, or when other conditions determine that automatic control is now possible, the operator will switch the master demand into service. The difficulty is that up to that instant, the system cannot be made aware of which mill group is about to be transferred to respond to the master signal, and each group may be operating at a very different throughput from any other.

While a loop is being transferred from manual to automatic control (or vice versa), it is important that the plant is not subjected to a sudden disturbance. At the moment of changeover, the 'manual' and 'automatic' signals must be equal. This is called 'bumpless transfer', and it can be achieved by providing the operator with indications of both signals so that they can be made equal before changeover is initiated. However, such a system would not be acceptable in most cases, since the process of changing from one mode of control to another should be as quick and simple as possible, and should not require the operator to unduly disturb the operation of the plant.

To achieve what is known as 'procedureless, bumpless transfer' from manual to automatic control, a common technique is to make the controller output follow (or 'track') the manual demand, so that when the system is switched to automatic the signal to the actuator is not subjected to a sudden change.

This is easy enough with a single controller positioning a single actuator, but what happens when one controller commands several subloops as shown in Figure 5.17. It is clearly impossible to force the master controller output to adopt a value that cannot be known ahead of time, or to change the output of the controller if it is already modulating one or more mills.

This problem is frequently not recognised by DCS vendors who have little or no experience of boiler control, and it can be quite difficult to explain it to them. But understanding it and resolving it are absolutely essential if the system is to be expected to operate smoothly and with minimal operator intervention. Various solutions have been developed, such as 'freezing' the master demand while the transfer is achieved and gradually ramping one signal up or down to match the other. It is important, however, that the DCS vendor should be able to demonstrate the solution offered within their system, and that they should be able to demonstrate its use on an existing power plant. Most DCS suppliers have their own solution to this problem. For example, Emerson has a balancer module that can share the load between multiple mills.

5.4 Air distribution and pressure control

In Chapter 3 we saw that, in a fired boiler, the air required for combustion is provided by FD fans and the exhaust gases are drawn out of the combustion chamber by an additional set of fans. On boilers with retro-fitted flue gas desulphurisation plant, additional booster fans may also be provided. The control of all these fans must ensure that an adequate supply of air is available for the combustion of the fuel and that the combustion chamber operates at the pressure determined by the boiler designer. In a fluidised-bed boiler the air must also provide the pressure required to maintain the bed in a fluid state.

The FD and ID fans also contribute to the provision of another important function, purging of the furnace in all conditions when a collection of unburned fuel or combustible gases could otherwise be accidentally ignited. Such operations are required prior to light-off of the first burner when the boiler is being started, or after a trip.

The control systems for the fans have to be designed to meet the requirements of start-up, normal operation and shut-down, and to do so in the most efficient manner possible, because the fans may be physically large and require a large amount of power for their operation (up to 14 MW in some cases). In addition, as we saw in Chapter 3, the performance constraints of the fans, such as surge and stall, have to be recognised, if necessary, by the provision of special control functions or interlocks. Chapter 3 also described the methods of controlling the throughput of the fans, that is, pitch control, dampers, vanes or speed adjustment. In this chapter, we shall examine how these elements are adjusted to address the operational requirements of the boiler.

5.5 NO_x control

The low NO_x burner provides no challenge to the control engineer. The overall air to fuel ratio logic remains unchanged. The distribution of air streams within the burner is entirely mechanical though additional controls could be introduced to alter damper or even swirler positions if desired.

Subsequently the staging was improved by reducing the total air to the burner zone with the balance of air being supplied as over fire air (OFA). Typically, the air required for stoichiometric combustion is introduced via the burner zone and the excess air enters the furnace at a higher level. Depending upon the burner design deeper staging may be possible where the burner is fired sub-stoichiometrically and the balance of air added as OFA. It is possible to have more than one stage of OFA (Figure 5.19).

This works best if the furnace has sufficient volume and therefore residence time to allow for significant primary combustion before the OFA is added.

- OFA flow = total secondary air – air at the burners.
- Total air = fuel × theoretical air × economiser exit λ.
- Theoretical air to fuel ratio = ratio of air to coal required to burn all the coal with no excess air. Typically, 8:1.
- If $\lambda = 1$, there is no excess air.
- $BZ\lambda$ = burner zone lambda.
- $WB\lambda$ = working burner lambda.

All lambdas change with boiler or mill load though it may be desirable to maintain $WB\lambda$ or even $BZ\lambda$ over a load range for NO_x control. The 8:1 is typical

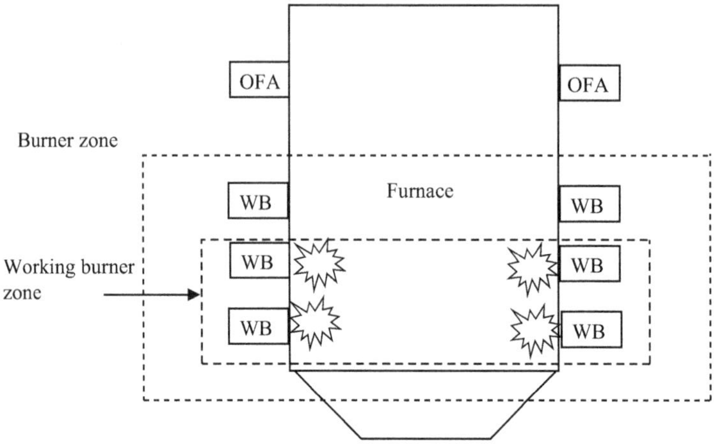

Figure 5.19 Definition of terms. Based on Doosan Babcock technology and published with permission

for some coals, but will be lower if firing brown coal lignite. Similarly, it may be 16:1 if firing fuel oil. The C&I engineer is expected to be aware of these ratios so he can challenge the 4:1 or 40:1 that are likely to be typing errors. However, the actual ratio 8:1 or 8.6:1 seconds determined by the combustion engineer for each project.

The logic can be developed in several ways. The following is the Doosan Babcock approach.

The combustion or process engineer determines the following:

1. Theoretical air to fuel ratio. This is a constant and based on the design coal.
2. The economiser exit λ over the boiler operating range.
3. The BZ λ over the boiler operating range.

The logic calculates the total air required by multiplying fuel demand by theoretical air to fuel ratio and (2) by economiser exit λ.

It also calculates the air to the burner zone by multiplying fuel demand by theoretical air to fuel ratio and (2) by BZ λ (3).

The latter is then subtracted from the total air to give the required OFA (Figure 5.20).

The calculated result changes only with fuel demand. That is, we could replace these six logic blocks with a single $f(x)$ of OFA against fuel demand. The advantage of the six blocks is that the actual calculation is visible and two crucial (fx) functions are used elsewhere to determine the O_2 setpoint and the WBλ. The O_2 setpoint is determined by the combustion engineer. However, the C&I engineer can estimate the value using the formula:

$$O_2 = 21 - \frac{21}{\lambda} \tag{5.1}$$

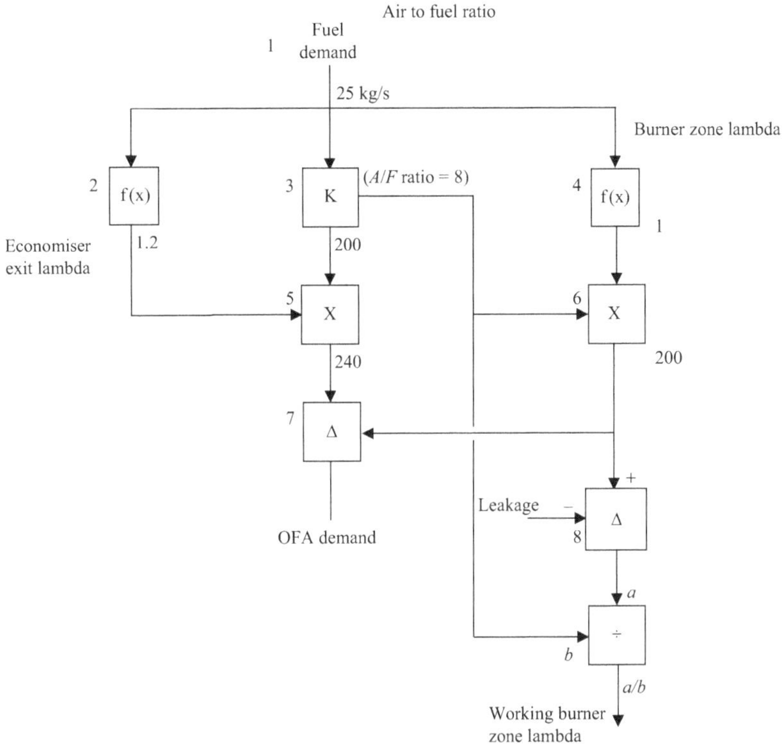

Figure 5.20 OFA logic. Based on Doosan Babcock technology and published with permission

For example, if lambda = 1. $O_2 = 21 - 21/1 = 0$, that is, no excess air
If lambda is 1.19.

$$\begin{aligned} O_2 &= 21 - 21/1.19 \\ &= 21 - 17.647 \\ &= 3.353 \end{aligned} \tag{5.2}$$

At low loads OFA is not usually needed.

However, a cooling flow of air is always required via the burner and OFA ports to prevent damage to the ports. Note the cooling air may well be at 350 °C.

Metal temperature thermocouples may be attached to the burner and OFA ports, to set up the correct cooling flows during commissioning. Sometimes these are retained and used to provide operator alarms.

An alternative configuration is to use boosted OFA which is delivered at higher pressure and velocity. This provides better penetration of the OFA resulting in better mixing and more complete combustion.

At low loads BOFA is not usually required but it is still necessary to purge the BOFA ductwork as part of the furnace purge. In order to maintain a cooling flow at low loads the BOFA fan is bypassed.

A HAZOP is likely to identify the risk of a runaway BOFA fan starving the firing burners and so logic is added to compare the measured flow with the expected flow and trip the BOFA fans if the flow is excessive.

To determine the OFA flow rate we need to know the working BZλ (WBZ λ) so we can set up the correct air to fuel ratio at the burners. Typically, the combustion engineer specifies the (total) BZ λ, say, 0.9 or 1.0 and the control engineer sets the logic to determine the actual Lambda for the firing burners. These are the same if all the burners are firing. However, most coal-fired plants operate with say five of six mills at full load. The air registers on the non-firing burners are opened to allow cooling air to pass. This cooling air contributes to the total available air for combustion. Hence the controlled air to the remaining burners must be reduced to maintain the required total air.

The principle of the calculation is as follows: Assume economiser exit lambda of 1.2 and OFA flow of 20 units. The corresponding BZ λ is 1.0 and requires 100 units of air. Then if all the mills/burners are firing each will have a lambda of 1.0. However, if only five mills are firing and four units of air are used to cool the non-firing burners, then only 96 units of air are needed for the firing burners and their lambda is 0.96. The furnace exit lambda changes depending upon the fuel being fired.

If the load reduces to 80% BMCR then there are 16 units of OFA and 80 units going to the furnace. If a mill is taken out of service and assume the two non-firing mills each require 4 units of cooling flow. Then the remaining firing burners need $80 - 8 = 72$ units, instead of the 80 units required for $\lambda = 1$. Hence their stoichiometry reduces to $72/80 = 0.9$.

So while the BZ λ is maintained at 1 the WBZ λ has reduced to 0.9. While this may not be a problem in this example if deeper staging is used for the OFA, with say 30% OFA then the same scenario at 80% BMCR results in a WBZ λ of 0.825, which maybe a problem for some burners with some fuels.

The logic can be improved to prevent the WBZ λ being driven too low. In its simplest form by an overall limit or in a more sophisticated way by also applying a load related limit to each mill.

5.6 Maintaining the furnace draught

The heavy solid line of Figure 5.21 shows the pressure profile through the various sections of a typical balanced draught boiler system. It shows the pressure from the point where air is drawn in, to the point where the flue gases are exhausted to the chimney, and demonstrates how the combustion chamber operates at a slightly negative pressure, which is maintained by keeping the FD and ID fans in balance with each other.

If that balance is disturbed the results can be extremely serious. Such an imbalance can be brought about by the accidental closure of a damper or by the sudden loss of all flames. It can also be caused by maloperation of the FD and ID fans. If the combustion process and fan fault conditions combine to significantly reduce the furnace pressure it may implode. The results of an implosion are

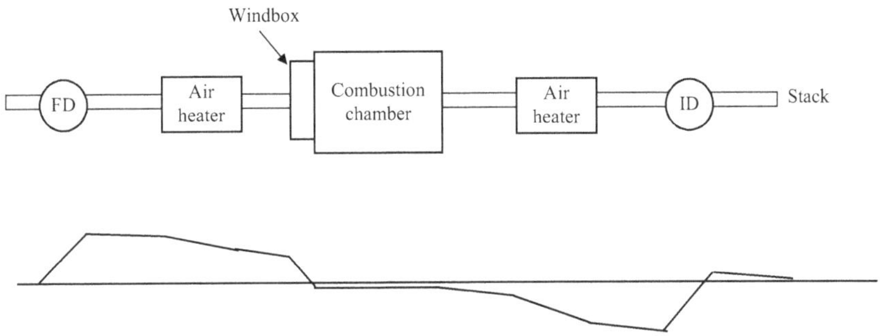

Figure 5.21 Draught profile of a boiler and its auxiliary plant

extremely serious because, even though the pressures involved may be small, the surfaces over which they are applied are very large and the forces exerted become enormous. Such an event would almost certainly result in major structural damage to the plant.

The logic starts really simply with the two ID fans being modulated in parallel to control the pressure in the furnace. However, if following NFPA 85 [1], it becomes one of the most regulated loops on the boiler,

NFPA 85 asks for the following:

- Three furnace pressure transmitters. Their median is used as the measured value.
- Directional blocking to prevent both sets of FD and ID fans moving on the wrong direction when an excess pressure deviation is detected. For example, if the furnace pressure is too high, reducing the ID fan throughput and increasing the FD fan throughput are blocked. Similarly, if the furnace pressure is too low, increasing the ID fan throughput and decreasing the FD fan throughput are blocked. The block affects both auto and manual control.
- High- and low-pressure alarms.
- A feedforward but not from measured air flow.
- A reduction in ID fan throughput following an MFT to minimise the sudden collapse in furnace pressure. The scale of the reduction increases with load. Some suppliers take this further and reduce the ID fan throughput on loss of any mill even without an MFT.

Various companies might add one or more of the following.

- Reduction is controller gain at low deviations to reduce wear on the fan linkages.
- Gain change depending upon the number of fans in auto.
- Gain change depending if the error is positive or negative.

Associated with the modulating control logic is the boiler protection logic initiating MFT or boiler trip. These are detailed in Chapter 3. This logic is applicable to oil-, gas-, coal- and biomass-fired boilers (Figure 5.22).

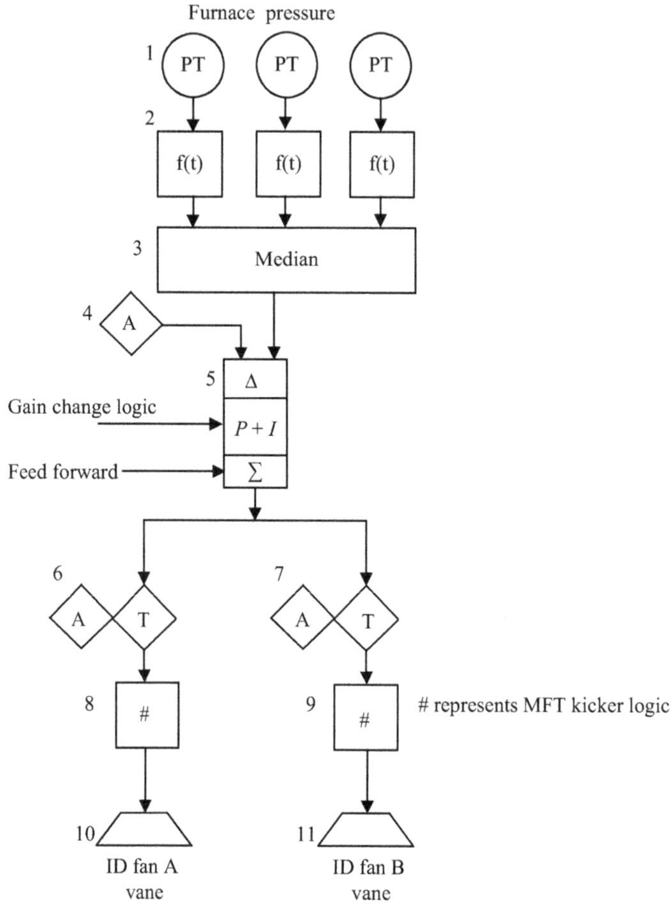

Figure 5.22 Furnace pressure control

5.7 Cross-over pressure control

The throughput of say two FD fans operating together can be regulated by a common controller or by individual controllers for each fan. Although a single controller cannot ensure that each fan delivers the same flow as its partner, this configuration is much simpler to tune than the alternative, where the two controllers can interact with each other and make optimisation extremely difficult. Whichever option is used, the control system must be designed to provide sufficient air to support combustion. One common approach is to use the single controller to move both fan vanes in parallel and then use a balancing loop to make small adjustments to vane position to ensure that both fans pass the same flow.

If we consider secondary air cross-over pressure control then Figure 5.10 shows how the air to a particular mill group is controlled by regulating the secondary

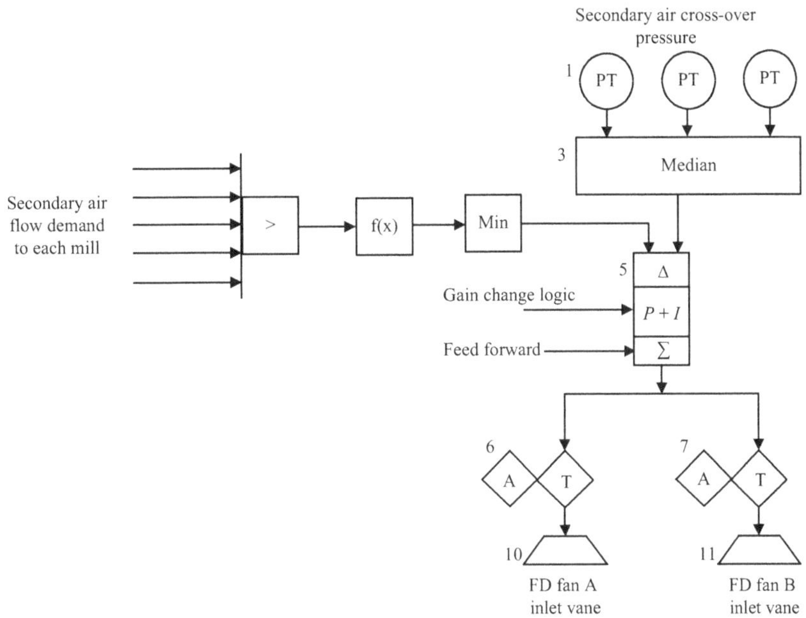

Figure 5.23 Secondary air cross-over duct pressure control

air flow to each windbox. In such cases this air supply is drawn from a common windbox which is maintained at a pressure which may be fixed or varying with boiler throughput.

Figure 5.23 shows how such a control system can be implemented. The desired value signal for the pressure controller is derived from load, so that the pressure in the windbox will change over the boiler load range to a characteristic that will be defined by the process engineer.

The minimum limit ensures that the pressure demand signal cannot fall below a predetermined minimum value. The measured value for the controller can be based on a measurement of the windbox pressure or the windbox-to-furnace differential pressure.

This arrangement works well if all the mills are in auto. However, if firing at say 60% BMCR a boiler that normally needs five mills of full load could have all five or just four mills firing but turned down, or three mills at maximum load. Any of these three combinations would achieve the 60% BMCR load. The three-mill combination will require more secondary air per mill than the five-mill combination and so need a higher cross-over pressure. To achieve this the demands for each mill are compared and the highest demand used to set the cross-over pressure. This tends to keep the secondary air control dampers on each windbox in a reasonable control position. Additional logic can be added to look at the damper positions and trim the cross-over pressure to give optimum damper control. This is often a trade-off between the control engineers who want the dampers, say, 50%–60% open to improve control and the process engineer who wants them as wide open as possible to reduce pressure loss and improve fan efficiency. Similar logic can be used for the hot PA cross-over pressure.

5.8 Binary control of the combustion system

So far, we have considered only the modulating systems involved with the combustion plant. In practice, these systems have to operate in concert with binary control systems such as interlocks and sequences. The purpose of an interlock is to coordinate the operation of different, but interrelated plant items: tripping one set of fans if another set trips, etc. The purpose of a sequence system is to provide automatic start-up or shutdown of the plant, or of some part of it.

The logic for interlock operations will be defined by the boiler designer and will probably have to comply with some local, national or international standard. The systems are very specific to the particular plant, and no attempt will therefore be made in this book to define these because the objective here is to provide a general overview of boiler control systems.

However, one topic that we shall look at is burner management since, like modulating loops, this type of system is very dependent on the correct operation of input and output transducers.

5.8.1 Burner management systems (BMS) and plant safety

The design of the BMS will aim to address critical safety issues, and the sequences for a given type of boiler or burner will be defined in conjunction with the plant designer, bearing in mind the requirements of applicable codes such as NFPA 85 [1]. In fact, the NFPA standard defines in some detail the exact sequences involved in lighting-off, monitoring and running-down operations of burners, and shows how these are to be linked with the plant interlock systems (e.g. ensuring that the furnace has been purged before any attempt can be made to initiate a burner light off sequence). For these reasons, the sequences will not be described here. However, attention will be paid to certain safety-related aspects of BMSs. Safety requirements are very comprehensively defined in every applicable standard. For example, NFPA 85 describes the events and failures which should be recognised in the design of the system.

5.8.2 Sequence controls

So far, we have considered modulating control logic for combustion efficiency furnace pressure control, etc., and discrete logic in the BMS to protect the boiler. In addition, sequence logic is required to ensure the safe start-up and shutdown of the plant. This is predominately Boolean logic and it is used to control everything from drain valves, air heaters, fans, mills circulating pumps and the overall sequence in which these are started and shutdown. This logic is not considered as challenging as the other two, because the actual requirement is usually defined by the equipment Supplier. They might say start the lube oil and hydraulic oil pumps before starting the fan motor.

There are significant challenges in stitching together the individual plant requirements into an overall boiler start-up logic, and determining the amount of operator intervention to be allowed. This requires close cooperation between the control and sequence engineers creating the logic and the process and operations engineers defining the requirements.

Part 2: Instrumentation

5.9 Furnace instrumentation

Only boiler-specific applications such as air flow measurement, flue gas analysis, furnace conventional closed-circuit television (CCTV), coal level and flame monitors are considered. The concepts are outlined later. There is more than one way to measure each parameter. The best choice is not always obvious and may be overridden by commercial decisions or the owner's specification but most often because of one engineers particularly good or bad experience. There is no logic to counter the last and you will soon find that a belief in extractive rather than in situ analysers, or Venturi rather than Pitot tubes is as ingrained as their choice in football teams! The following reflects experience gained on a range of predominately coal-fired boilers throughout China, Korea, the United States, and Europe.

5.9.1 CCTV systems

CCTV systems are used in several places around a power station. They are used with displays in the gate house as part of a security systems to monitor gateways, and perimeter fences and to monitor hazardous operations, for example, tanker unloading. They can be used to monitor a drum-level gauge and give the remote display in the control room to satisfy safety standards and to monitor the stack so that any smoke plumes can be seen.

Specialised CCTV are used to monitor inside the furnace. They are used to monitor the ash hopper and separate cameras are used to monitor the flames. Modern solid-state cameras are exceptionally reliable (up to 10 years) providing the camera's environmental control is maintained. There are two options for primary cooling of the camera housing that protects the electronics, water or air. Secondary cooling for the lens is always by air purge, as this allows the optical surfaces to be kept clear of dust and slag particles. The air system uses a vortex cooler.

While water cooling is more efficient at heat transfer from the camera housing, the additional complication of providing a clean water supply and drain will normally balance things in favour of air cooling. Extra precautions are needed on outdoor boilers to stop the water freezing. Air-only cooling requires a larger quantity of air. The secondary cooling needs clean instrument air to keep the camera lens clean.

At the temperatures typically associated with large boilers, loss of cooling will result in almost immediate camera damage or destruction. Because of this, most installations will mount the camera housing on a mechanism which will retract the

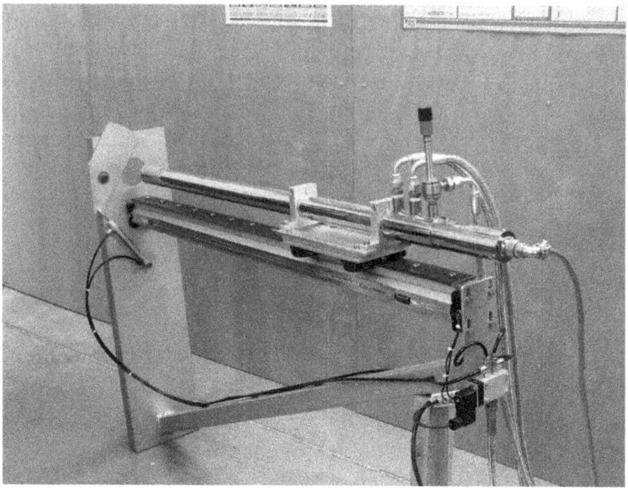

Figure 5.24 Camera with retraction unit. © Video Technology Ltd., reproduced with permission

housing from the boiler if it becomes too hot or on loss of services. A local control panel houses the camera power supplies and environmental monitoring systems. Transmission of the pictures to the plant operators may be via dedicated coaxial cable, optical fibre or Ethernet.

Amorphous ceramic coatings can be applied to the external surface of the camera housing to resist acid attack from high sulphur and chlorine content fuels. These are most likely to condense on water-cooled camera housings (Figure 5.24).

During the design stage, the layout engineer can check how much of the boiler interior can be seen from the camera position. Ideally with a 60° viewing field, although wider angle lenses are available greater than 90° is likely to be unusable.

5.9.2 Acoustic leak detection

While boilers are usually reliable, they sometimes have tube leaks. A small leak of high-pressure steam can quickly cut through adjoining pipes and eventually lead to a catastrophic steam leak. In a survey carried out in Europe and the United States, the Electric Power Research Institute (EPRI) found that around 6% of lost availability for power generation plants were caused by boiler tube leaks.

Originally developed for CEGB by Lintvalve (now Procon Engineering) a system was developed in the 1970s to alert the operator to potential steam leaks. The acoustic leak detection system consists of several acoustic microphones around the boiler casing that listen for leaks. The quantity of microphones is usually specified by the end user and might range from 16–20 on a 300 MW unit through to 24–32 on a 600 MW unit to 48 on a 1000 MW or above unit.

The sound levels from each device are displayed in a bar graph format on the screen and monitored as green orange or red, depending upon the DB reading. There are also elevation and plan displays to help locate the source of the leak. The system is nominally standalone and because it is not actually controlling the boiler may be shared between multiple boilers. The Procon Engineering system described here can accept up to 96 inputs. Other companies offer similar technology without the PC display allowing the end user to create his own displays on the DCS. In addition to the sound levels it is usual to display some indication of load.

The microphone needs protecting from the heat and ash from the furnace. See Figure 5.25 for a typical assembly. The tube distances the microphone from the

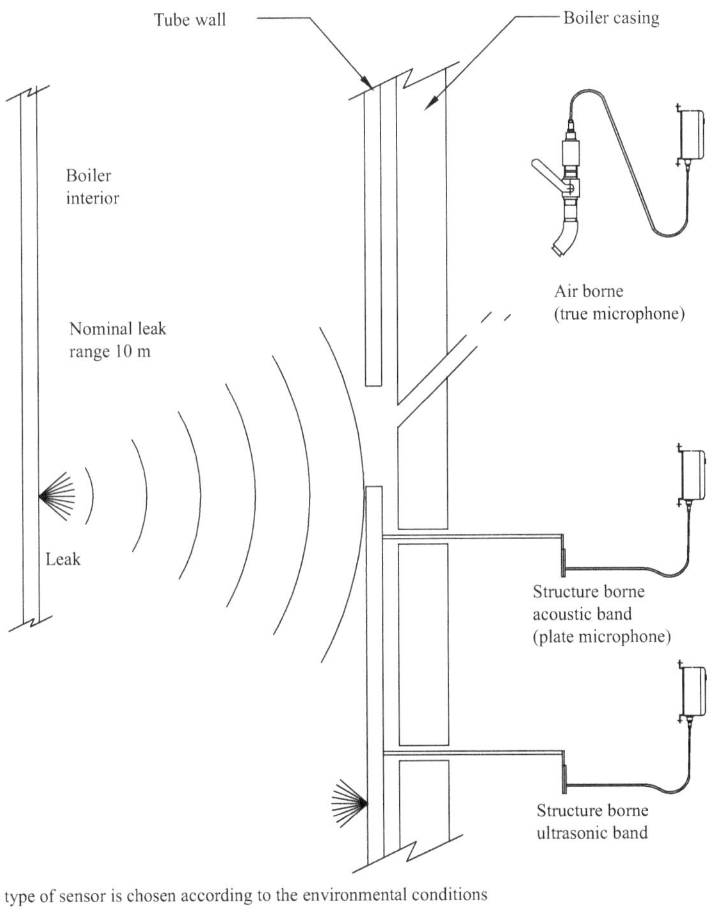

Figure 5.25 Acoustic probe assembly. © Procon Engineering, reproduced with permission

heat. Its angle encourages the ash to fall back into the furnace. In addition, an intermittent air purge of between one and five seconds is used to blow out the carrier tube. The tube penetrates the membrane between the boiler tubes.

In certain parts of the boiler where the ambient noise levels are high an alternative to the acoustic transducer is a structure-borne sensor. This comprises of a waveguide that is welded to the boiler wall that is connected to a piezo electric plate that detects vibrations. The same technology is used for down shot boilers where the inside of the furnace wall is refractory lined which blocks the sound waves.

Precautions are taken with all types to prevent any covering from rubbing on the probe. The location of a tube leak is found by interpreting the increasing sound levels from nearby microphones. The leak being closer to the noisiest detector. As a soot blower releases steam or air into the furnace this noise will be detected by the microphones. An experienced operator will recognise the noise profile that sootblowers make as they traverse the tube banks. The noise decreases as the sootblower moves away and increasing as it returns. Alternatively, the alarms can be masked during sootblowing. Apart from the obvious leak detection, there is anecdotal evidence to suggest that the system has detected open hatches in duct work.

5.9.3 Flame monitoring and purge air requirements

5.9.3.1 Flame monitors

The requirements for a comprehensive BMS have already been discussed in Chapter 3 and attention was drawn there to the importance of flame monitoring.

Monitoring the status of a flame is not easy. The detector must be able to discriminate between the flame that it is meant to observe and any other in the vicinity, and between that flame and the hot surfaces within the furnace. The detector must also be able to provide reliable detection in the presence of the smoke and steam that may be swirling around the flame. To add to the problems, the detector will be required to operate in the hot and dirty environment of the burner front, and it will be subjected to additional heat radiated from the furnace into which it is looking.

With their attendant BMSs, flame scanners of a boiler are vital to the safety and protection of the plant. If insufficient attention is paid to their selection, or if they are badly installed or commissioned, or if their maintenance is neglected, the results can be, at best, annoying. The problems will include nuisance trips, protracted start-up of the boiler and the creation of hazardous conditions that could have serious safety implications. The scanner system diagnostics alarm and trip the flame fail relay if any error is detected.

Figure 5.26 shows a typical flame detector and the swivel mounting that enables its sighting angle to be adjusted for optimum performance. A flame scanner is a complex opto-electronic assembly and modern scanners incorporate sophisticated technologies to improve flame recognition and discrimination. Although the

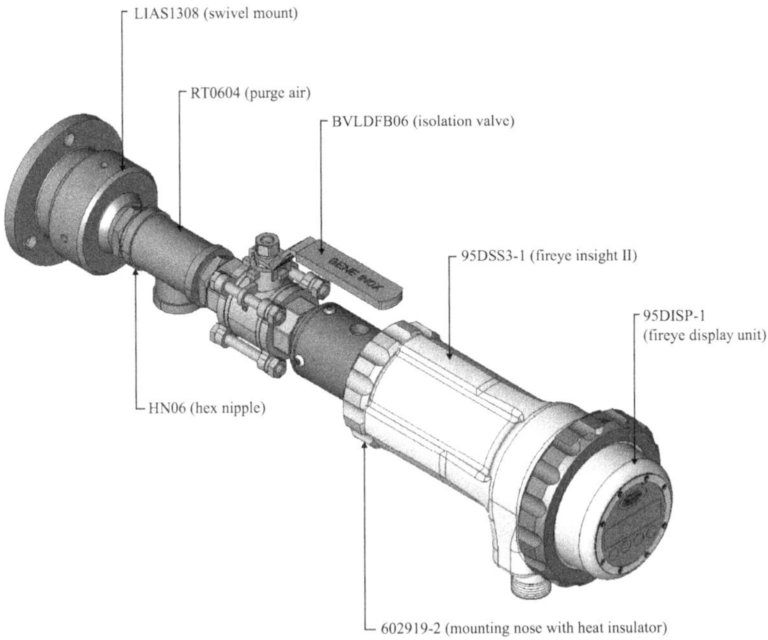

Figure 5.26 Flame monitor complete assembly. © Lias Industrial Ltd., reproduced with permission

electronics assembly will be designed to operate over a wide temperature range (typically –40 °C to +85 °C), unless great care is taken this value could easily be exceeded and it is therefore important to take all possible precautions to reduce heat conduction and radiation onto the electronic components. The illustration shows how a heat insulator is used to prevent undue heat being conducted from the boiler structure to the electronics enclosure.

The spectrum of radiation from a flame is determined by many factors, including the type of fuel being burned and the design of the burner. The intensity of the flame tends to be low for gas and high for coal and oil. The flame will also flicker and, in general, low-NO_x burners will demonstrate a lower flicker frequency than gun-type burners. Oil and coal flames tend to produce a higher degree of infrared radiation, whereas a gas flame is rich in ultraviolet radiation. Radiation in the visible part of the spectrum will also depend on these factors, but these days the tendency is to use detectors whose response is biased towards either the infrared or the ultraviolet end of the spectrum since emissions in these ranges provide better indication of a flame than visible radiation, which can be plentiful and misleading. See Figure 5.27 for definition of spectrum parameters.

See Figure 5.28 for flame emissivity curves and ranges of different detectors. Each type of fuel also produces by-products of combustion, which affect the

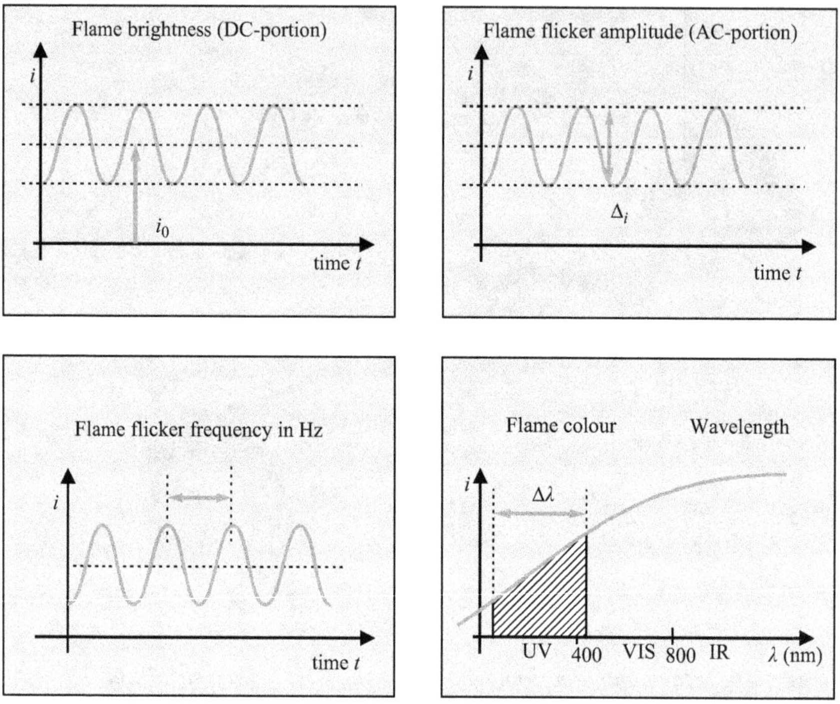

Figure 5.27 Definitions of flame spectrum parameters. © Durag Group, reproduced with permission

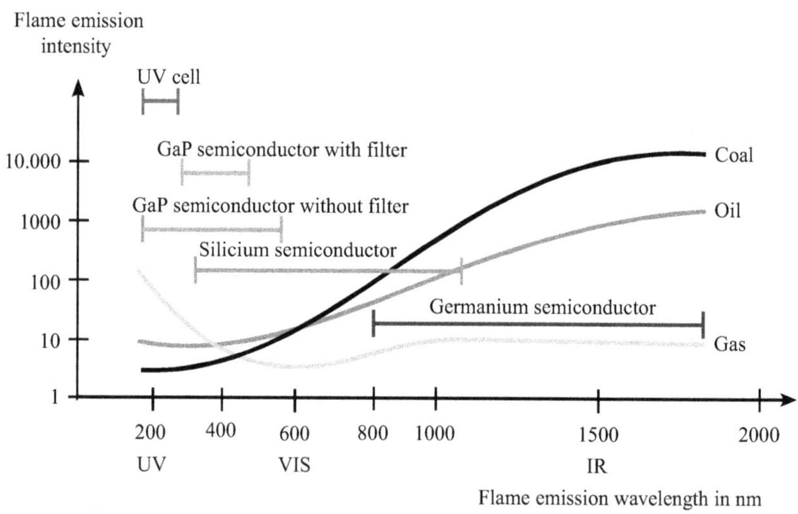

Figure 5.28 Flame emissivity curves and ranges of different detectors. © Durag Group, reproduced with permission

transparency of the flame and therefore the blanking effect it has on adjacent flames or on any flames on the opposite side of the furnace. Oil and coal flames tend to obscure infrared radiation, while gas flames produce water vapour which obscures ultraviolet radiation.

Choosing the scanner type used to be easy, UV for gas IR for coal and some discussion over the best choice for oil. Sometimes IR for heavy fuel oil and UV for diesel. Now as detector technology has improved there are additional choices of gallium and silicium semiconductors. See Figure 5.35 for flame emissivity curves and ranges of different detectors.

Reputable manufacturers will be pleased to provide application-specific guidance. At the design stage, this advice will be based on previous experience of similar installations. For a retrofit on an existing plant, the manufacturer should be asked to carry out a comprehensive site survey, using various types of scanner, while the burners are started, operated under various loads and stopped. Several tests may be required, and a survey may last for several days. The greater the attention that is paid to this study, the better will be the performance of the final installation. As a minimum for a new build the specification to the flame monitor Supplier should include the fuel specifications and sufficient furnace drawings for them to determine burner separation and to then guarantee flame discrimination.

While line of site flame monitors with limited adjustment via the swivel are excellent for viewing oil or gas burners on a front-wall-fired furnace there are times when fibre optics are required. They have been used to look sideways at the root of the flame, to avoid pick up from opposite burners and they are essential when used on a corner-fired furnace with tilting burners. See Figure 5.29 for the fibre optics used in a tilting burner.

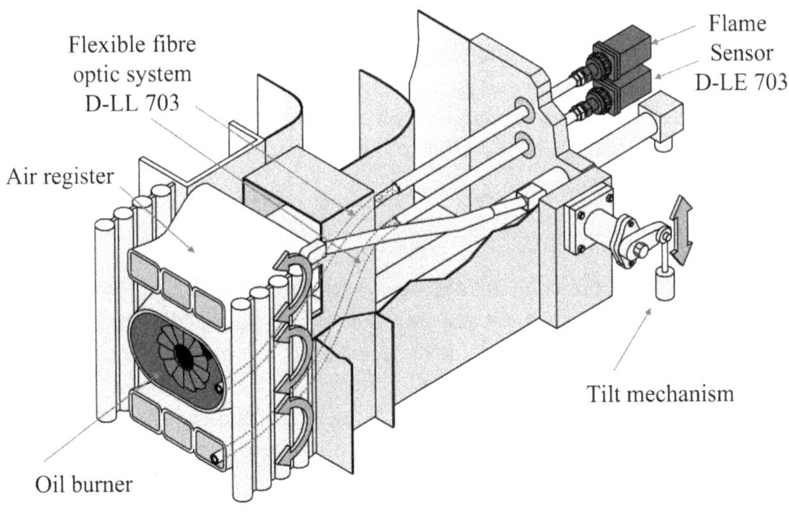

Figure 5.29 Flexible fibre optics used in a tilting burner. © Durag Group, reproduced with permission

Figure 5.30 Video-based thermography system. © Durag Group, reproduced with permission

The fibre optics offer the best solution for difficult access or customised view. They have two drawbacks. They are more expensive, doubling the cost of the flame monitor and 'glue' holding the fibre optic bundle together melts at 350 °C and requires an enhanced purge air system.

Companies are looking at unusual ways to monitor the flame. One possibility is the video-based thermography system, which combines the benefits of a CCTV with temperature profiling. Although this has not been used in place of conventional flame monitors it is a welcome addition to the conventional CCTV system allowing temperature distribution to be monitored particularly on large scale burner test rigs and as an indication of complete burn out on travelling grate incinerators (see Figure 5.30).

5.9.4 The requirements for purge air

The purge air that is supplied to the scanner serves two purposes: it provides a degree of cooling and it prevents dust, oil and soot from being deposited on the optical parts of the unit. The air should be available at each burner, even if the burner itself is not operating. The same supply is often used to keep the observation ports clean.

It should therefore be obvious that the air used for purging should be cool, dry and clean, and that it should be available at all times. But in many cases these requirements are ignored, and the performance of the instrument is thereby inevitably degraded.

Purge air can be obtained from the instrument air supply or it can be provided by dedicated blowers. In some cases, it is taken from the FD fan discharge.

Each of these is viable, provided the requirements outlined earlier have been thoroughly considered. It is also important that the presence of the purge air supply should be monitored and its loss transmitted to the DCS because a long-time failure of the air supply could result in expensive and possibly irreparable damage to the scanners. There is some debate on tripping the furnace on loss of purge air. A long-term loss will lead to the optics becoming dirty and eventually all the burners tripping when the flames are no longer detected. With a properly installed thermal barrier the 'cooling air' can be tripped without the monitor overheating. If this is accepted by the Client then the loss can be alarmed and he can schedule a shutdown rather than immediately lose generation.

If fibre optic flame monitors are used cooling air is always required to keep the fibre optic bundle cool. This requires around twice the amount of purge or cooling air used on a conventional flame detector. Loss of air results in loss of all the flame monitors, which requires two or more redundant fans, and at least one must be powered from a secure back up, usually it has a DC motor.

5.10 Oil (and gas) flow instrumentation

Where oil is used as a support fuel the orifice plate is the simplest solution. It is well understood, low cost, usually able to buy in a carrier tube and so completely assembled and can be calibrated before delivery. Because oil is not burned all the time and there is no requirement for fiscal measurement. Most end users will accept duplicating the tapping points and transmitters with a single orifice plate as providing independent measurement. SIL 2 DP transmitters are readily available.

They have a limited turndown with a maximum control range of, say, 6:1 and indication range of 10:1. At start-up approximately 10% of the fuel is recycled to the tank, so the 10:1 indication range is never exceeded. During start-up if only a few oil burners are in service the excess air is so high that inaccuracies do not matter. During starting and stopping of a mill, all of its oil burners are lit and close to maximum firing, so say a fifth or a sixth of the maximum flow. That is, the turn down is satisfactory when oil is used as a support fuel.

Where the oil is also capable of 50%–100% BMCR firing then owners ask for a more accurate measurement. Traditionally oval wheel meters were used. This is a mechanical device, very accurate but expensive. Turbine meters have also been used but are not recommend as a dirty fuel can clog the mechanism, stop it turning. The control system then drives the valve wide open!

A popular compromise is to use Coriolis meters. More expensive than an orifice but more accurate and with a good turndown and SIL capable. Two small cautions:

If used on heavy fuel oil, where the lines may be steam cleaned, you need to rate them for the steam temperature.

If redundancy is specified the only way to achieve it is to duplicate the meters.

Figure 5.31 shows a U-shaped Coriolis meter used for gas flow measurement.

Figure 5.31 Coriolis meter used for gas flow. © Emerson, reproduced with permission

5.11 Air flow

The most common measurement techniques with their advantages and disadvantages are listed later. They have all been used successfully but sometimes commissioning has been difficult.

5.11.1 Venturi

The duct is collapsed usually on all four sides and the resultant pressure drop measured. The principle is well documented and understood; see BS EN ISO 5167 (see Figure 5.32). They can be manufactured from the same material as the duct work. They are relatively long so do not fit in short secondary air ducts. They may need extra stiffening to withstand maximum duct design pressure. If the ductwork can accommodate them and their pressure loss they are my preference. They can be used in the traditional sense with a DP transmitter but this has limited turndown or with any of the following devices (except the aerofoil) which can be mounted in the Venturi throat. In this arrangement, the Venturi acts a flow conditioner, and its reduced throat makes most of the other devices cheaper as less probes or sensors are required.

5.11.2 Aerofoil

A series of wing-shaped sections in parallel across the duct (see Figure 5.33). They require a shorter duct length than Venturi. Parallel sections lend themselves to splitting the duct and providing two independent measurements. They are not as well understood as Venturi and need better precision in manufacturing. They are prone to impulse line blockages but this can be reduced if manufactured as a seam welded pressure vessel. Any leak in the welding will jeopardise the flow reading.

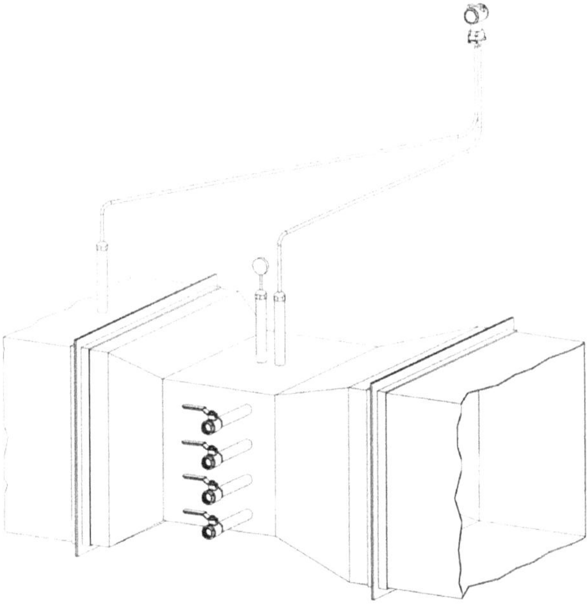

Figure 5.32 Typical Venturi. © Storm Technologies Inc, reproduced with permission

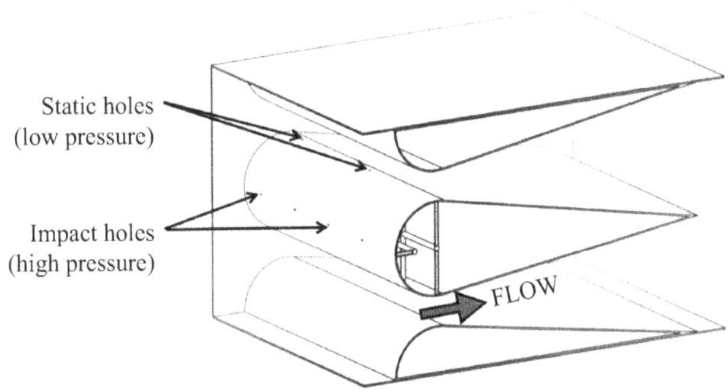

Figure 5.33 Typical aerofoil. © Storm Technologies Inc, reproduced with permission

5.11.3 Thermal mass flow

Sometimes called hot wire anemometer. Relies on the cooling effect of the fluid over the RTD. A thermal sensor measures the difference in temperature between a reference RTD (ambient temp) and the active RTD which is preferentially heated with a constant power heater. At zero flow delta T is maximum and at highest flow Delta T is at minimum. See Figure 5.34 for the flow meter with constant power heater or alternatively with a variable power heater as recently launched by FCI.

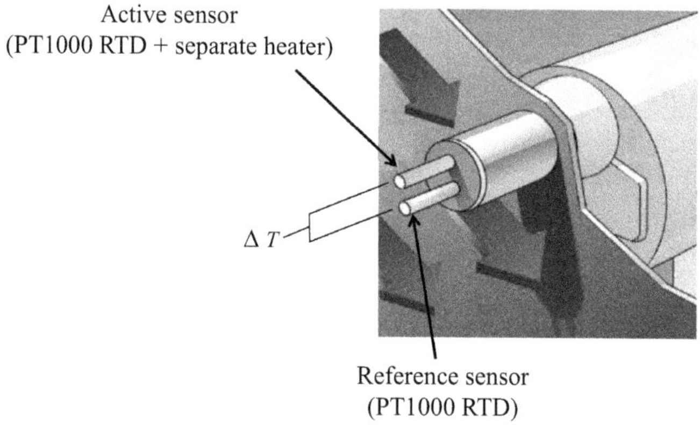

Figure 5.34 Flow meter with constant or variable power heater. © FCI, reproduced with permission

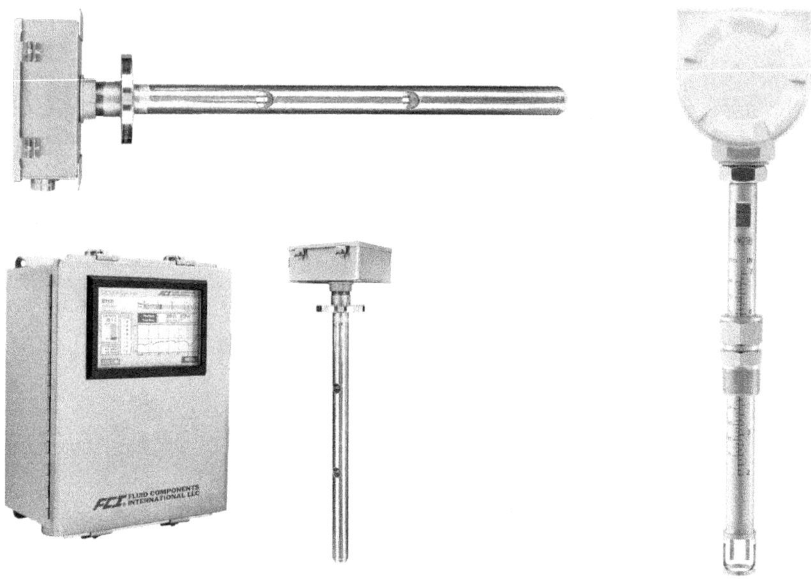

Figure 5.35 Alternative probe arrangements. © FCI, reproduced with permission

The figures showing FCI probes were provided by Allison Engineering who distribute these probes in the UK.

The Delta T is directly proportional to the mass flow because the cooling effect of the air on the heated RTD is a function of the velocity *and* the density of the air. Hence if the pressure and therefore gas density changes then for the same velocity the temperature difference will change and the new flow will be calculated without the need for external density compensation. They come in many designs form single point to multipoint (see Figure 5.35).

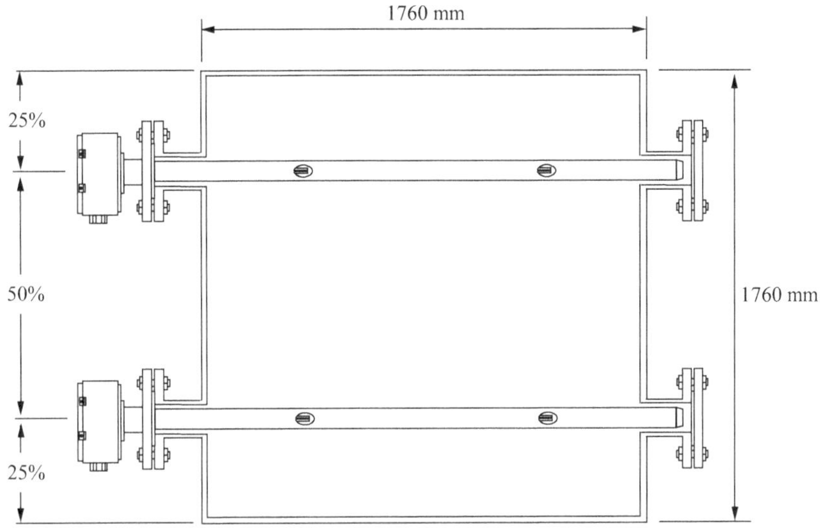

Figure 5.36 Typical arrangement as used for OFA systems.
© FCI, reproduced with permission

They are often used on retrofit projects, for example, when adding an OFA system. Rugged design with no moving parts, suitable for power plant application. Standard electronics can average up to eight-point measurements. They can be used in gas up to 450 °C and has minimal pressure drop. Can be rated for SIL 1 and /or ATEX Exd certified for use in hazardous areas.

Note the constant power heater design takes a short time to follow step changes in temperature. This is not normally a problem in combustion air and OFA systems, but might cause concern on PA systems where a low PA flow is used to trip the mill. In these circumstances, a variable power heater with a faster response can be used.

See Figure 5.36 shows two carrier tubes each with two thermal mass flow sensors, as used on OFA projects. The four points are averaged in the common electronics unit.

5.11.4 Capacitance/cross-correlation

The Promecon system uses two probes installed one behind the other and spaced by 350 mm; as the dust passes the first probe it changes the capacitance between the probe and the pipe. This change in capacitance is repeated as it passes the second probe. Cross-correlation techniques are used to monitor the changing capacitance and record the time lapse between seeing the same pattern on the two probes. By knowing the time taken and the distance between probes the velocity is accurately calculated. The probes are made of a hard-wearing tungsten carbide material. The probes are chosen from two standard lengths. The system uses the capacitance between the whole probe and the wall so different speeds along the probe are averaged out. The main advantages of this system are that it is not affected by the dust, which might block DP devices, and the long life of the probes. The system can be SIL 2 rated. Using an emitter upstream of the probes allows it to work with clean air and increases the turn down range to between 3 and 20 m/s so that cooling flows to burners can be measured (see Figure 5.37).

This provides a user-friendly solution on regular flow measurements as might be found in a Venturi throat without the usual risk of impulse tube blockage. It can be used on wider ducts by paralleling up a second set of probes to the same electronics to average over a wider area. While this is excellent for small ducts or in a Venturi throat it is effectively measuring in one plane over part of the duct and so may not be as accurate in larger ducts. Where the advantage of a Venturi or straightening device to condition the duct is not available a second independent system can be used in each half of the duct. The positioning of the probes is determined by CFD studies.

On the extremely large air heater ducts a multi-zone sensor system is available to measure the air flow at different locations in the duct. This system also measures the O_2 and so can be used to determine air heater leakage (see Figure 5.38).

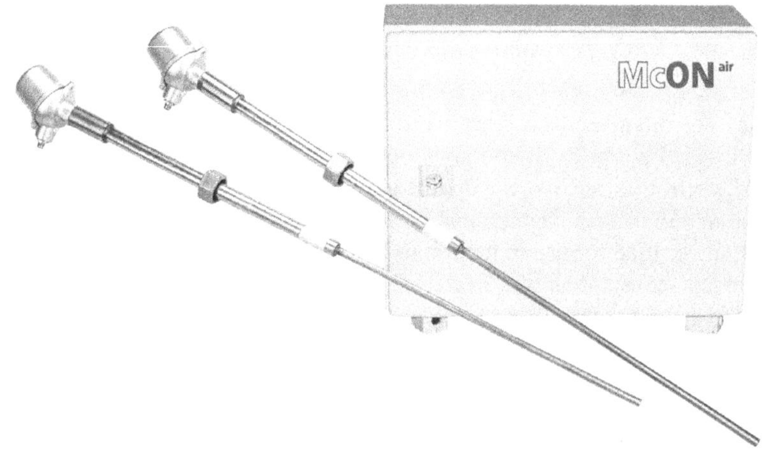

Figure 5.37 Typical capacitance probes and electronics for transmitter. © Promecon, reproduced with permission

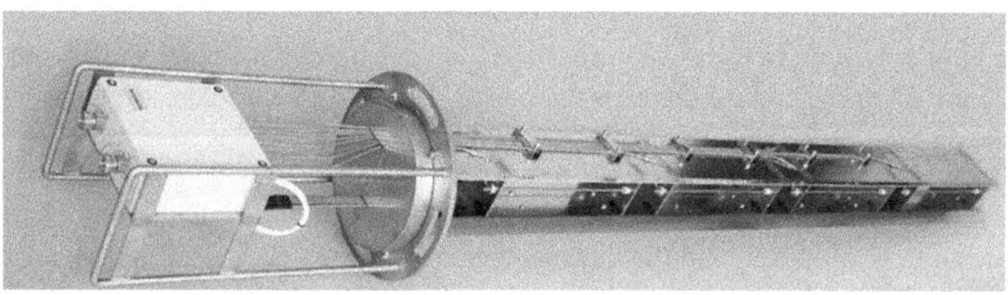

Figure 5.38 Multizone sensor. © Promecon, reproduced with permission

Figure 5.39 Coal flow system. © Promecon, reproduced with permission

The same system can be used to measure the velocity of PA with coal suspended in it. To be effective one flow measurement is required for each PF line. In addition to the flow the density is required and this is determined from microwave resonance. The measurements from each PF line are wired back to a common cubicle where the flow calculations are carried out. These measurements can be used during commissioning to balance the flow to all the burners supplied from one mill. If motorised variable orifices are used then the operator can also balance the flow during normal operation (see Figure 5.39).

5.11.5 Averaging Pitot tube system

A single-averaging Pitot measures the velocity across the probe. It gives a true average along the probe but does not allow for poor distribution across other parts of the ductwork. Figure 5.40 shows three single-averaging Pitots. These are low cost and have minimal pressure loss.

Figure 5.40 Averaging Pitots. © Air Monitor, reproduced with permission

To give a more accurate measurement multiple averaging Pitot tubes can be used. This could be multiple sets of single Pitots each with their own transmitter or it could be a set of single-averaging Pitots that are then piped together outside of the ductwork using a single transmitter. This overcomes measurement errors caused by poor distribution across the duct.

The Pitot installation requires some care. If the air flow hits the Pitot at a significant angle it distorts the pressure reading. This can be overcome by using an upstream flow straightener (see Figure 5.41). This arrangement with multiple probes provides a comprehensive flow measurement system.

While all DP flow measurements may suffer from blockages or impulse pipe leaks this technique is most vulnerable because of the very low DPs. Suppliers of multiple Pitot systems offer automatic purge systems and bespoke DP transmitter designed for low DPs. See Figure 5.42 for the transmitter and purge unit. They can be re-zeroed during the purge period to overcome zero drift. The flow signal is frozen during the purge period. It is usual to freeze the control system at the same time. This is usually not a problem in normal boiler operation as the HAZOP team may look for alternative trip signals on critical PA flows and as the transmitter is a bespoke design it may not be SIL rated.

Figure 5.41 *Integral flow straightener and six averaging Pitots supplied as a single assembly. © Air Monitor, reproduced with permission*

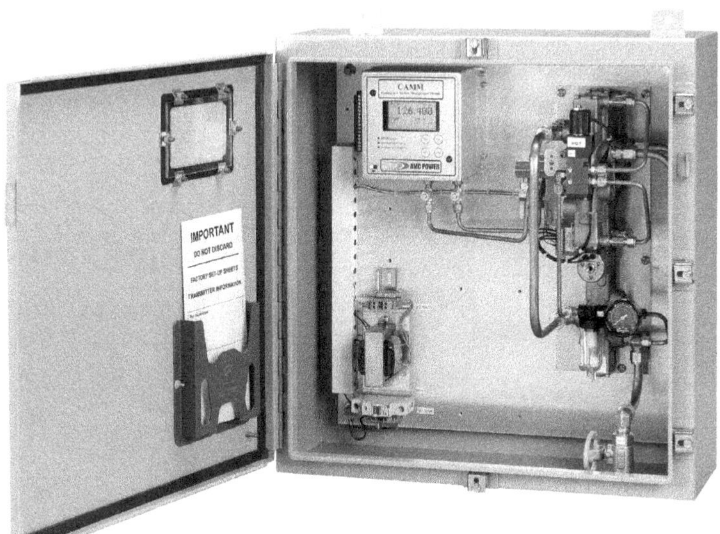

Figure 5.42 *CAMS. © Air Monitor, reproduced with permission*

5.12 Flue gas analysis

5.12.1 Oxygen and CO analysers

The transmitters used for measuring flue gas oxygen are usually based on the use of zirconium probes, whose conductivity is affected by the oxygen content of the atmosphere in which they are installed. These have proven to be reliable and are the de facto standard. They are accurate to ±1% of reading or 0.1% Oxygen. The zirconia analyser measures the raw sample, that is, on a wet basis and care must be taken to make the correction in logic if the O_2 setpoint is given on dry basis. They can be used in parallel with a cross-duct CO analyser. Some cross-duct CO analysers are affected by glowing dust particles common at the economiser exit. Under these circumstances the analyser could only be used to indicate CO break through and is not accurate enough for combustion control.

Recently a combustibles sensor has been added to a zirconia probe system to monitor the CO equivalent (COe). This sensor detects combustibles including CO, CH_4 and H_2. It uses a thick film calorimeter monitors and COe levels with a precision of ±25ppm. Combustible analysers are recommended by NFPA 85 [1] for use in reburn applications, where it is used to ensure complete combustion of the reburn fuel.

Although originally supplied as in situ analysers, suppliers now recommend a close extractive analyser system. (see Figure 5.43).

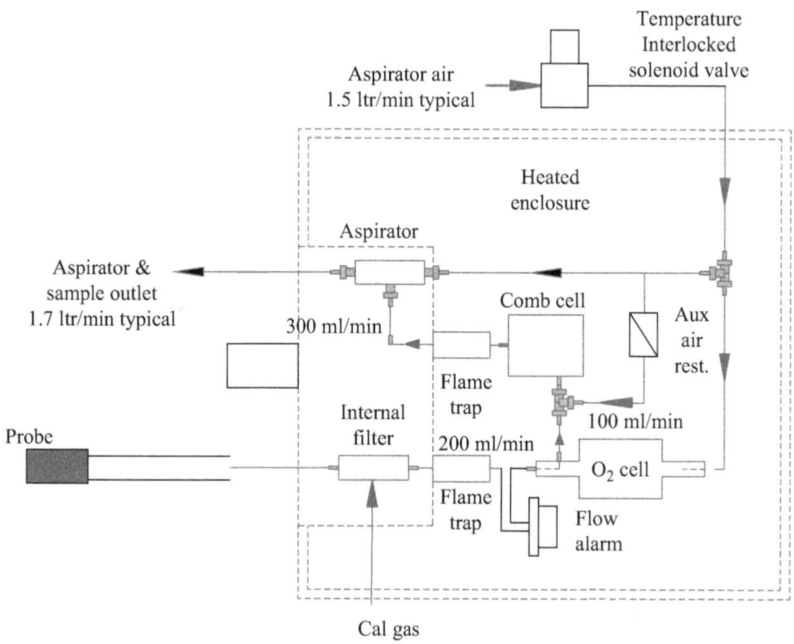

Figure 5.43 Heated enclosure for O_2 and COe analyser. © Servomex, reproduced with permission

A minimum of two O_2 analysers are provided per boiler side to provide some redundancy. As boiler size has increased three per side is the norm and some clients ask for six per side.

Although zirconia O_2 probes and COe are ideal for control of the boiler, Tuneable diode lasers (TDL) are now available for both O_2 and CO. They are more expensive than conventional analysers but if O_2 is deemed part of a safety loop they can be certified suitable for use in SIL2 SIS. CO analysers are recommended as a safety measure for furnaces in the petrochemical sector by The American Petroleum Institute (AP RP 556). If this requirement migrates to power station boilers then TDL will be more appropriate for COe because of their SIL2 potential.

Infrared cross-duct CO systems would pick up glowing ash particles as CO and were not accurate. Lasers provide a more accurate CO measurement and can be certified for use in a SIL2 loop if needed.

5.12.2 NO_x measurements

NO_x was originally measured as part of a continuous emissions monitoring system (see 5.10.3). More recently they have been used to measure the NO_x into and out of the selective catalytic reduction (SCR) systems. The inlet NO_x measurement multiplied by the gas flow indicates the total amount of NO_x present and allows the required ammonia flow to be calculated. This feedforward control injects approximately the correct amount of ammonia. A feedback loop is also provided based on the actual exit NO_x level that trims the ammonia demand.

In situ analysers can be provided to measure the NO_x, but multiple analysers are required at both the inlet and outlet, which becomes expensive. SICK developed their SCR bypass system to measure a representative sample. See Figure 5.44 for the probes.

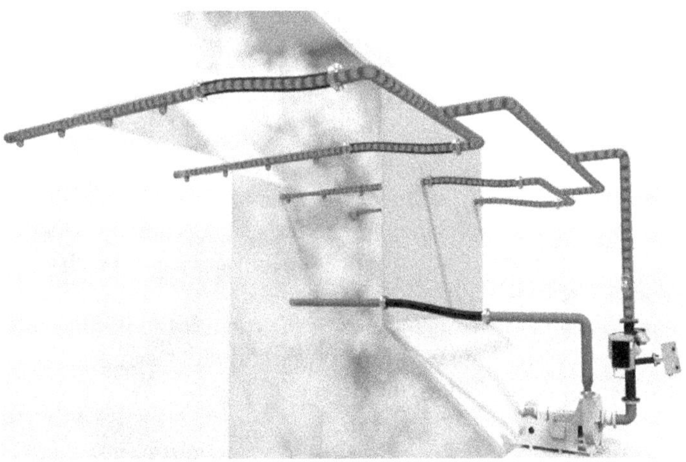

Figure 5.44 Multiprobe extractive NO_x system. © SICK, reproduced with permission

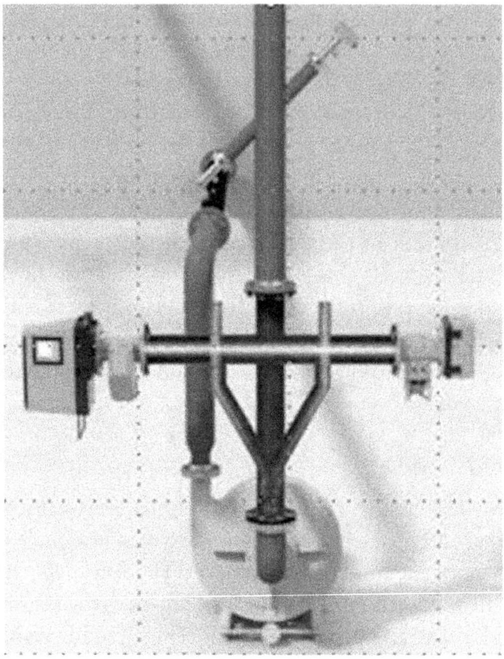

Figure 5.45 Cuvette for NO_x measurement. © SICK, reproduced with permission

The system has a number of probes typically four per reactor. Each probe has a number of holes sized so that each draws the same amount of gas into the system. This representative sample is then passed through a cuvette across which a cross-duct UV analyser is installed. The transmitter and receiver for these are kept clean by a purge air system. Additional instruments are provided to measure oxygen, pressure, temperature and humidity. The bypass flow is also measured so that loss of sample can be alarmed. The sampled gas is then returned via a fan to the reactor. In designing this system, the C&I and process engineers need to consider the overall reliability. This can be improved by completely duplicating the system or just installing a standby fan. Some redundancy is possible by using the analyser on the other side of the boiler. If the boiler is in steady state the NO_x reading at the inlet to both SCRs should be the same unless there is a combustion problem. The outlet readings are more likely to differ. Some possible strategies are to switch to just feedforward control with a less onerous NO_x target. Some clients will also ask for ammonia slip to be measured. This provides an alarm if too much ammonia is injected or the catalyst life is almost over so the ammonia is not consumed. Figure 5.45 shows the cuvette to carry the NO_x analyser. The gas flow measurement is above the cuvette.

5.12.3 Continuous emission monitoring systems

The local environmental protection agency requires power station to monitor the stack emissions on a continuous basis to prove that the emissions do not exceed the

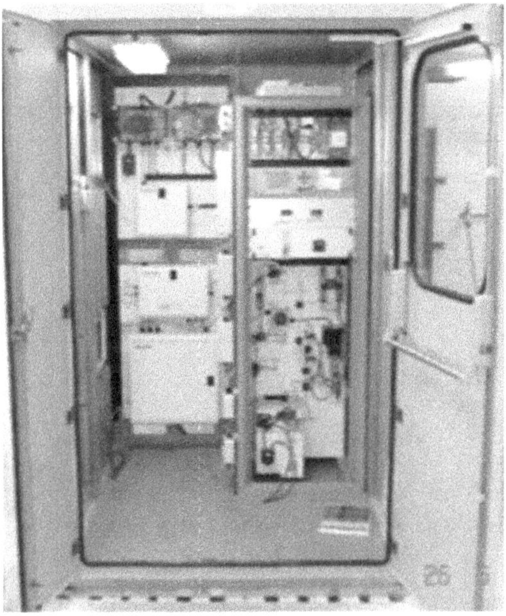

Figure 5.46 Typical extractive CEM. © Servomex, reproduced with permission

national regulations. They may require to reports to be automatically sent to them each month. There are three types of systems: extractive, cross-duct and in situ. While each has its advantages, environmental protection agencies recommend the extractive system.

A typical system will measure O_2, NO_x, CO_2 flow and temperature. As these do not influence the actual boiler performance it is normal to provide the type requested by the owner. See Figure 5.46 for a typical extractive system.

5.13 Mill and silo fire detection and penthouse monitoring

Coal can self-combust as it starts to oxidise it gives off CO and hence CO monitoring is a good early warning of a coal fire. This approach can be used for coal bunkers and for coal mills. NFPA 850 Recommended Practice for Fire Protection for Electric Generating Plants and High Voltage Direct Current Converter Stations recommends monitoring coal silos each day [1]. It suggests that this can give up to 24 hours advanced warning. Carbon monoxide detection is strongly recommended for coal mills. The same techniques are also applied to biomass silos and mills. An additional CO_2 measurement is sometimes used for biomass silos because bacterial action such as fermentation produces CO_2. This is also a source of heat and can lead to spontaneous combustion. For an out-of-service mill CO detection will work in a comparable way. While there is no mandatory requirement for mill CO detection, competent

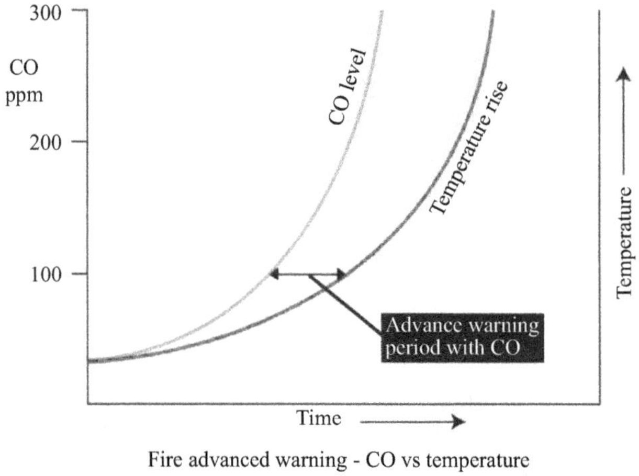

Figure 5.47 Graph showing advanced warning gained by using CO instead of temperature. © AMETEK Land, reproduced with permission

engineers will provide this system or carry out a risk assessment to demonstrate that other safeguards are adequate. See Figure 5.47 for CO and temperature against time.

The original Mill Watch® was a CO detection system using a single probe and analyser for each mill. Tube mills were supplied with two, one per end. Now there are several systems available. All draw a sample of air from the mill and pass it through an analyser. It is used to alarm on high CO levels giving an advanced warning of mill fires. The system is relatively expensive as 'just an alarm' and some suppliers offer a multiplexed system where samples are drawn from several mills or coal bunkers and fed to a single analyser. The analyser can be a pellistor or electrochemical sensor. The latter is more sensitive. Both have a shelf life of around 2–3 years meaning an in-service life of 12–18 months. If a multiplexed system is used, a backing pump should be provided to continuously draw a sample from each mill to the multiplexer. Then local to the analyser the correct sample is connected overcoming the large transport times had the sample been drawn from one end of the mill bay. As with all instrumentation measuring PF area there is a risk of PF dust blocking the system and so a regular purge is used to keep the probe clean. The multiplexing and purge sequencers are controlled by a local programmable-logic controller. While this multiplexed approach is adequate for detecting spontaneous combustion, it is no longer considered fast enough for use on in-service mills where fires form rapidly. Mill fires often originate in the underbowl area which suggests that static fuel has been allowed to accumulate or an in-service mill might catch fire due to burning coal being passed via the feeder.

Figure 5.48 shows a Land Mill Watch system and the CO probe. The analyser cabinet includes sample conditioning, gas measurement and calibration as well as the control electronics. A twin-stream version is available which continuously measures samples from two measurement points.

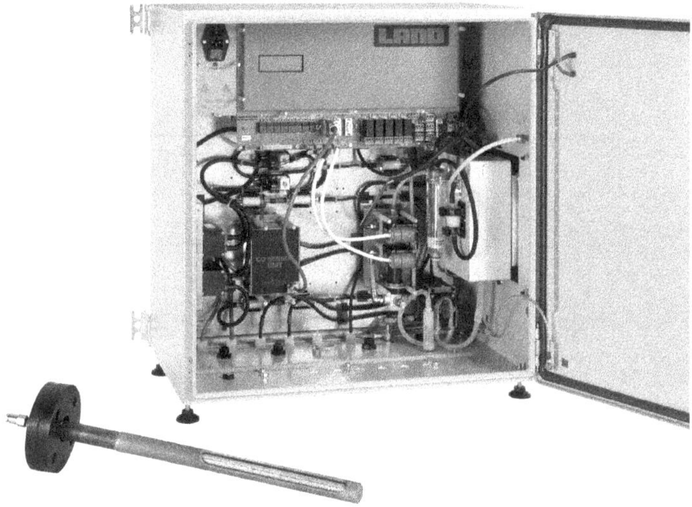

Figure 5.48 Land Mill Watch system. © AMETEK Land, reproduced with permission

Following a mill fire it is normal to inert the mill with steam or nitrogen or CO_2 to smother the fire. They are then kept inerted until the mill cools to ambient, and it is reasonable to assume that there are no fires in the mill. The inerting medium reduces the oxygen in the air to safe level. NFPA 69 states 60% of limiting oxidant concentration (LOC) if calculated or a higher level if the O_2 is not actually measured [1]. The additional cost of the O_2 meter means they are not normally supplied.

Probes are usually mounted in the classifier area. For indoor applications, a plain stainless steel tube can be used to connect the sample probe to the analyser. A heat-traced sample line is required in most outdoor locations. A typical probe with an abrasion shield is also shown in Figure 5.48.

A similar system but looking for methane may be used in the penthouse to detect unburnt gas leaks on gas-fired boilers.

5.14 CIA measurements

If the fuel is not properly burned some carbon will be carried over into the fly ash. This means that the combustion process is not efficient and so fuel is wasted. Also, they cannot sell the ash if the carbon content is too high. Figure 5.49 shows a carbon in ash analyser with an Archimedes screw that tacks the sample into the analyser. Both screw and analyser are at precipitator hopper temperature, typically 80 °C to avoid condensation. The signal is transmitted to a central cabinet in the equipment room. The CIA signal can be used in the DCS to slowly trim the excess air in the furnace to optimise combustion.

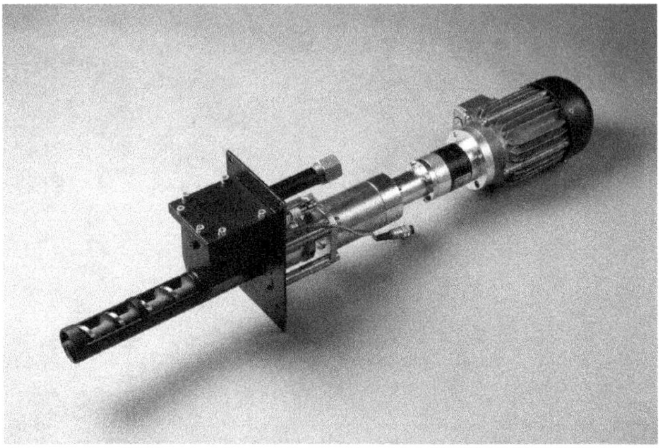

Figure 5.49 Carbon-in-ash analyser. © Promecon, reproduced with permission

5.15 Summary

Having looked at the control systems applying to the combustion and draught plant, with their associated instrumentation, in the next chapter we shall turn our attention to the feedwater systems.

References

[1] NFPA 85 'Standard for the prevention of furnace explosions/implosions in multiple burner boilers'. National Fire Protection Association, Batterymarch Park, Quincy, MA, USA.

[2] Anson D, Clarke W H N, Cunningham A T S, and Todd P. 'Carbon monoxide as a combustion control parameter'. *Inst Fuel*, 1971, **xliv** (363).

[3] Dukelow SG. 'The control of boilers'. Instrument Society of America, Research Triangle Park, NC, USA, 1991.

[4] Eskenazi DV, D'Agostini M, Levy, EK *et al.* 'On-line measurement of unburned carbon'. *EPRI Heat Rate Improvement Conference*, EPRI, Palo Alto, CA, 1989.

[5] Hurt R H, Lunden M M, Brehob E G, and Maloney J. 'Statistical kinetics for pulverized coal combustion'. *Proceedings of the 26th International Symposium on Combustion*, Naples, Italy, 1996.

Chapter 6
Feedwater control and instrumentation

David Lindsley[1] *and John Grist*[2]

Part 1: controls

6.1 Drum level and feedwater control

The objective of a feedwater control system may seem simple: it is to supply enough water to the boiler to match the evaporation rate. But as is so often the case with boilers, this turns out to be a surprisingly complex mission to accomplish. There are difficulties even in making the basic drum-level measurement on which the control system depends. The design of the control system is then further complicated by the many interactions that occur within the boiler system and by the fact that the effects of some of these interactions are greater or smaller at various points in the boiler's load range.

The control system designer's task is to develop a scheme that provides adequate control under the widest practicable range of operational conditions and to do so in a manner that is both safe and cost-effective. To do this it is necessary to understand the detailed mechanisms of the feedwater and steam systems and to be fully aware of the operational requirements. In all but the smallest and simplest boilers, each of the interrelated factors has to be taken into account, and it is insufficient to rely on simple responses to the three parameters which seem to be relevant to the supply of feedwater: steam flow, feedwater flow and the level of water in the drum. The following logic descriptions are applicable to oil, gas coal and biomass-fired boilers. The logic is used on boilers with conventional burners and on circulating fluidised bed boilers.

6.2 One, two, three and four-element control

The level of water in the drum provides an immediate indication of the water contained by the boiler. If the mass flow of water into the system is greater than the mass flow of steam out of it, the level of water in the drum will rise. Conversely, if the steam output is greater than the feed inflow, the level will fall.

[1]Retired
[2]Consulting Engineer

As stated in Chapter 2, the purpose of the drum is not only to separate the steam from the water but also to provide a storage reservoir that allows short-term imbalances between feedwater supply and steam production to be handled without risk to the plant. As the level of water in the drum rises, the risk increases of water being carried over into the steam circuits. The results of such 'carry-over' can be catastrophic: cool water impinging on hot pipework will cause extreme and localised stresses in the metal and, conversely, if the level of water falls there is a possibility of the boiler being damaged, partly because of the loss of essential cooling of the furnace water walls. Therefore, the target of the feedwater control system is to keep the level of water in the drum at approximately the midpoint of the vessel. Given this objective, it would appear that the simplest solution would appear to be to measure the level of water in the drum and to adjust the delivery of water to keep this at the desired value – feeding more water into the drum if the level is falling, and less if the level is rising (see Figure 6.1). Unfortunately, the level of water is affected by transient changes of the pressure within the drum and the sense in which the level varies is not necessarily related to the sense in which the feed flow must be adjusted. In other words, it is not sufficient to assume that simply because the level is increasing the feedwater flow must be decreased, and vice versa.

This strange situation is due to effects known as 'swell' and 'shrinkage'. Boiling water comprises a turbulent mass of fluid containing many steam bubbles, and as the boiling rate increases the quantity of bubbles that is generated also increases. The mixture of water and bubbles resembles foam, and the volume it occupies is dictated both by the quantity of water and by the amount of the steam bubbles within it. If the pressure within the system is decreased, the saturation temperature is also lowered and the boiling rate therefore increases (because the temperature of the mixture is now higher in relation to the saturation temperature than it was before the pressure change occurred). As the boiling rate increases, the density of the water decreases, but since the mass of steam and water has not changed the decrease in density must be accompanied by an increase in the volume of the mixture.

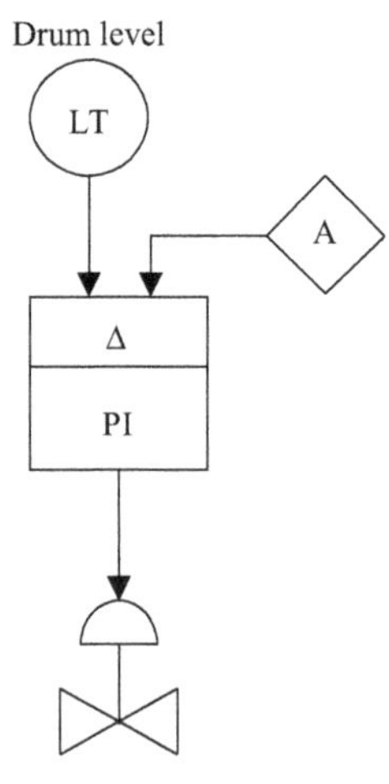

Figure 6.1 Single-element level control

By this mechanism, the level of water in the drum appears to rise, a phenomenon referred to as 'swell'. The rise of level is misleading: it is not indicative of a real increase in the mass of water in the system, which would require the supply of water to be cut back to maintain the status quo. In fact, if the drop in pressure is the result of the steam demand suddenly increasing, the water supply will need to be increased to match the increased steam flow.

Swell also occurs at start up, but this is because of a genuine expansion of the water as the temperature increases as well as the sudden expansion of bubbles at 100 °C as the water boils. An analogy is how milk boils over.

'Shrinkage' is the opposite of swell: it occurs when the pressure rises. The mechanism is exactly the same as that for swell, but in the reverse direction. Shrinkage causes the level of water in the drum to fall when the steam flow decreases, and once again the delivery of water to the boiler must be related to the actual need rather than to the possibly misleading indication provided by the drum-level transmitter.

If a slow change of steam flow occurs, all is well because the pressure within the system can be controlled. It is when rapid steam flow changes happen that problems occur since, due to swell or shrinkage, the drum-level indication provides a contrary indication of the water demand.

Following a sudden increase in steam demand, which causes the pressure to drop (and therefore the drum level to rise), a simple level controller would respond by reducing the flow of feedwater. Equally, a sudden decrease in steam flow, which would be accompanied by a rise in pressure and an attendant fall in the drum level, would cause a level controller to increase the flow of water. Both actions are, of course, in the incorrect sense.

The effects of swell and shrinkage, in addition to being determined by the rate of change of pressure, also depend on the relative size of the drum and the pressure at which it operates. If the volume of the drum is large in relation to the volume of the whole system the effect will be smaller than otherwise. If the system pressure is low the effect will be larger than with a boiler operating at a higher pressure since the effect of a given pressure change on the density of the water will be greater in the low-pressure boiler than it would if the same pressure change were to occur in a boiler operating at a higher pressure.

Faced with this situation, designers of control systems have responded by implementing a variety of solutions. The simplest of these is a 'two-element' system since it is based on the use of two process measurements in place of the single drum-level measurement used earlier.

6.2.1 Two-element feedwater control

Remembering that the basic requirement of a feedwater control system is to maintain a constant quantity of water in the boiler, it is apparent that one way of addressing the problem would be to maintain the flow of water *into* the system at a value which matches the flow of steam *out* of it. One version of this system is shown in Figure 6.2. Here, the flow is controlled by an easily recognised device, a valve. We shall look at valves in more depth later, but for the moment assume that the version used in the diagram maintains the rate of water flowing through the valve at a figure which is directly proportional to the demand signal from the

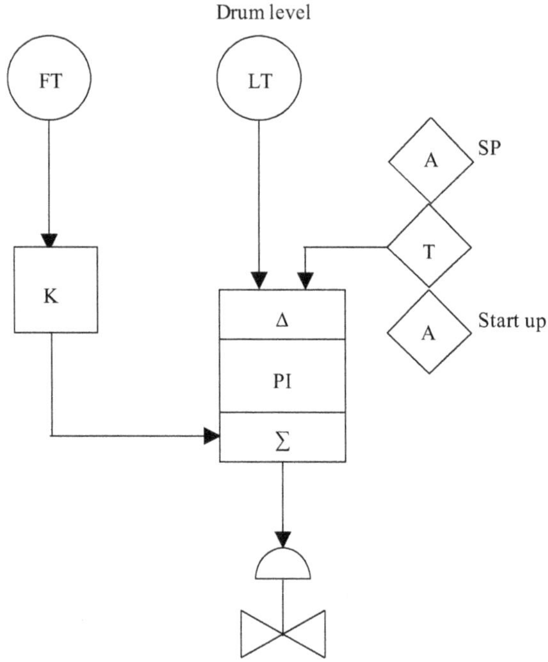

Figure 6.2 Basic two-element feedwater control

controller (i.e. if the demand signal varies linearly from 0% to 100%, the flow rate also changes linearly between 0% and 100%). Such a valve is said to have a 'linear characteristic' and in the system shown this is employed in conjunction with a transmitter that produces a signal proportional to steam flow. Used together, these two devices keep the parameters in step. If the transmitter produces a signal which is equal to the steam flow at all loads and if the flow through the valve is matched with this signal at every point in the flow range, a controller gain of unity will ensure that, throughout the dynamic range of the system, the flow of water will always be equal to the flow of steam.

Naturally, scaling factors of the transmitter and the valve must be taken into account. If the range of the flow transmitter is different from the valve's flow-control range, the controller gain will need to be adjusted accordingly, and in practical systems this is always necessary.

In order to provide an adequate operational margin of confidence, the range of the control valve is always designed to be greater than the flow range of the boiler. For example, in a boiler producing 200 kg/s of steam, the valve may be sized to deliver 220 kg/s of water when it is fully open. In this example, with a linear valve characteristic, an opening of approximately 91% will be needed to pass a flow of 200 kg/s.

In this case, if the steam-flow transmitter produces an output of 100% at 200 kg/s flow, the controller gain must be such that a measured value of 100% produces an output of 91%. This is a proportional band of 110 (i.e. a gain of 200/220), and if this gain is assigned to the controller, the feed flow will match the steam flow over the entire range of boiler load (assuming that the valve characteristic is linear, that the flow transmitter output is 4 mA at zero flow, and that zero flow of water occurs with a valve signal of 0%). The calculated valve position is only an estimate and any error will eventually fill or empty the drum. In normal operation this long-term error is corrected by the level controller. However, during a load change the mismatch between steam flow and expected feed flow, together with the swell or shrinkage error gives poor control.

To counter these effects, it is necessary to add a feedback element, consisting of a feed flow controller which will act to correct for any mismatch between the actual and desired feed flow.

6.2.2 Three-element feedwater control

Throughout the aforementioned analysis, reference has been made to the feedwater valve characteristic being linear and the valve being sized to produce a fixed flow when it is 100% open. However, the flow through a valve depends both on its opening and on the pressure drop across it. In a feedwater system, the pressure drop across the valve varies from instant to instant, and the flow through it at any given opening will therefore vary. For reasons given earlier, in a simple two-element system based on drum level, the inclusion of an integration element in the level controller is undesirable. Therefore, the varying flow results in the level control becoming offset, to restore the steam flow/feedflow balance. This offset is undesirable since it needlessly erodes the safety margin provided by the presence of the drum.

One method of correcting for the error produced by the feed valve is the addition of a third element to the system – a measurement of feedwater flow.

There are various ways of implementing such a system, one of which is shown in Figure 6.3. Here, the output of the drum-level controller is trimmed by a signal representing the difference between the feed-flow and steam-flow signals. A gain block (11) is introduced to compensate for any difference between the ranges of the two transmitters. This is not required if engineering units are used. In most cases the steam flow and feed-flow signals will cancel out, and the drum-level controller will be modulating the feed flow to keep the level at the setpoint. In this case, it is reasonable to apply an integral term in this controller, as shown.

In another implementation of this familiar system shown in Figure 6.4, a 'cascade control' technique is applied. The drum-level controller (item 5) compares the measured level signal with a set value and produces a bidirectional output proportional to any error. This trims a modified steam-flow signal, which is acting as the desired value for a closed-loop feedwater controller (7). A gain block (11) adjusts for any range difference between the steam-flow and feed-flow transmitters. Note the gain block can be in the steam-flow or feed-flow line. As before, it is not required if engineering units are used.

These are not the only ways of implementing three-element control. Several variants of the system are in common use, each with its own advantages and

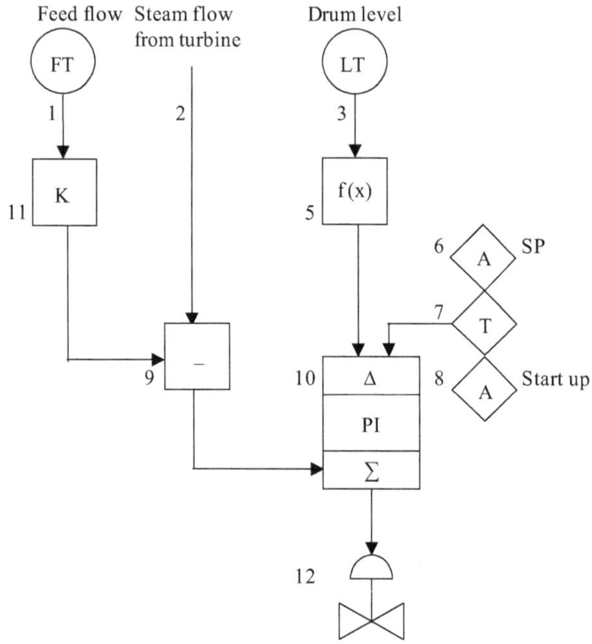

Figure 6.3 One type of three-element feedwater control system

disadvantages. However, each system has one factor in common, the use of steam-flow, feed-flow and drum-level measurements. The application of the feed-flow measuring element compensates for any variations in feed flow, whether these are due to the pump characteristics or other factors, and the three-element system is therefore recommended wherever accuracy of control is required.

As might be expected, a three-element system is more expensive than a single-element or two-element system. The feed-flow measurement in the system does necessarily add a significant cost burden. Typically requiring triplicated differential pressure (DP) transmitters, a primary element (such as a Venturi, or flow nozzle) which will add further cost and require the provision of adequate lengths of straight pipework upstream and downstream of the device.

Large boilers required for flexible operation will always need three element feedwater control or an equivalent. Siemens offers a two-loop control system for drum-level control, but has additional proprietary features so cannot be included here. Their two-loop concept for steam temperature is described in Chapter 7.

Having concluded that we need three-element feedwater control, we immediately have a problem that the flow measurements are inaccurate at low loads and could drive the control valve and drum level in the wrong direction. Standard practice is to start with single element control and transfer to three element control at say 25% on an increasing load. Then transfer back to three-element control, on a decreasing load at 20% Boiler Maximum Continuous Rating (BMCR). The logic to decide when to transfer is simple, but ensuring bumpless transfer is more difficult.

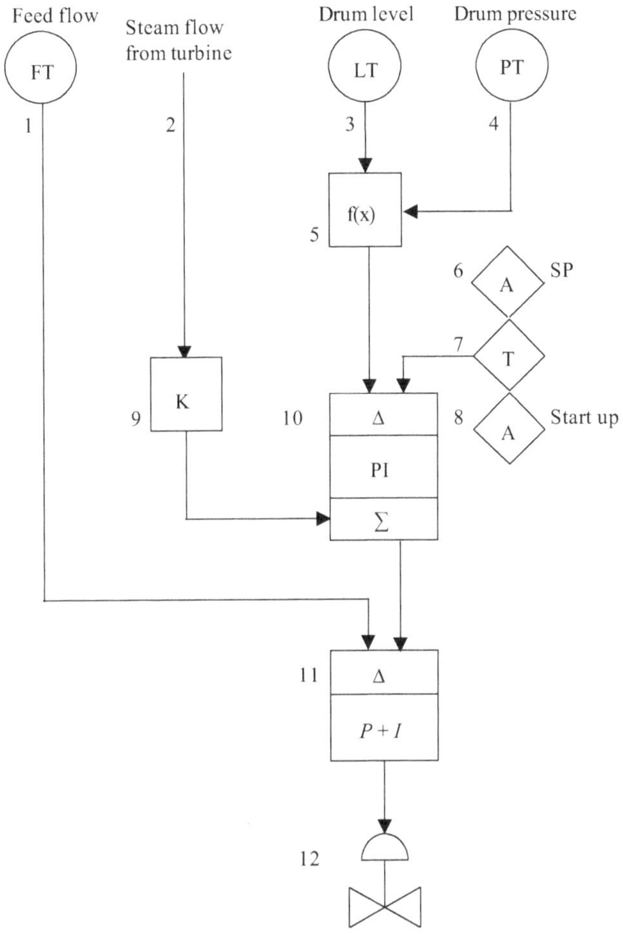

Figure 6.4 Alternative (cascade) three-element feedwater control system

In both the two-element and three-element systems, an assumption made in the aforementioned examples is that the steam output by the boiler eventually returns to the inlet in the form of water. This is not true where there is significant blow-down or if any steam is extracted for applications such as soot blowing. Here the steam is effectively lost. The amount being used will not be included in the flow measurement. This will result in the drum level being offset from the desired value since the flow of feedwater into the system should be equal to the steam taken by the load (plant or turbine) plus the steam used for soot blowing. This is corrected by the integral action term in the level controller.

6.2.3 Four-element level control

This is really an attention-grabbing title. The fourth element does not help in the dynamic stability of the loop. There are two instances where fourth measurement might be considered. Both are dictated by the plant arrangement.

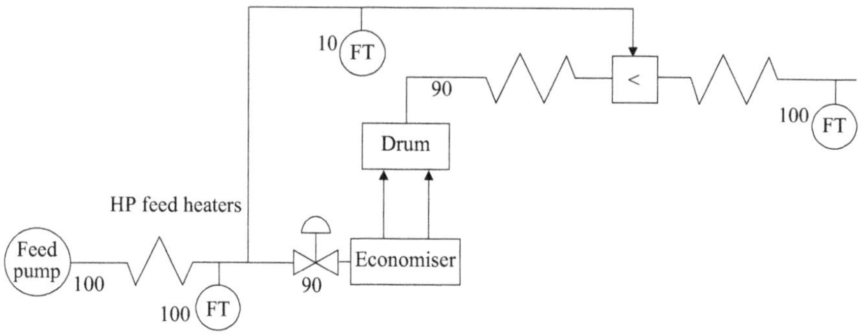

Figure 6.5 Hot spray

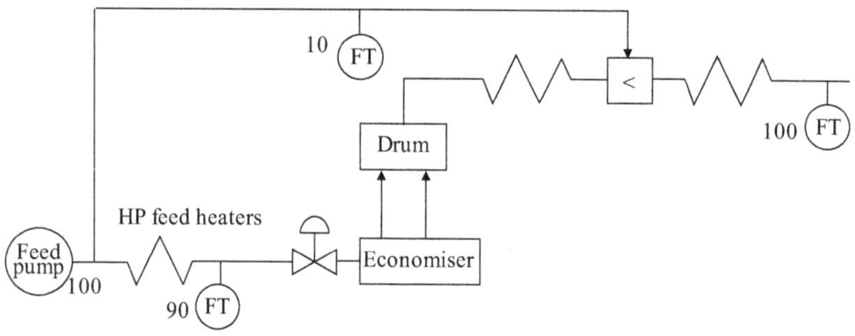

Figure 6.6 Cold spray

The first depends upon the arrangement of the spray water take off with respect to the feedwater flow measurement. Traditionally boilers used cold spray. That is, spray taken before the HP feed heaters. The inherent pressure drop over the heaters ensures that the resultant steam is at a lower pressure than the spray water and so a 'pinch valve' is not required. However, hot sprays are overall more efficient and most boilers now use hot sprays.

Assume a hot spray system as shown in Figure 6.5. Then each 100 units of feedwater from the pump is measured by both the feed-flow transmitter and the steam flow transmitter. The spray, although shown as 10 units, has no effect on these measurements and the logic of Figure 6.4 works as expected.

However, with a cold spray system as shown in Figure 6.6 the same 100 units of feedwater from the pump does not all pass via the feed-flow measurement. In this instance 10 units are shown as spray. The steam flow/feed-flow feedforward now shows an error equivalent to the spray flow, which is corrected by the drum-level controller. Every time the spray flow changes it introduces another error. While this effect is attributed to hot or cold spray the real problem is the location of the feed-flow measurement with respect to the spray water take off.

This is overcome by correcting the logic as shown in Figure 6.7.

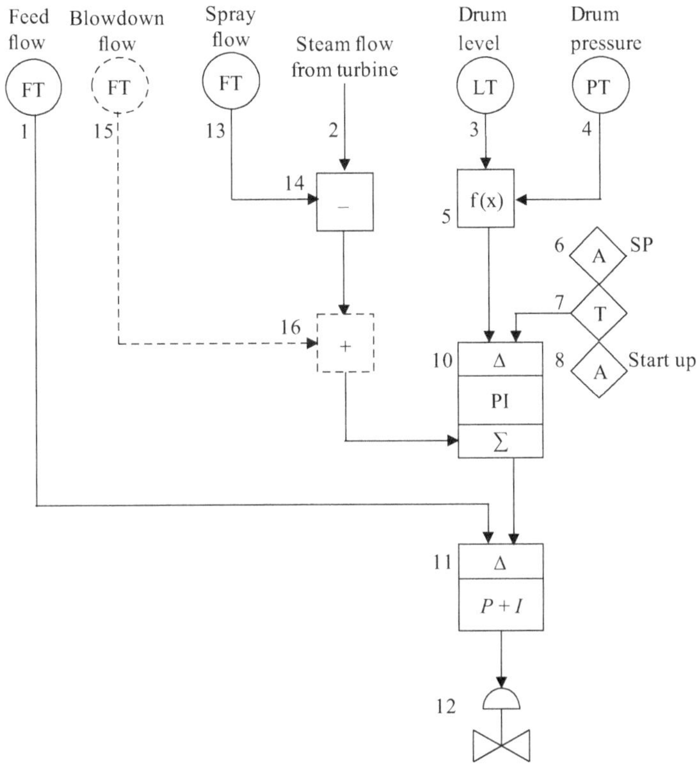

Figure 6.7 Four-element level control

The other reason that four-element control has been suggested is to allow for the blowdown. During a cold start the water in the boiler swells and sometimes the drum level is dangerously high. The excess water is removed via drum-level-lowering valves. These are often restricted to use below 30 bar to prevent a major loss of water. At start up the boiler is in single-element control and so blowdown measurement is not required. Blowdown is also used to help control the water chemistry and be adjusted daily by the station chemist. However, this quantity of blowdown is usually less than 0.5% of BMCR and has no significant effect on the level control. It is often not measured and dumped to the boiler drains vessel. Some boilers will have a condensate recovery system where the blowdown passes via an intermediate flash tank. Some of the pressurised water flashes off as steam but the rest is recovered. In these instances, it is normal to include a blowdown flow measurement. Theoretically this could be used in the drum-level control system as shown by dashed lines in Figure 6.7, but is not standard.

In addition to these variables used for control drum pressure is also required to compensate the drum-level measurement, and feedwater temperature is used to correct the feed-flow measurement.

6.3 Feedwater valve control

In the simplified logic, we have shown for three-element-level control we have only considered a single control valve. The final control is achieved by a combination of feed valves and feed pumps. In the 1980s power station had feed valves capable of controlling over the complete range. Some would have two 100% duty valves and a 50% start-up valve, others might have three 50% duty valves. They would also have say one 50% electric feed pump and two 100% steam-driven feed pumps. The multiple valves helped achieve the necessary turn down and provided redundancy. Sometimes at loads >90% the valves would be driven wide open to minimise pressure loss and the feed flow controlled by varying the pump speed.

Today most new plants have a single start-up valve and transfer to pump speed control between 20% and 40% BMCR. One arrangement is shown in Figure 6.8. In this arrangement, the main line might be 600 mm diameter and isolated by a motorised valve (2), the control valve (1) is in a smaller 300 or 400 mm bypass line. There are additional isolating valves around the start-up valve used for maintenance.

At start-up the isolating valve (2) is closed, and the DP (3) across the control valve (1) controlled to say 5 bar by the controller (4) that regulates the speed of the pump.

The control valve itself is controlled by a single-element-level controller or the start-up logic in a once-through boiler. It will initially be closed. The feed pump will be at minimum speed and protected by a leak off valve.

When the level system is released to auto control the valve (1) moves to control the level or feedflow, and the pump speed changes to keep the DP across the valve constant. As the load increases the valve opens and the pump speeds up. At some load the control valve is almost fully open and the system switches to pump speed control only. To enable this to happen the isolating valve (2) is slowly opened. The control valve can be left open or closed.

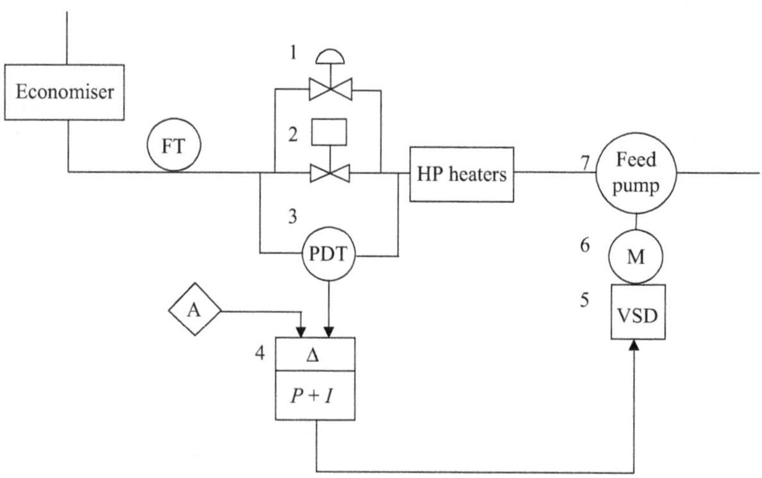

Figure 6.8 Start-up valve arrangement

Between 20% and 40% BMCR several logic transfers occur:

- The system transfers from single-element to three-element control if a drum boiler.
- The system transfers from recirculation mode to once-through mode if a once-through boiler.
- The system transfers from feed valve DP to pump speed control, the main isolating valve opens.
- The pump control moves from the start-up pump to the main pump control.

As one of these might disturb the boiler operation, some try to stop them happening at the same time. One that is easy to move is the transfer from start-up valve (1) to fully open valve (2). To achieve this the start-up valve must be used at a higher load. That is, it must pass a higher flow. The flow through the start-up valve may be increased by increasing the DP from say 5 bar to 10 or 15 bar, but only if the valve was originally specified for a 5 bar DP. This also increases the velocity through the valve and its associated pipework. Ideally on say a 60 cm feed line the bypass line should be 30 cm but if only 20 cm, the supplier and piping engineer may accept the higher velocity as it only happens during start-up and shutdown.

6.4 Storage vessel-level control in a once-through boiler

The control engineer needs to understand the Benson process before trying to understand the control logic. The once-through concept was introduced in Chapter 2, but further explanation is included here. Note that the water evaporates in the tubes that form the walls of the furnace. Hence the correct term when describing the steam and water circuit is evaporator. However, the description furnace is in common use, particularly when describing the walls and tubes.

In a drum boiler, the volume of water circulating is up to three times the volume of steam being produced. In a once-through at high loads they are the same. The water keeps the furnace tubes cool, lack of water will cause overheating and damage to the tubes. In a once-through boiler this is a problem at low loads where a disturbance in combustion say from starting a mill may cause overheating.

To prevent this, the once-through reverts to a forced circulation system below a critical load or flow. This is known as the Benson load and is around 30%–35% BMCR. Below this the water from the evaporator walls passes to the storage vessel instead of a drum. From the storage vessel, it is pumped by the circulation pump back to the economiser inlet (see Figure 6.9).

The associated control logic is shown in Figure 6.10.

There are two main control loops shown here. The storage vessel-level control and part of the overall boiler feedwater control which is shown by dashed lines.

From start-up to Benson load the boiler feedwater control (1–5) ensures a minimum flow of 35% BMCR flow passes via the economiser and evaporator. The storage vessel-level control system controls the flow via the circulation pump in a relation to the level in the storage vessel (6–10) and during start-up when the level swells it also controls the dump valves (11–13).

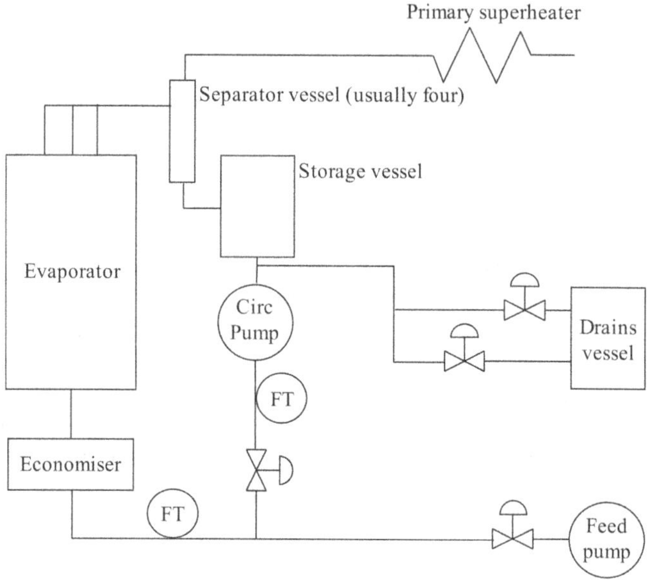

Figure 6.9 Once-through start-up system

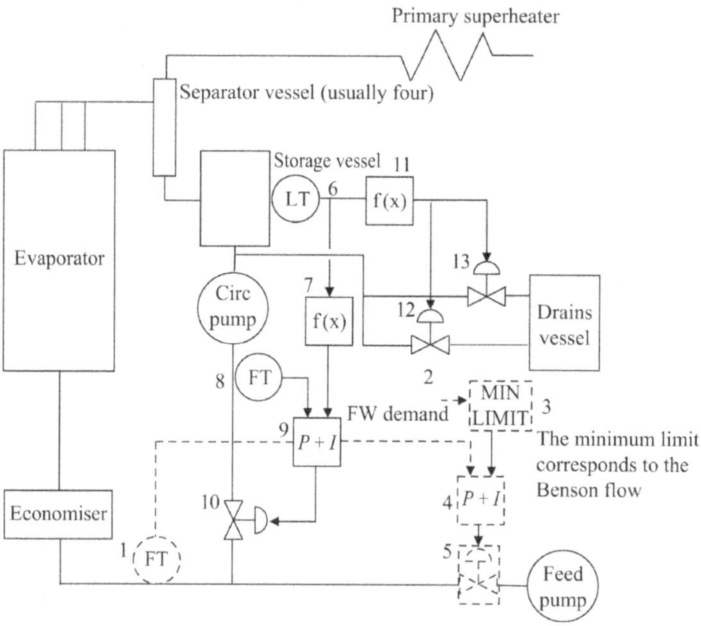

Figure 6.10 Storage vessel-level control

The action of these loops is described here and supported by Table 6.1.

Before firing for the first time the pumps are run to circulate the water through the condensate polishing system and so reduce the iron levels. This is not needed on every start-up. When the iron levels are OK the storage vessel which is around 10 m tall is filled to the 6 m level (all values are generic and must be designed for each project); the paragraph numbers correspond to those in Table 6.1.

1. The circulation pump running at a maximum of 35% BMCR the first burner is lit. The 35% flow corresponds to the Benson flow or minimum safe flow. The main feed pump is running on leak off mode.
2. Lighting the first burner introduces some logic to reduce the circulation pump flow by around 3% BMCR. This reduction is seen by the feedwater control loop and valve 5 opens so that FT(1) sees the 35% flow.
3. As more burners are lit the water swells and the level in the storage tank rises.
4. At around 6.3 m the first dump valve starts to open to control the level.
5. At 7.8 m both dump valves have opened and a maximum swell is dumped. Around 60 m^3 might be dumped from an 800 MWe boiler.
6. As the temperature passes through the boiling point the swell decreases and steam starts to leave the evaporator.
7. Extra oil burners are added and more steam is produced as the water is now being extracted.
8. At a minimum load of say 10% the turbine is rolled and MW produced.
9. As load is increased the first mill is fired.
10. The second mill is fired.
11. With two mills firing the load and steam flow continue to increase.
12. The load has increased to 30%, and the level in the vessel reduced to 3.8 m.
13. At around 35% BMCR load all of the fluid leaving the evaporator is steam and the feed flow entirely from the boiler feed pumps. This is the Benson point. The circulation flow valve is closed and the pump operates on leak off.
14. The third mill is started. The actual load at which mills start depends upon how fast a ramp is required and the relative capacities of the mills and oil support system.
15. At around 60% load the fourth mill is fired. Soon the circulation pump will be automatically stopped on high load or tripped by low level in the storage vessel.
16. As load continues to increase the fifth mill is started. Usually there are six mills and the last is a spare.
17. Load continues to increase until full load is reached. From steps 13 to 17 the feed flow loop (1–5) dominates. The method of control is known as enthalpy control and is described in Chapter 7.

The table appears to show a simple relationship between the feedwater and circulation flow at low loads: they add up to 35%. Some people have assumed that the feed flow is driven as a function of load. This is not the case. The feed-flow logic follows the boiler demand or a minimum of 35%. When the load is 25%, the recirculation is 10% and the feed-pump controller set point is 35, hence it passes an extra 25% to satisfy the setpoint. Note this is compared with the measured economiser flow. If there was an error in the circulation flow measurement, so it passed

Table 6.1 Table of parameters changing with time

State	Feedwater flow	Circulation pump flow	Economiser flow (%)	Steam flow	Storage vessel level	Activity
Preparing to fire	0	35	35	0	6	
First burner lit	3	32	35	0	6	
Several burners lit, water starts to expand	3	32	35	0	6.2	
Several burners lit, water continues to expand	3	32	35	0	6.3	Dump valve opens a little
Water starts to boil with large swell	3	32	35	0	6.4	Dump valve(s) open to remove swell
Water temperature increases	3	32	35	1	6.2	
Extra burners added	5	30	35	5	5.9	
Turbine roll	10	25	35	10	5.9	
First mill added	15	20	35	15	5.9	
Two mills firing	20	15	35	20	5.9	
	25	10	35	25	5.9	
	30	5	35	30		
Three mills firing	35	0	35	35	2	Benson point
	40	0	40	40		Once through operation no circulation
Four mills firing	60	0	60	60		
Five mills firing	80	0	80	80		
Five mills firing	100	0	100	100		

only 8% then the feed controller would add more feedwater until the economiser inlet flow measured 35%. As the Benson flow is critical to protect the boiler, it follows that the boiler must be tripped on low economiser flow. Typically, at 90% of the flow an alarm is raised and at 90% of that the boiler is tripped. Sometimes a small time delay is included to allow the standby feed pump to start.

As the load increases the circulation pump is tripped either on low storage vessel level or if the load rises above 45% BMCR. Ideally the vessel now stays at its low level until the load falls below 35% when the process is reversed. While this is the ideal case it does not actually happen. The steam from the boiler passes via the separator vessels to the superheater. At low loads this is feedwater, as load increases it is a mix of steam and water and above the Benson point it is all steam. Some of this steam will still pass via the storage vessel and may evaporate any

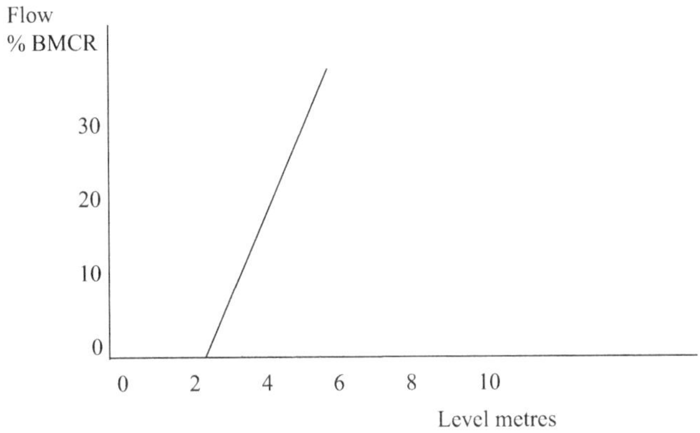

Figure 6.11 A typical function curve of circulation flow against storage vessel level

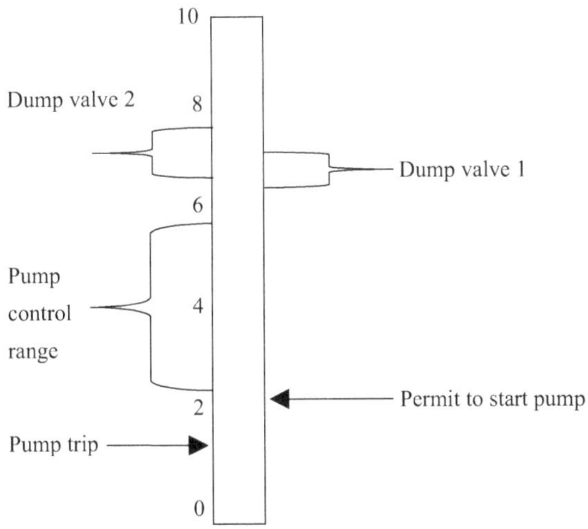

Figure 6.12 Typical storage vessel pump start, stop and control ranges

water in the vessel maintaining the level. However, this is not an exact design, also the warming flow described later is added to the vessel and it is expected that the level will continuously rise. Depending upon the boiler supplier the level may be allowed to completely fill the vessel where the top layer is evaporated by the passing steam flow or removed. The flow of water to be removed is much smaller than the dump flow, and the removal valve is either piped to the drains vessel or the first stage attemperator where it is injected into the steam line.

In normal operation below the Benson point the circulation flow (8) is controlled as a function of the storage vessel level (6) by the $P + I$ controller (9). A typical function curve is shown in Figure 6.11.

Typical storage vessel pump start/stop and control ranges are shown in Figure 6.12.

6.5 Once-through super critical additional systems

The once-through boiler has some extra systems and equipment.

6.5.1 Sub-cooling line

Cavitation can damage the pump. Cavitation occurs when the net positive suction head (NPSH) is less than the saturation pressure.

The pump sits directly below the storage vessel, sometime just below it and sometimes at ground level. The column of water above the pump determines the suction head and in normal operation no cavitation happens. However, as the boiler load increases the water temperature increases and its density decreases, effectively reducing the suction head. This can be prevented by adding colder water from the economiser inlet to the pump suction at the storage vessel. The sub-cooling valve may be opened if the water temperature approaches saturation or whenever the pump is running as advised by the boiler supplier. The sub-cooling circuit is shown in Figure 6.13 and the associated logic in Figure 6.14.

While this circuit is designed to prevent the pump from overheating another circuit is required to keep it warm. This is to avoid thermal shock when the boiler is suddenly reduced in load after a period at high load. In the high load period the pump, its control valve and the storage vessel dump valves cool down to ambient

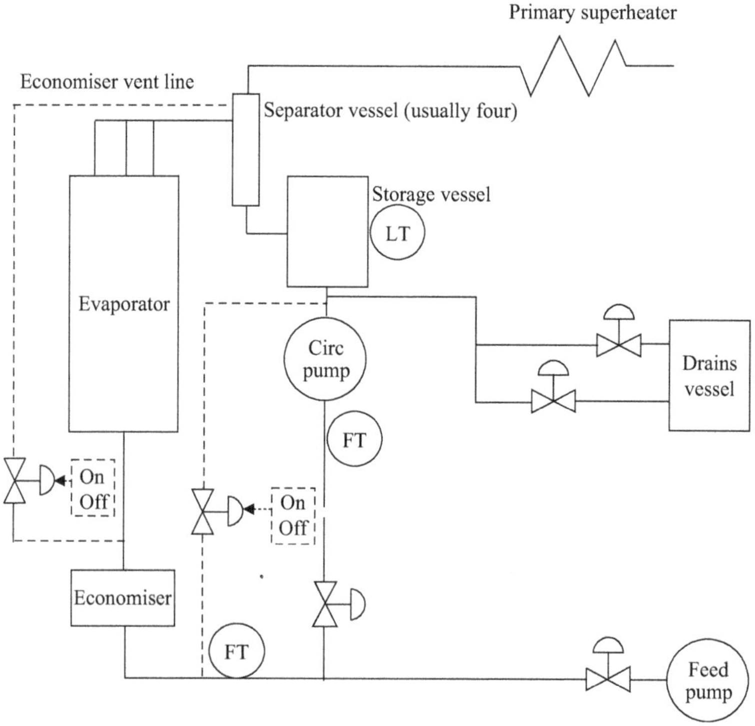

Figure 6.13 Sub-cooling and vent line

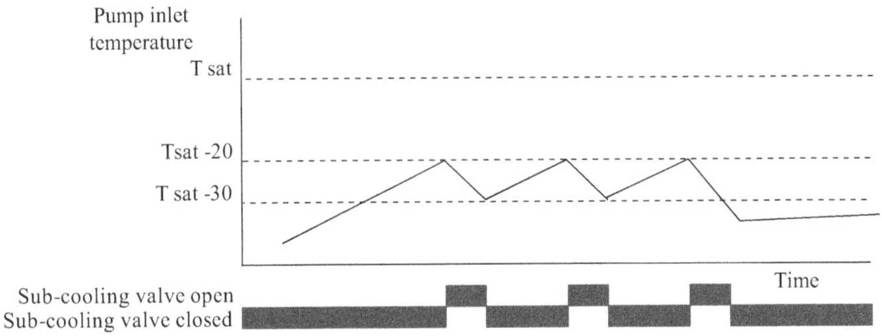

Figure 6.14 Sub-cooling logic

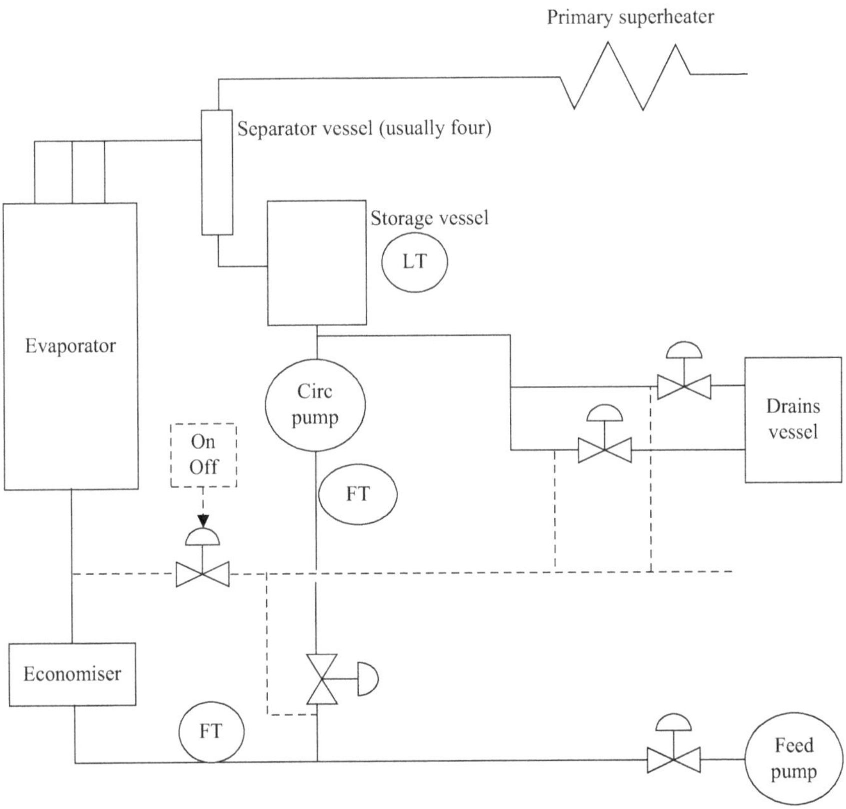

Figure 6.15 Pump and dump valve warming

temperature, then as load is reduced or the boiler tripped they see high-temperature water and suffer thermal shock. This is prevented by passing water from the economiser outlet backwards through the pump and the dump valves (see Figure 6.15). The extra pipework is shown by dashed lines. The method of control is generally on/off and a regulating valve is fitted in each supply line. This is set and locked during commissioning.

6.5.2 Economiser vent line

Because of the relatively low circulation flows at low loads, it is important to prevent steaming in the furnace wall tubes. The temperature of the steam is not the problem but the risk that the steam bubbles will prevent the proper flow of water and then the tubes will overheat. On a drum boiler, there is a similar concern if the economiser steams. The minimum feed flow prevents steaming in normal operation. However, following a master fuel trip (MFT) the post-trip purge carries the hot gases from the furnace over the tubes in the rear of the furnace. If the feed pump has been tripped as part of the MFT then there is no flow in the economiser and it will steam. Vent lines are installed between the economiser and the separator vessels to allow this steam to bypass the furnace. A solenoid operated valve is fitted that is open if no burners are firing but closes whenever any burner is firing. This line is shown in Figure 6.13.

6.6 Basic enthalpy control

6.6.1 Enthalpy control process considerations

With a once-through boiler the ratio of feedwater to fuel is controlled to maintain the correct outlet steam temperature.

The concept is shown in Figure 6.16. This shows a simplified steam and water circuit and both the temperature and feedwater loops. The temperature loop is explained in Chapter 7, but here is referred to as 'two-loop control' (1 and 2). The first two-loop controller trims the final superheater outlet temperature and sends a signal to the second two-loop controller (2) as the setpoint for the second superheater outlet temperature. This is derived to produce a load-dependent temperature drop across the second-stage attemperator (3).

The second two-loop controller (2) acts in a similar way to control the outlet of the secondary superheater by using the first-stage attemperator (4). This concept can be applied to both once-through and drum boilers. On a drum boiler, the drum temperature corresponds to the saturation temperature for the operating pressure, and the pick up across the primary superheater is a function of load. Hence its outlet temperature cannot be influenced by the control engineer.

In a once-through boiler the evaporation point in the furnace walls is not fixed. For example, if the feedwater flow is increased the same amount of heat will result in a lower temperature increase and the evaporation point will move up the furnace wall. The temperature will continue to increase above this evaporation point but as it exits the evaporator it will be lower than it was before the feed flow increased. This gives the control engineer the opportunity to vary the evaporator outlet/primary superheater inlet, and hence the temperature drop, across the first-stage attemperator.

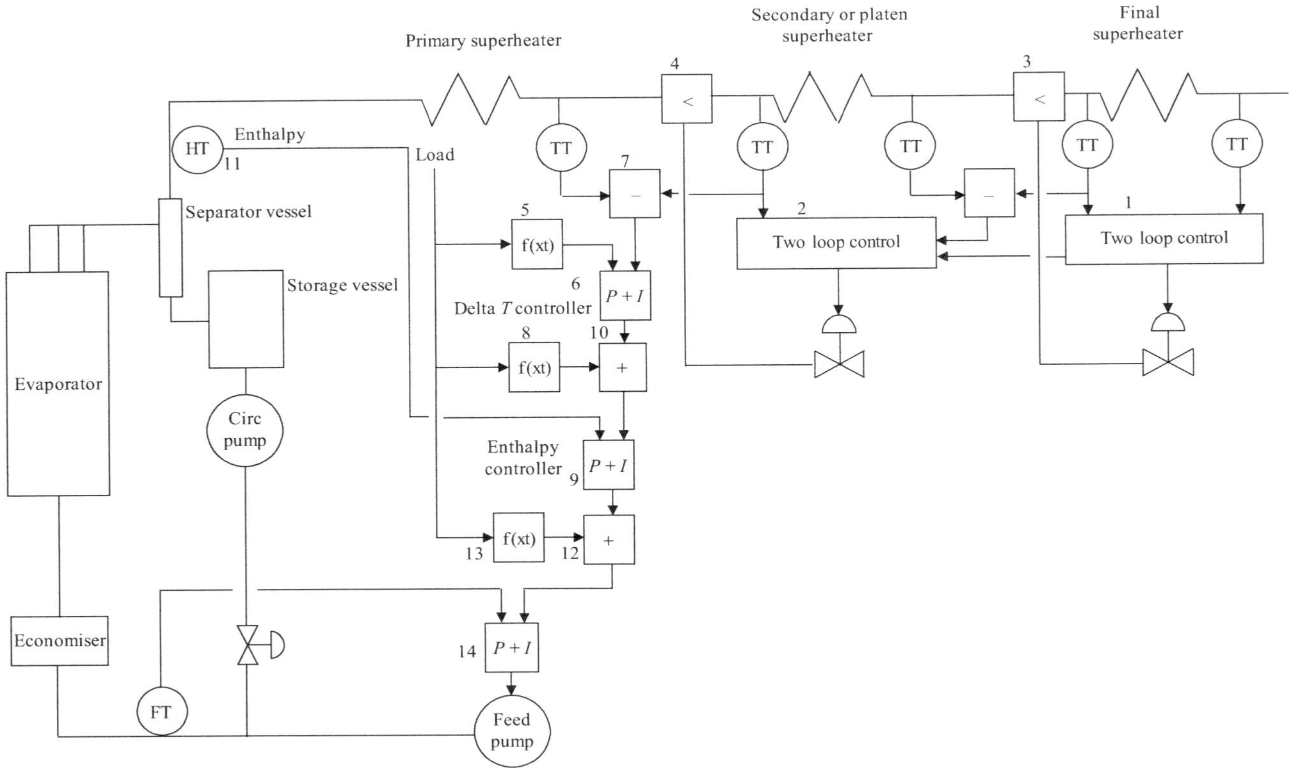

Figure 6.16 Basic enthalpy logic

6.6.2 Enthalpy control logic considerations

The expected temperature drop, across the first-stage attemperator, is set in an $f(x)$ (5) and compared in controller (6) with the temperature drop derived from the subtractor (7). This (6) is called the delta T controller.

A separate controller (9) called the enthalpy controller compares the design enthalpy set in $f(x)$ (8) with the furnace outlet enthalpy (11). The set point for the enthalpy loop is modified by the delta T controller outlet (10) and the enthalpy controller outlet is added (12) to the load-dependent feedwater flow signal derived from $f(x)$ (13). This modified feed flow demand passed to the feedwater controller (14) which modulates the feed valves or feed pump speed to achieve the required flow.

This is a simplified description and in practice the feed flow will be pressure and temperature compensated, and the enthalpy is not measured directly but derived from pressure and temperature and a look-up table. The minimum Benson feed-flow limit needs to be added. Siemens has developed a more advanced enthalpy control solution but it is only available if you have a Benson licence.

6.7 Deaerator control

Strictly speaking, control of the deaerator is not a function of feedwater control which is the subject of this chapter. However, as the deaerator is an essential link in the feedwater supply system it is appropriate to consider its control systems here.

In Figure 2.5 we saw how steam admitted to the deaerator rises upwards past metal trays over which the water is simultaneously cascading downwards. As the water and steam mix and become agitated, entrained gases are released. The dissolved gases are vented to the atmosphere because the vessel is pressurised by the steam. The deaerator is situated in the water circuit between the discharge of the condenser extraction pump and the inlet of the feed pumps, as shown in Figure 6.17.

It will be evident that two control functions are required by the deaerator: one to maintain the steam pressure at the optimum value, the other to keep the storage vessel roughly half-full of water.

6.7.1 Steam pressure control

The pressure of the steam entering the deaerator is maintained by a simple controller whose measured value signal is obtained from a transmitter measuring the steam pressure in the deaerator. The set value of the controller is normally fixed.

It has already been explained that the steam supply may be obtained either from the boiler or from an extraction point on the turbine. If the latter source is used, special consideration is given to ensuring that an event such as a turbine trip does not deprive the deaerator of the steam supply it needs for its operation. This purpose is served by taking a steam supply from the boiler and passing it to the deaerator via a pressure reducing and desuperheating system (PRDS). This steam supply is referred to as 'pegging steam'. Judicious adjustment of the PRDS controller setpoint will ensure that an adequate steam supply for the deaerator is always obtained. However, to ensure rapid response to a turbine trip, a system of interlocks

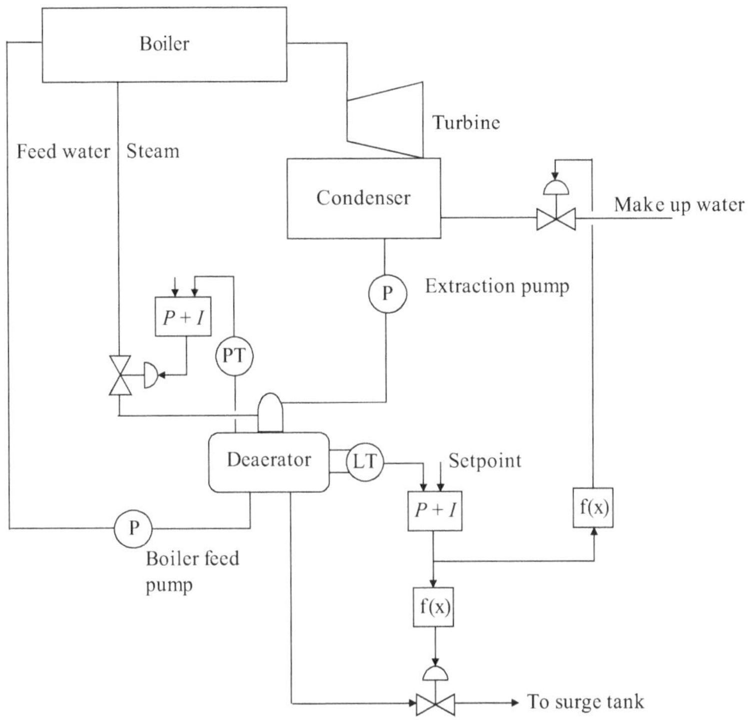

Figure 6.17 Principle of deaerator-level control system

should be provided, so that the pegging steam supply is brought into service immediately on detection of a trip.

6.7.2 Level control

The deaerator storage vessel provides a measure of reserve capacity for the plant. To achieve this function the level of water in it must be maintained at roughly the midpoint. This is achieved by means of a level controller whose measured value signal is obtained from a DP transmitter or from capacitive probes which would normally be connected to tappings of an external water column which is in turn connected to the top and bottom of the deaerator storage vessel. If there were no losses in the system, the amount of water would be constant and the level in the deaerator storage vessel would remain at the correct value set during commissioning. However, losses are inevitable (e.g. due to leakages at pump glands or during soot blowing or blowdown operations), and a supply of treated water must therefore be made available. The deaerator level controller output adjusts the opening of a valve that admits this make-up water to the condenser, as shown in Figure 6.17.

The make-up supply is conventionally fed into the system at the condenser. Figure 6.17 shows that interaction between the level controllers of the deaerator and condenser is inevitable. The situation is made more complex because the condenser extraction pump is provided with a bypass arrangement to always maintain a minimum flow through the pump.

In fact, the conditions which cause the deaerator level controller to call for more water to be added to the system will also cause the condenser level to fall, and so the two systems do not act in opposite senses. Nevertheless, they do interact, and care must be taken to minimise the instability that is likely to arise.

6.7.2.1 Spill control

In addition to the level control system described earlier, a system must be provided to drain excess water from the deaerator storage vessel in the event of an oversupply of water and steam. This function is normally achieved on large power station boilers by using the output of the level controller to operate the make-up and spill valves in split range. At mid-level both the spill and make-up valves are closed as the level continues to fall the make-up valve will start to open, while the spill valve will only be opened if the level increases significantly. There is a large dead band between these two actions. Smaller boilers will be limited to having a spill valve whose opening is normally controlled on an on/off basis from a high-level alarm.

6.7.3 Integrated-level control

The long dead time and long-time constant of the deaerator-level system complicates the design and commissioning of this control loop. This situation has been worsened by the commercial pressure to operate power station boilers at partial load and the increased need for two-shifting operations. [The term 'two shifting' refers to operations where the plant runs for two eight-hour periods (or shifts) in a working day and is shut down for the remaining eight-hour shift.] There is also a drive to reduce the consumption of controllable make-up water, although this necessarily increases with the amount of load changing to which the boiler is subjected. A typical scenario is that if the boiler experiences a large load reduction (say, 50%) the deaerator level would tend to rise because as the steam flow from the boiler reduces so must the feed flow. During this time, the condensate flowing to the deaerator would not be required by the feed pump, causing the level in the storage vessel to rise. The level controls would react by dumping the excess water that is filling the deaerator. If there is a load increase *to* the original load, the deaerator level would drop due to the increased flow from the feed pump, and the level controls would tend to add make-up water to the condenser. This cycle of making-up and spilling can occur every time the load changes.

The requirement to reduce the wasted make-up water has resulted in the use of advanced control strategies for controlling the make-up and spill on larger boilers. For example, a fuzzy logic control algorithm embedded in the distributed control system (DCS) software can be used. The inputs to the controller are the steam flow and the deaerator level, the outputs are to the make-up and spill valves. The system uses rule-based logic to take decisions on making up and spilling. The goal of the scheme is to minimise make-up and spill by retaining water within the condensate system until it is required. These alternative algorithms are specific to the DCS supplier and so cannot be described here.

Part 2: feedwater instrumentation and control mechanisms

6.8 Measuring and displaying the drum level, steam and feed flows

From the aforementioned text, it is apparent that the primary objective of the feedwater control system of a drum-type boiler or heat-recovery steam generator (HRSG) is to maintain the drum level at the correct value. We shall now look at how this parameter is measured. It is an area where the problems are unexpectedly complex.

Figure 6.18 shows one method of measuring the drum level. This connects the DP transmitter directly to the drum via isolating valves. Note the 'constant-head' reservoir connected to the upper tapping point. Because the impulse pipework to the transmitter is outside the heated zone of the boiler, any steam within it will tend to condense, and the pressure applied to the HP port of the transmitter will therefore be the steam and water pressure plus the pressure due to the weight of this condensate. Note that some suppliers connect the condensate leg to the LP connection of the DP transmitter. Refer to supplier's drawing for details of lagging and orientation of isolating valves. Globe valves are mounted with their stems horizontal to not impede the flow of condensate.

The latter will depend on the volume of condensate that has collected, and this will be time-dependent. The pipe will be full of steam after the transmitter impulse lines have been 'blown down' on start-up, but it will afterwards start to fill with condensate. The constant-head vessel is left deliberately unlagged, so that the steam in it condenses, maintaining the pipe to the transmitter full of condensate.

The DP appearing at the transmitter ports is a function of several variables: the water level in the drum, the densities of this water and the steam above it and the density of the water in the pipes to the transmitter. In addition, the derivation of a level signal needs to take into account the density of the fluid used during the initial calibration of the transmitter. Finally, it is also necessary to recognise the

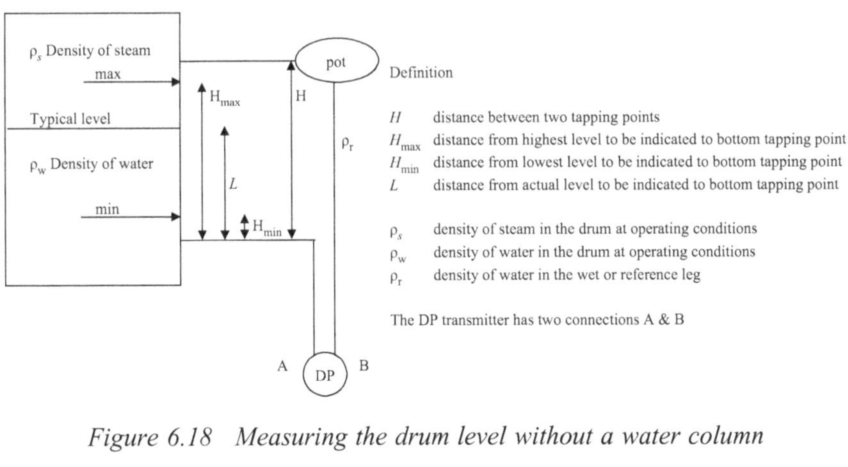

Figure 6.18 Measuring the drum level without a water column

requirement that the lowest drum water level at which the system will be allowed to operate will be at some point above the level of the lower tapping point.

The DP at the transmitter is defined by:

$$DP = (1/\rho_t) \times [(H \times \rho_\tau) - [(L \times \rho_w) + (H - L) \times \rho_s]] \tag{6.1}$$

where DP is the differential pressure at the transmitter (at 20 °C); H is the distance between the bottom and top tappings; L is the height of water in the drum above the bottom tapping; ρ_t is the density of water used for calibrating the transmitter (at 20 °C); ρ_w is the density of the water in the drum at the operating pressure; ρ_s is the density of the steam in the drum at the operating pressure and ρ_r is the density of the water in the reference leg.

It should be noted that the DP at the transmitter will be highest when the separation between the water level and the top tapping is the greatest, that is, at low levels. In other words, a 4–20 mA transmitter will produce 4 mA at the highest level and 20 mA at the lowest. This apparently reversed 'sense' must be corrected by the DCS before the density calculation is applied.

Since the densities of the steam and water in the drum will both depend on the conditions that exist in the drum, the pressure needs to be taken into account when calculating the actual level based on the differential head produced at the transmitter. At one time, various proprietary devices were available for performing the required calculation, but it is most economical these days to perform the calculation in the DCS, based on the DP and pressure signals. In this case the aforementioned equation will enable the level to be calculated for any combination of signals from the two transmitters.

However, an important point to bear in mind when using the DCS to perform the pressure compensation in this way is that the corrected signal will be available to the operator *only while the DCS is operational*. If a major failure should occur in the computer system, it is important that the operator can still be able to monitor the drum level by other means. If based on DP measurements these must be compensated in a similar way to the above, so that the indication is relatively unaffected by pressure. Electronic drum-level monitors (EDLMs) or gauge glasses with CCTV or fibre optic do not require extra compensation.

Most DCS suppliers have a standard code for this compensation. The control engineer only needs a specimen calculation to test during the FAT. An associated but more important consideration is how to range the transmitter itself. Because of the density effects described earlier there can be a 3:1 difference between the actual DP (in inches or millimetre) seen by the transmitter and the physical difference (in inches or millimetre).

6.8.1 Some words of caution

All the measurements are from the lower tapping on the drum. The transmitter is often mounted several metres below this point and so sees an additional pressure due to this extra head. This extra head is applied to both sides of the transmitter and so ignored in these calculations.

Avoid confusion between 40 °C, the assumed temperature in the reference leg, and 20 °C, the assumed temperature of the water used to calibrate the transmitter.

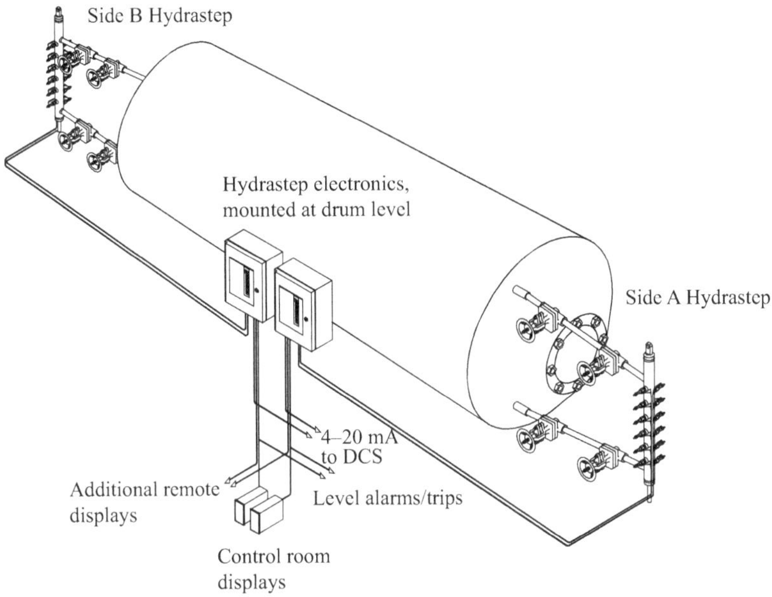

Figure 6.19 The 'Hydrastep' level indicating system. © Rosemount Measurement Ltd, reproduced with permission

The DP connections are shown as A&B rather than high and low as some suppliers connect A to the high side and others connect it to the low side.

Figure 6.19 shows a proprietary system which generates a drum-level signal for display locally and in the control room. This technology is based on the fact that the conductivity of water is different from that of steam, and a series of electrodes mounted in a water column attached to the drum uses this fact to detect the interface between water and steam. (The use of an external water column is necessary because of the number of penetrations that would otherwise be necessary in the boiler drum.) The detection circuit is divided into two groups of 'even' and 'odd' electrodes, so that failure of a single drive circuit cannot disable the entire system.

Although such a system will be affected by the difference in the densities of the water in the drum and the column, careful design of the installation will minimise any errors.

At present, although such devices provide an excellent indication of the drum level, they are not suitable for control because of the transient disturbance that occurs as the level moves from the position of one electrode to another. These step changes can produce unpredictable effects in the control loop.

6.8.2 Using an external water column

Although the method of connection shown in Figure 6.18 is viable, it has the disadvantage of being sensitive to errors during sudden reductions in the boiler pressure, caused by the condensate 'flashing off'-boiling as the temperature of the fluid suddenly finds itself above the saturation temperature.

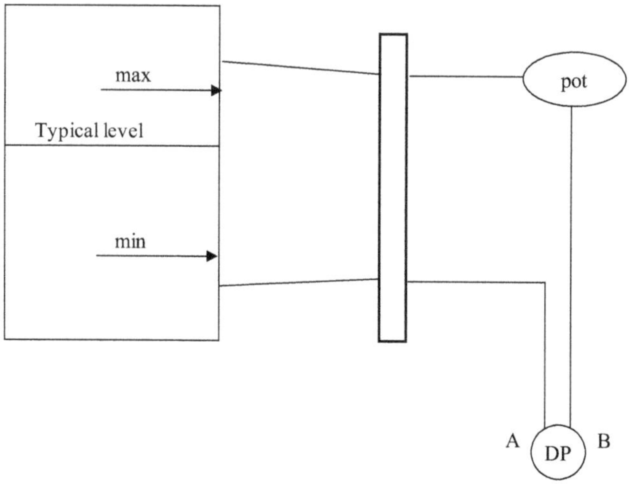

Figure 6.20 Drum-level measurement with a column

An arrangement that minimises this problem is shown in Figure 6.20, where an external water column is connected to the drum, so that the level of water in column is (theoretically) the same as the level within the drum. The column uses a volume of stored fluid which is larger than the volume of condensate in the small-bore HP leg of Figure 6.18, and the system is therefore less vulnerable to flashing off.

As with the detection column shown in Figure 6.19, great care must be taken to avoid errors caused by the temperature of the fluid in the measuring system being very different from that of the steam and water in the drum. This leads to a density error since the water column at the gauge will balance with the level in the drum, although its length is less than the distance between the lower tapping and the level of the water within the drum.

Considering the drum and the column of water in the column to be a 'U' tube, it will be seen that balance will occur when the weight of fluid in the left-hand leg equals the weight of fluid in the right-hand leg.

By keeping the temperature conditions within the column as close as possible to those in the drum, the density error will be minimised. This is done by arranging the pipework so that fluid flows through the water column. Unlike the drum itself and all the other pipework, the water column and the section of pipe connecting it to the tapping point isolation valve are left unlagged so that the steam condenses and the impulse line remains full of condensate. A circulation of fluid is then established through the column, with steam leaving the drum at the upper tapping point, condensing in the pipe and water column, and returning to the drum via the lower tapping point. This flow tends to maintain the temperature within the water column at a value which is as close as possible to the condition within the drum. Nevertheless, some temperature difference will still exist, and this will have the effect of increasing the density of the fluid in the cooler parts of the system.

6.8.3 Statutory requirements

In many countries, there is a legal requirement to provide separate systems to make the boiler operator aware of the level of the water in the drum. EN 12952-7 and ASME 1 both allow the use of EDLM but also require at least one gauge glass. If the EDLMs or other devices are providing a permanent indication to the boiler operator then the gauge glass may be isolated from the boiler.

The Indian Boiler Regulations also require a minimum of one level gauge. It is important that a specific installation is designed to meet the standards required at the actual point of use since the standards set by authorities in the relevant country, or by the insurers, may differ substantially from those indicated by this diagram.

The size of connecting pipework is also governed by some standards, and so must be checked on a case-by-case basis.

While electronic drum level monitors (EDLM) are common and specifically designed for their application they are not all Safety Integrity Level (SIL) rated. If the HAZOP calls for a SIF for the low-level trip then a SIL device such as the Hydratect must be used. This is a simpler version of a Hydrastep with no local and remote indication. In this application, it would be used only for the trip function; however, it can be used as standalone for drain pot-level control. Hydrastep and Hydratect are Emerson™ products.

6.8.4 Discrepancies between drum-level indications

It sometimes happens that various instruments connected to the same boiler drum display level measurements that are significantly different from each other. Since it is unlikely that the actual drum is anything but horizontal (except for installations on submarines during diving operations) such discrepancies must be due to some error or other. The following list summarises the factors that can cause errors:

- *Density errors*: Differences between installations can cause one instrument to be more affected by density factors than another. (One possibility is that, inadvertently, lagging has been applied to one of the condensation reservoirs.)
- *Turbulence*: The surface of the boiling water inside the drum is anything but still. It has been known for 'standing waves' to exist around the downcomers, affecting some measurement points more than others.

  Some boiler suppliers also avoid sampling in the rounded drum ends to avoid end effects, while others prefer this location to avoid the downcomers.
- *Flashing-off*: Differences in the geometry of the measuring systems can cause some measurements to be more affected than others by flashing off during pressure changes.
- *Calibration*: It is vital that all transmitters are carefully and accurately calibrated, and that any density compensation is correctly set-up.
- *Installation*: As stated earlier, errors or sluggish response can be the result of partial or complete plugging of impulse lines, or imperfect blowdown operations. As the measuring device is some distance from the drum, it is possible that two devices at opposite ends are actually at different levels and so will always read differently.

6.8.5 Drum-level gauges

While the electronic drum-level measurement system uses a specialised measuring technique, more conventional instruments can be used for local level indication. The most basic is a conventional-level gauge. On atmospheric tanks these can be a single see-through tube, but these tubes would rupture at boiler drum pressures and so a series of small gauges or bull's eyes are used (Figures 6.21 and 6.22).

You will see that each gauge is illuminated from behind to help viewing. Where bull's eyes are used a prism, together with the different refractive indices of steam and water, ensures that only a red or green bull's eye is seen depending upon steam or water being present.

On low-pressure drums, say less than 60 bar, float-operated magnetic gauges can be used. However, they are usually used on steam condensate tanks, where a single device can be used to give very clear local indication, level switches and remote indication. Originally potentiometric retransmission was used but can now be replaced by guided wave radar. This is independent of the float, the only common point being the pressure vessel and connections to the vessel (see Figure 6.23).

Figure 6.21 Drum-level gauge. © TC Fluid Control, a part of WIKA Instruments, reproduced with permission

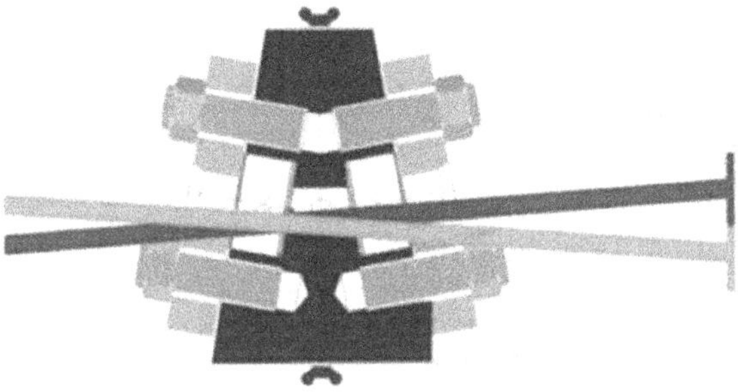

Figure 6.22 *Refractive effect used in level gauge. © TC Fluid Control, a part of WIKA Instruments, reproduced with permission*

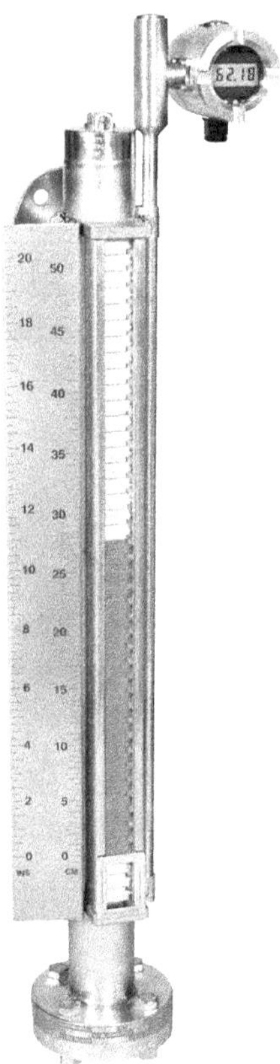

Figure 6.23 *Magnetic-level gauge with magnostrictive transmitter. © TC Fluid Control, a part of WIKA Instruments, reproduced with permission*

6.9 Feed-flow measurement

This is usually achieved by using a flow nozzle. Flow nozzles are well understood; see ISO 5167-3:2003 or ASME MFC-3M-2004. However there are some special considerations for feed-flow measurements. These are as follows: The feedwater is at high pressure, close to 180 bar on a drum boiler and 250 or above on a once-through. An all-welded construction is the norm. The impulse pipe work specification needs to be checked to ensure that it has an adequate pressure rating. Main feed pipe is rated for full-feed pump pressure. Because of the high pressure and quantities required off the shelf, main pipe is not always available and controlled bore pipework may be used. You cannot easily buy small quantities of this size pipework, so normal practice is to free issue a length to the flow nozzle manufacturer. Because of the size and weight, it is unlikely that a 'test house' will be provided with say 15–20 diameter straight lengths and a 4 m length weighing over 13 tonnes will be free issued.

The ASME standard expects the eccentricity to be within controlled limits. If it is not the expectation is that the pipe will be bored out. Boring out carries a risk that it will make the wall thickness too small. One way to avoid this is to calibrate the nozzle, so the result is accurate even if not concentric. It is custom and practice to have at least three sets of tappings for DP measurement on this element. The DP device measures volumetric flow so a density correction is necessary as the feed-flow temperature changes. The pressure also affects the density but to a lesser extent. Pressure compensation is optional for a drum boiler, but would be expected on a once-through super critical boiler maybe operating at 250 bar.

6.10 Steam flow measurement

On small boilers such as HRSG a flow nozzle may be used. However, on large power station boilers the Engineer Procure Construct (EPC) contractor must guarantee a maximum pressure drop between the boiler and the turbine. This has associated liquidated damages (LDs). These are punitive and so most EPCs do not provide any flow measurements, but rely on a calculated steam flow derived from the pressure after the first set of turbine blades, or an ellipse law using the pressure before and after the HP turbine. The turbine supplier provides the calibration curve.

On a drum boiler the steam flow signal is not needed at low loads as only single-element-level control is used. This turbine-derived flow is accurate at higher loads unless a turbine bypass valve opens, when the steam through the bypass does not pass through the turbine. To overcome this some suppliers have tried to derive steam flow from the DP across the superheater steam path. A better approach is to 'measure' the bypass flow and add it to the turbine-derived steam flow. The bypass flow may be a real measurement with a flow nozzle or a flow derived by the bypass valve supplier from his knowledge of valve CV with valve opening and the prevailing pressure drop.

6.11 The mechanisms used for feedwater control

In this analysis, we have looked at the principles of control and seen that because of various problems a variety of control methods has evolved, the selection of which depends on a variety of engineering and economic considerations relating to each application. In the discussion, some reference had to be made to the mechanism for controlling the flow and for simplicity it was assumed that this was by means of a familiar device, a valve. Now we shall look at the nature of valves in greater depth, and then we shall examine other methods of controlling the flow.

6.11.1 Valves

What follows is merely a practical overview of valve designs in general. It is not intended to be a deep analysis of what is in itself a specialised subject. If more detailed information is needed, it can be obtained from the many textbooks on valve design or from the publications produced by various valve manufacturers. (See Reference section of this book for information on such books.)

A control valve consists of many components which may conveniently be considered as falling into one of two groups: the valve body and the actuator. The former is the part through which the water flows and this flow is controlled by adjusting the resistance offered to the water. This is done by moving the position of a plug-in relation to its seat. The position of the plug is controlled by an actuator which acts via the stem.

Figure 6.24 shows an IMI CCI DRAG® control valve for feedwater control. The body has welded end connections and the DRAG trim provides the required rangeability (the 'trim' being the part of the valve which is in flowing contact with the water).

In addition, the trim design will determine the valve *characteristic,* which is the curve relating the stem position to the rate of flow of water through the valve. This is an important feature since the characteristic determines the gain of the valve system, which forms part of the overall loop gain.

For a given opening, the flow through the valve will be determined by the delivery pressure of the feed pump and the resistance that the boiler pipework offers to the flow: to simplify the task of selecting the correct valve size and definable set of conditions. This is achieved by determining what the flow through the valve would be if a fixed DP were to be maintained across it. This is termed the *inherent characteristic* of the valve.

Once the valve is operating on the actual plant, the position/flow relationship achieved in practice will not match the inherent characteristic because in the real world the inlet pressure and system resistance will vary, producing a pressure drop which is different from the value that was used to define the inherent characteristic. The pressure/flow relationship achieved in actual operation is called the *installed characteristic.*

As stated earlier, the gain of the valve is initially defined by the inherent characteristic, three types of which are commonly available, as shown in Figure 6.25. The operation of these different characteristics is now examined.

Figure 6.24 *A typical feedwater control valve. Image reproduced with permission of IMI CCI. Copyright © IMI CCI. A trading brand of Control Components, Inc., an IMI Critical Engineering company*

6.11.2 Quick opening

With a quick-opening valve, the flow rate through the valve changes very rapidly at low openings, with a slope that is fairly linear. Once the valve has opened about half way the rate of change of flow diminishes. This type of characteristic is usually applied only to shut-off valves.

6.11.3 Linear

When a valve has the linear characteristic, the flow rate through it at any given opening (in terms of percentage of maximum flow) is directly equal to the valve stem position (as a percentage of its full travel). With this type of characteristic, the gain of the valve system is constant for all openings. However, the flow through the valve at any given opening depends on the pressure drop across it and the linear characteristic

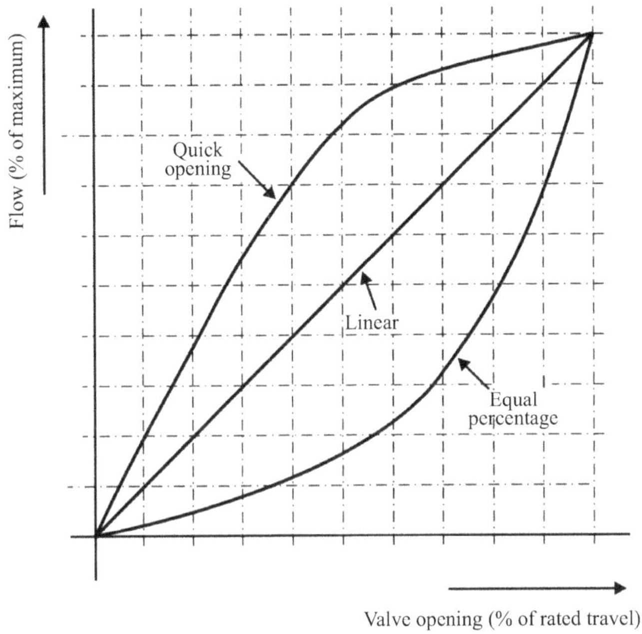

Figure 6.25 Inherent characteristics of valves

applies *only if the pressure drop across the valve is constant for each opening,* a condition that may not exist in practice unless special attention is paid to achieving it.

As a general rule, a linear characteristic is preferred for feedwater control applications since it simplifies tuning of the loop and enables good performance to be achieved over the widest possible range of flows.

Where optimum performance and efficiency is required the additional cost of providing a constant pressure drop across the valve should be considered. Such a solution will add cost, but this will be offset by improved control performance and plant life and by the savings achieved by not running the feed pump or pumps at a higher speed than necessary at reduced loads.

6.11.4 Equal percentage

The third characteristic is called equal percentage. Here, for all positions of the stem, the flow change achieved by moving the stem by a given amount is a constant proportion of the previous flow. What this means is that a given stem movement will change the flow by the same ratio as the *previous flow,* at any point in the valve travel. Therefore, the larger the opening, the greater will be the change of flow produced by a given stem movement.

This is shown by the curve in Figure 6.25, where moving the stem from 20% of full travel to 30% of full travel changes the flow from 5% to 7.5%, while moving the stem from 80% of full travel to 90% of full travel increases the flow from 50%

to 75%. In both cases, a stem movement of 10% of full travel results in a 50% increase of the previous flow.

From this it is apparent that, with this type of characteristic, the change of flow for a given stem movement is smallest at low openings and greatest at high openings. In other words, the gain is lower at low flows than it is at high flows. Equal-percentage valves are used where the mechanical plant design is such that there is only a small pressure drop available for the valve or where the pressure drop across the valve is likely to fluctuate over a wide range.

6.11.5 The valve sizing coefficient

Feed valves are designed to pass a flow that corresponds to the maximum flow requirement of the boiler plus a safety margin, and the capacity of the valve is related to a factor known as the valve sizing coefficient. The factor used widely across the world is based on US units and relates the capacity of the valve to the pressure drop by the following formula:

$$Q = Cv\ \sqrt{(\Delta P/G)} \tag{6.2}$$

where Cv is the sizing coefficient, Q is the capacity of valve in US gallons per minute (1 US gallon = 3.785 l), ΔP is the pressure differential across valve in pounds per square inch (psi) (1 psi = 0.069 bar) and G is the specific gravity of the flowing fluid.

The sizing coefficient is determined by experiment for each type and size of valve using water as the test fluid. It is equal to the volume of water (in US gallons) that will flow through the wide-open valve in one minute when the pressure drop across the valve is 1 psi. The European equivalent of the sizing coefficient is known as Kv. To convert between these units, multiply the Kv figure by 1.66 to obtain the Cv.

Given the pressure, temperature, flow and line-size characteristics for a given feedwater application, valve manufacturers will be able to provide detailed guidance on the correct valve size for a given application and from this will be able to predict the pressure drop across the valve.

6.11.6 Fail-safe operation

In the course of designing a feedwater control system, another matter that must be considered is that of selecting the 'fail-safe' position of the plug, the state that will arise if the actuator fails, or if the command signal to the valve is lost. The actual selection will depend on a range of factors, but in determining the safest option it is important to consider the effects of the flow on the valve itself since the applied pressure may tend to force the plug open or closed. Once again, valve manufacturers will be able to provide advice on this matter in relation to the actual installation being designed.

6.11.7 Selecting the valve size

The size of valve to use in a given application will be determined by many factors, only one of which is the physical size of the line in which it is fitted. Clearly, the valve must be large enough to pass the required flow with ease. Oversized valves will be unnecessarily expensive and should be avoided as unable to effectively control small flows.

For a valve to maintain any control over the process there must be some pressure drop across it, but if the pressure drop is too great several undesirable events start to occur. The lowest pressure occurs a short distance downstream from the restriction (the vena contracta) and the pressure rises past this point although it obviously never regains the initial value. If the outlet pressure drops below the vapour pressure, the fluid will start to boil and bubbles will form in it at the point where this occurs.

One of two things will occur as the pressure rises again after the restriction. If it rises above the vapour pressure, the bubbles will collapse, a process known as cavitation, which is accompanied by acoustic noise, the degree of which depends on the scale of the cavitation. Apart from being undesirable on environmental grounds, this noise represents an expenditure of energy. Furthermore, if the bubbles collapse close to metal surfaces, the localised energy release can damage the metal. In some cases, this will become apparent as severe pitting of the valve plug and cage, or it could appear as pitting of the pipe itself if the point where the vapour pressure threshold is passed occurs some way down from the valve. High recovery valves are more likely to experience cavitation because the downstream pressure is more likely to exceed the vapour pressure.

If, on the other hand, the pressure does not rise above the vapour pressure the bubbles will remain suspended in the fluid, to be carried downstream of the restriction. This is known as 'flashing', and it can result in erosion damage to the valve internals at the point of maximum velocity (usually at or near the point where the plug seats against the ring).

These bubbles reduce the valve's ability to pass fluid and eventually it 'chokes'. When this occurs no more flow can occur irrespective of how much water pressure is applied at the valve inlet. From these considerations, it is apparent that the production of bubbles may cause noise and damage to the valve, and possibly the pipework, and it will be correctly surmised that an optimum set of conditions will exist for a given design of valve. In other words, the valve will work best at one pressure-drop point. To summarise, if the supply of water to a boiler is controlled by throttling the flow through a valve, this can cause erosion and noise, and although control can only be effected by maintaining some pressure drop across the valve, this loss represents a loss of energy which should be reduced to the minimum

6.11.8 Specialised valve trims

The feed valve station has a particularly difficult set of operating conditions to cope with. At start-up, it must drop the feed-pump discharge pressure to the pressure needed to overcome the head of the drum. This requires a small CV and must be able to withstand damage caused by the high-pressure drop. At full load, the drum pressure is at maximum and the upstream pressure has been reduced by pipeline losses. This requires a large CV. Traditionally this was overcome by using a start-up valve and one or more duty valves.

CCI (Now IMI CCI) originally developed a Drag® valve. Since then Weir have released their Xtreme® trim and SPX Flow their Raven® trim (see Figure 6.26). While each has its own special features they all drop the pressure over a number of stages.

In this example, the flow is from the outside of the stack to the inside. You will see that the flow changes direction 14 times while passing through the labyrinth.

An analogy is to consider something being dropped out of window and hitting the ground with great force and then comparing it to the same object being rolled down a flight of stairs, where it has a series of smaller falls (see Figure 6.27) the flow is from inside to outside. Each section has two entries to minimise blockages and multiple outlets to reduce noise.

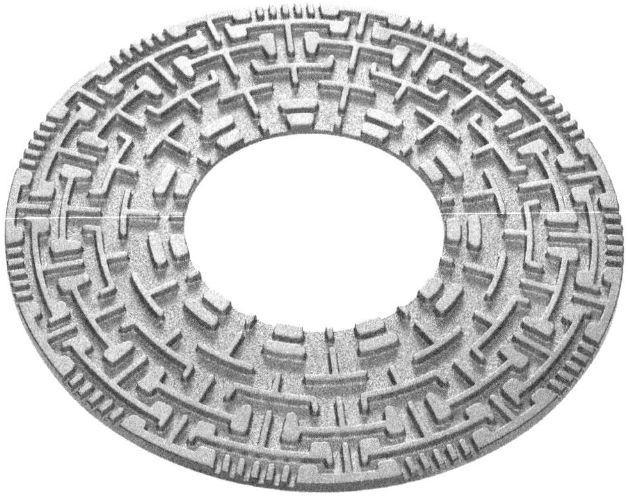

Figure 6.26 Raven trim disc. © SPX Flow, reproduced with permission

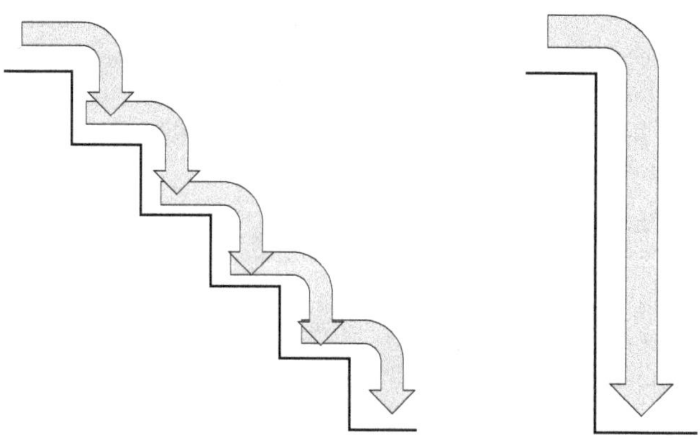

Figure 6.27 Concept of a multiturn valve giving multiple pressure drops

The valve trim is assembled by many of these discs stacked on top of each other. If the design of the discs is varied the CV can be contoured with valve opening.

6.12 Pumps

The water flowing through the feed valve into the boiler is delivered at pressure by one or more feed pumps. These produce a head of water which is related to the flow through the pump by a characteristic that will be similar to Figure 6.28, which shows that although the discharge pressure remains relatively constant as the flow rises from zero to (in this example) about 50%, beyond this value the pressure tends to decay as the flow increases. From this it will be apparent that the feed valve has to produce a greater pressure drop at low loads than at high loads.

Figure 6.29 shows the flow/pressure characteristic of a pump delivering water into a fixed system. In practice, as the flow through the system increases the resistance offered to it also increases, as shown by the dotted line in the diagram.

At any flow, in order to deliver water into the boiler, the pressure drop across the control valve will be the difference between the pump delivery pressure and the system resistance. For much of the flow range the pressure drop will inevitably be greater or less than the optimum for the valve design. This can be overcome by changing the inlet pressure so that the pressure drop always remains at the optimum value. This is achieved using variable-speed pumps.

6.12.1 Variable-speed pumps

Although in many cases the feed pumps operate at a fixed speed, at the design stage consideration should be given to the option of using variable speed pumps

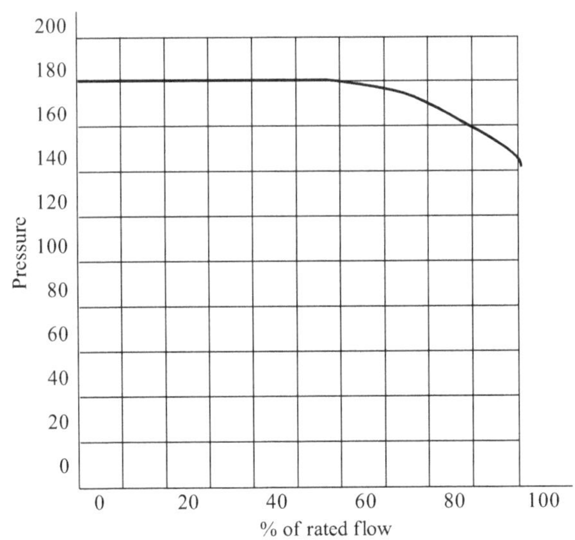

Figure 6.28 Typical flow curve for a feed pump at maximum speed

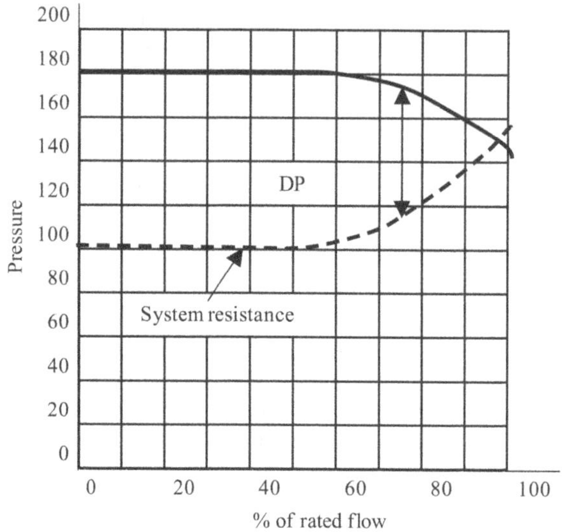

Figure 6.29 Pump delivery and system resistance

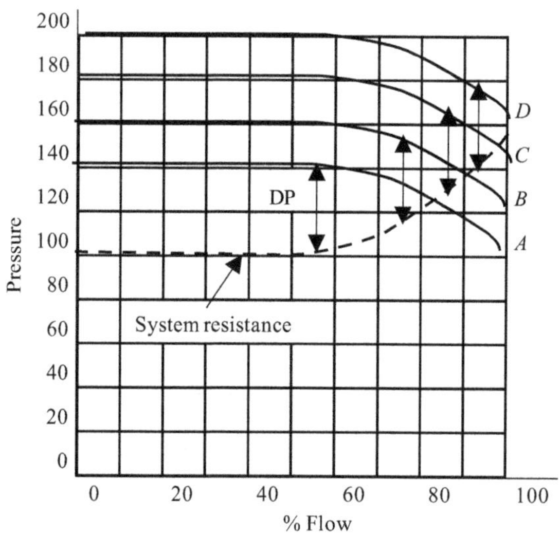

Figure 6.30 Variable-speed pump operation

(sometimes known as 'controllable-speed' pumps) because these will enable the feed valve to operate at the optimum pressure conditions for all loads.

The characteristics of a variable-speed pump are shown in Figure 6.30. The four curves show the pressure/flow characteristics at four different operating speeds (A–D), with A being the slowest and D the fastest speed. From this set of curves it will be appreciated that by adjusting the speed of the pump it will be possible to maintain a fixed DP across the feed regulating valve at all boiler loads. In the example shown the pressure drop across the valve (DP) achieved by operating the pump at speed A at 70% flow is maintained at the same value by operating the pump at speed B at 80% flow, speed C at 90% flow, etc., as indicated by the lines with arrows at each end.

The advantages of using a variable-speed pump include:

- Improvement of efficiency because of reduced pressure loss
- Reduction of pumping power
- Reduction of feed-valve wear due to erosion when operating at low flows
- Improved control because the valve operates at its designed pressure drop
- Improved control offered by the ability to operate with constant loop gain

Variable-speed pumps are more expensive than fixed-speed ones, but the increase in capital cost is offset by the revenue savings that will be gained, particularly if the boiler operates at reduced throughputs for a significant time over its life.

Of course, the pumps, valves and boiler pipework are only part of the overall system. Where spray attemperators are used (see Chapter 7), the feed pumps must also be capable of delivering cool water to the nozzles. The design of attemperators requires the water to be at delivered at a pressure which exceeds the steam pressure by a minimum value. Bypassing the pressure drop across the system as discussed in Section 6.2.3 allows sufficient DP to be maintained at the spray nozzle. But, if a variable-speed pump is used and run down to a low delivery pressure, there is a risk that the required differential may not be available. In such situations, the decision to use fixed- or variable-speed pumps will be affected by the need to maintain an adequate water pressure at the attemperators.

6.12.2 Pump controls

Pumps suffer from a phenomenon similar to surge or stall on a fan. To avoid damage due to flashing and consequent cavitation of the water inside the pump body, the pump must always operate within specific bounds of inlet and discharge pressures and flow. A high flow–low discharge pressure condition is generally protected by a circuit known as Q–H, or runout, or right-hand side protection which forces a speed reduction and raises an alarm if the critical Q–H limit is exceeded. The circuit is generally a strong proportional bias based on the flow deviation beyond the envelope limits, but can also be an independent flow controller that is activated during the event.

It is normal to duplicate the transmitters and logic. Alarming if the transmitters do not match. The expected flow against pressure curve is programmed into the DCS and a small margin added.

6.12.3 Boiler circulation pumps

In many boilers the water circulates around the furnace walls by natural circulation, the water from the feed pump passes via the economiser to the drum. From the drum, it flows via the downcomers to the bottom of the furnace. In the furnace walls, it rises as it is heated to the drum where the steam leaves and the remaining water circulates again. In some boiler designs, there is not enough natural circulation and circulation pumps are used.

Once-through boilers also need a circulating pump below the Benson point to ensure sufficient flow to prevent the furnace tubes from overheating.

The circulation pump is of a special design sometimes called a canned motor or wet stator motor. Effectively the 6 kV windings sit in water!

See Figure 6.31 for a boiler circulating pump. This design of pump is particularly reliable because it hangs from the pipework and therefore has no pipework forces imposed on it. It has water lubricated journal and thrust bearings. No water seal is required at the impeller as the water is allowed to flood the rotor and stator.

An auxiliary impeller circulates the boiler water though an external heat exchanger where it is cooled by low-pressure water.

All pumps generate heat when they run; this is dissipated in the water flowing through them. If the pump is run with no flow the heat cannot escape and the

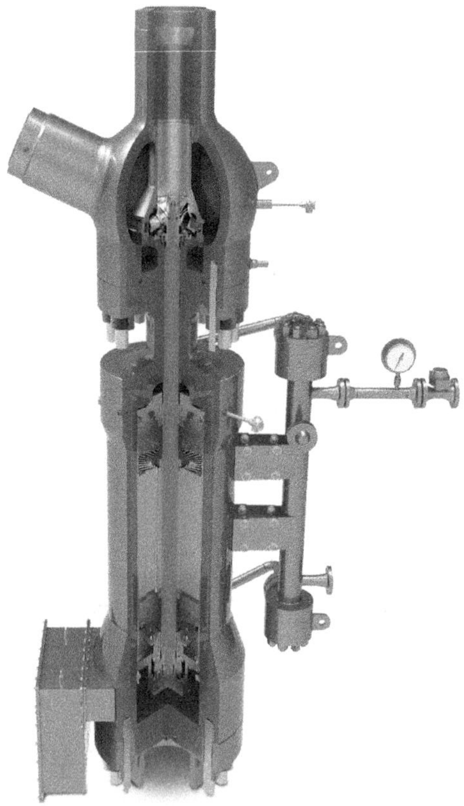

Figure 6.31 Wet stator boiler circulating pump. © Hayward Tyler Group, reproduced with permission

temperature of the pump increases, the water boils and cavitation occurs damaging the pump. Leak off is used to stop this happening. There are three ways of achieving leak off protection.

1. Having a permanent spillback around the pump. This wastes energy but may be used on a once-through boiler circulation pump if the standard pump size has the spare capacity. The pump is only run at low boiler loads, minimising the waste of energy. This arrangement is fool proof and requires no leak off valve, nor the associated logic.
2. Having a spillback controlled by logic. Simple logic is used to close the leak off valve above say 20% pump capacity and open it below say 15% pump capacity. The open and close flow rates are advised by the pump supplier.

6.13 Summary

Once the combustion process has occurred, the water has boiled and the steam has been generated, the next requirement is to ensure that the temperature of the steam that is delivered to the turbine or heat load is maintained at the correct value. In the next chapter we shall look at the control and instrumentation systems that are employed for this purpose.

Chapter 7

Steam temperature control and measurement

David Lindsley[1] and John Grist[2]

7.1 Why steam temperature control is needed

The rate at which heat is transferred to the fluid in the tube banks of a boiler or heat-recovery steam generator (HRSG) will depend on the rate of heat input from the fuel or exhaust from the gas turbine. This heat will be used to convert water to steam and then to increase the temperature of the steam in the superheat stages. In a boiler, the temperature of the steam will also be affected by the pattern in which the burners are fired since some banks of tubes pick up heat by direct radiation from the burners. In both types of plant the temperature of the steam will also be affected by the flow of fluid within the tubes and by the way in which the hot gases circulate within the boiler.

As the steam flow increases, the temperature of the steam in the banks of tubes that are directly influenced by the radiant heat of combustion starts to decrease as the increasing flow of fluid takes away more of the heat that falls on the metal. Therefore, the steam temperature/steam flow profile for this bank of tubes shows a decline as the steam flow increases. On the other hand, the temperature of the steam in the banks of tubes in the convection passes tends to increase because of the higher heat transfer brought about by the increased flow of gases, so that this temperature/flow profile shows a rise in temperature as the flow increases. By combining these two characteristics, the one rising, the other falling, the boiler designer will aim to achieve a fairly flat temperature/flow characteristic over a wide range of steam flows.

No matter how successfully this target is attained, it cannot yield an absolutely flat temperature/flow characteristic. Without any additional control, the temperature of the steam leaving the final superheater of the boiler or HRSG would vary with the rate of steam flow, following what is known as the 'natural characteristic' of the boiler. The shape of this will depend on the particular design of plant, but in general, the temperature will rise to a peak as the load increases, after which it will fall.

The steam turbine or the process plant that is to receive the steam usually requires the temperature to remain at a precise value over the entire load range, and it is mainly for this reason that some dedicated means of regulating the temperature

[1]Retired
[2]Consulting Engineer

must be provided. Since different banks of tubes are affected in different ways by the radiation from the burners and the flow of hot gases, an additional requirement is to provide some means of adjusting the temperature of the steam within different parts of the circuit, to prevent any one section from becoming overheated.

In theory, the design of the plant should be targeted on arranging for the natural characteristic to attain the correct steam temperature when the rate of steam flow is that at which the boiler will normally operate. If this is possible, it means that spray water is used only while the unit is being brought up to load or when it operates at off-design conditions. In practice this objective can be attained only to a limited extent because the boiler's natural characteristic changes with time due to factors such as fouling of the metal surfaces, which affects the heat transfer. In general, it is common to operate with continuous spraying, which has the advantage of allowing the steam temperature to be adjusted both upwards and downwards. If the required temperature were to be met solely by employing the natural characteristic as described, it would not be possible to produce temperature increases.

Before looking at the types of steam temperature control systems that are applied, it will be useful to examine some of the mechanisms which are employed to regulate the temperature according to the controller's commands. Depending on whether or not the temperature of the steam is lowered to below the saturation point the controlling devices are known as attemperators or desuperheaters. (Strictly speaking, the correct term to use for a device which reduces the steam temperature to a point which is still above the saturation point is an attemperator, while one that lowers it below the saturation point may be referred to either as an attemperator or a desuperheater. However, in common engineering usage both terms are applied somewhat indiscriminately.)

In general, the steam temperature control is independent of the fuel being fired.

7.2 The spray water attemperator

One way of adjusting the temperature of steam is to pump a fine spray of comparatively cool water droplets into the vapour. With the resulting intermixing of hot steam and cold water the coolant eventually evaporates so that the final mixture comprises an increased volume of steam at a temperature which is lower than that prior to the water injection point. This cooling function is achieved in the attemperator. The attemperator is an effective means of lowering the temperature of the steam, though in thermodynamic terms it results in a reduction in the performance of the plant because the steam temperature has to be raised to a higher value than is needed, only to be brought down to the correct value later, by injecting the spray water.

Although the inherent design of the attemperation system may, in theory, permit control to be achieved over a very wide range of steam flows, it should be understood that the curve of the boiler's natural characteristic will restrict the load range over which practical temperature control is possible, regardless of the type of attemperator in use. It is not unusual for the effective temperature control range of a drum boiler to be between only 75% and 100% of the Boiler Maximum Continuous

Rating (BMCR). The main steam temperature in a once through boiler may be controlled down to 40% BMCR.

7.2.1 The mechanically atomised attemperator

Various forms of spray attemperator are employed. Figure 7.1 shows a simple design where the high-pressure cooling water is mechanically atomised into small droplets at a nozzle, thereby maximising the area of contact between the steam and the water. With this type of attemperator the water droplets leave the nozzle at a high velocity and therefore travel for some distance before they mix with the steam and are absorbed. To avoid stress-inducing impingement of cold droplets on hot pipework, the length of straight pipe in which this type of attemperator needs to be installed is quite long, typically 6 m or more.

With spray attemperators, the flow of cooling water is related to the flow rate and the temperature of the steam, and this leads to a further limitation of a fixed-nozzle attemperator. Successful break-up of the water into atomised droplets requires the spray water to be at a pressure which exceeds the steam pressure at the nozzle by a certain amount (typically 4 bar). Because the nozzle presents a fixed-area orifice to the spray water, the pressure/flow characteristic has a square-law shape, resulting in a restricted range of flows over which it can be used (this is referred to as limited turndown or rangeability). The turndown of the mechanically atomised type of attemperator is around 1.5:1.

The temperature of the steam is adjusted by modulating a separate spray water control valve to admit more or less coolant into the steam.

Because of the limitations of the single nozzle, the accuracy of control that is possible with this type of attemperator is no better than ±8.5 °C.

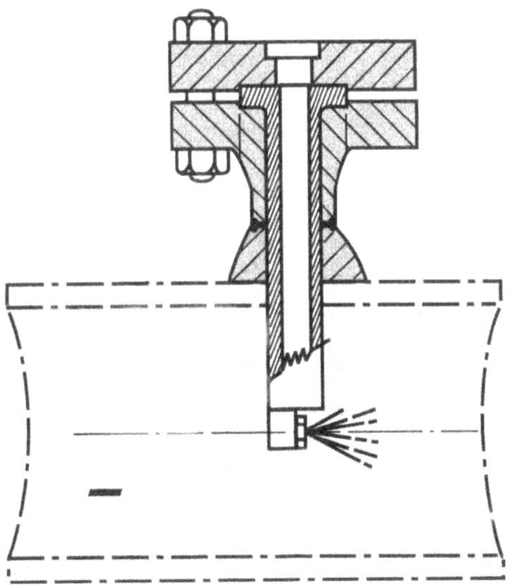

Figure 7.1 Mechanically atomised desuperheater. © SPX Flow, reproduced with permission

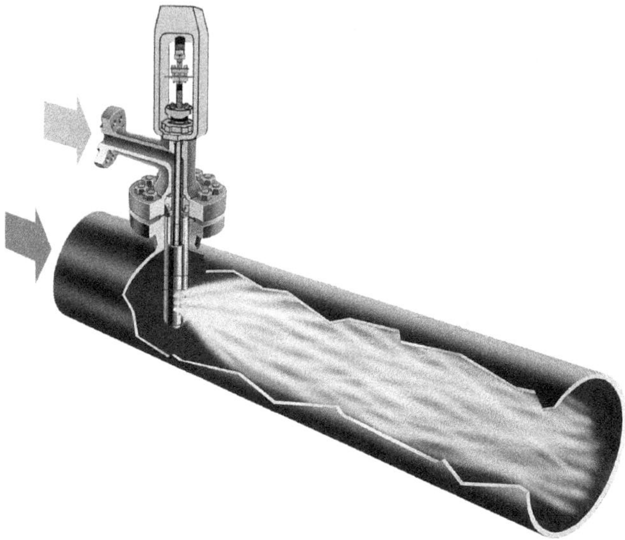

Figure 7.2 Principle of a multinozzle desuperheater. © SPX Flow, reproduced with permission

7.2.2 The variable area attemperator

One way of overcoming the limitations of a fixed nozzle in an attemperator is to use an arrangement which changes the profile as the throughput of spray water alters. Figure 7.2 shows the operating principle of a variable area, multinozzle attemperator. This employs a sliding plug which is moved by an actuator, allowing the water to be injected through a greater or smaller number of nozzles. With this type of device, the amount of water injected is regulated by the position of the sliding plug, a separate spray water control valve is therefore not needed.

Adequate performance of this type of attemperator depends on the velocity of the vapour at the nozzles being high enough to ensure that the coolant droplets remain in suspension for long enough to ensure their absorption by the steam. For this reason, and also to provide thermal protection for the pipework in the vicinity of the nozzles, a thermal liner is often included in the pipe extending from the plane of the nozzles to a point some distance downstream. This type of attemperator is often used on the reheater system.

The accuracy of control and the turndown range available from a multinozzle attemperator is considerably greater than that of a single-nozzle version, allowing the steam temperature to be controlled to $\pm$ 5.5 °C over a flow range of 40:1.

7.2.3 The variable annulus desuperheater

Another way of achieving accurate control of the steam temperature over the widest possible dynamic range is provided by the variable annulus desuperheater (VAD). Here, the approach contour of the VAD head is such that when the inlet steam flows through an annular ring between the spray head and the inner wall of the steam pipe its velocity is increased and the pressure slightly reduced. The coolant enters at this point and undergoes an instant increase in velocity and a decrease in pressure, causing it to vaporise into a micron-thin layer which is stripped off the edge of the spray head and propelled downstream. The stripping action acts as a barrier which prevents the coolant from impinging on the inner wall of the steam pipe. The principle is shown in Figure 7.3. The downstream portion of the VAD head is contoured, creating a vortex zone into which any unabsorbed coolant is drawn, exposing it to a zone of low pressure and high turbulence, which therefore causes additional evaporation. Due to the Venturi principle, the pressure of the cooled steam is quickly restored downstream of the vena contracta point, resulting in a very low overall loss of pressure.

An advantage of the VAD is that, due to the coolant injection occurring at a point where the steam pressure is lowered, the pressure of the spray water does not have to be significantly higher than that of the steam.

7.2.4 Other types of attemperator

At least two other designs of attemperator will be encountered in power station applications. The vapour atomising design mixes steam with the cooling water, thus ensuring more effective break-up of the water droplets and shrouding the atomised droplets in a sheath of steam to provide rapid attemperation.

Variable-orifice attemperators include a freely floating plug which is positioned above a fixed seat – a design that generates high turbulence and more efficient attemperation. The coolant velocity increases simultaneously with the pressure drop, instantly vaporising the liquid. Because of the movement of the plug, the pressure drop across the nozzle remains constant (at about 0.2 bar). The design of this type of attemperator is so efficient that complete mixing of the coolant and the steam is provided within 3–4 m of the coolant entry point, and the temperature can be controlled to ±2.5 °C, theoretically over a turndown range of 100:1.

Because the floating plug moves against gravity, this type of attemperator must be installed in a vertical section of pipe with the steam through it travelling in an upward direction. However, because of the efficient mixing of steam and coolant, it is permissible to provide a bend almost immediately after the device.

Figure 7.4(a)–(c) shows the principle of a variable orifice desuperheater as it moves from no load, through light load to full load. While this can be used on auxiliary steam systems, it is generally not suitable for power station boiler interstage attemperators that are normally mounted in a horizontal steam line.

7.2.5 Radial discharge mechanically atomised attemperators

Larger utility boilers have larger steam pipes or four instead of two steam paths. Multiple nozzles around the circumference of the pipe are used to ensure better penetration and distribution of the spray water. This approach is now well established among leading suppliers both for main-steam attemperators and in association with HP bypass systems. Each port is spring loaded which ensures that the port only opens when there is sufficient

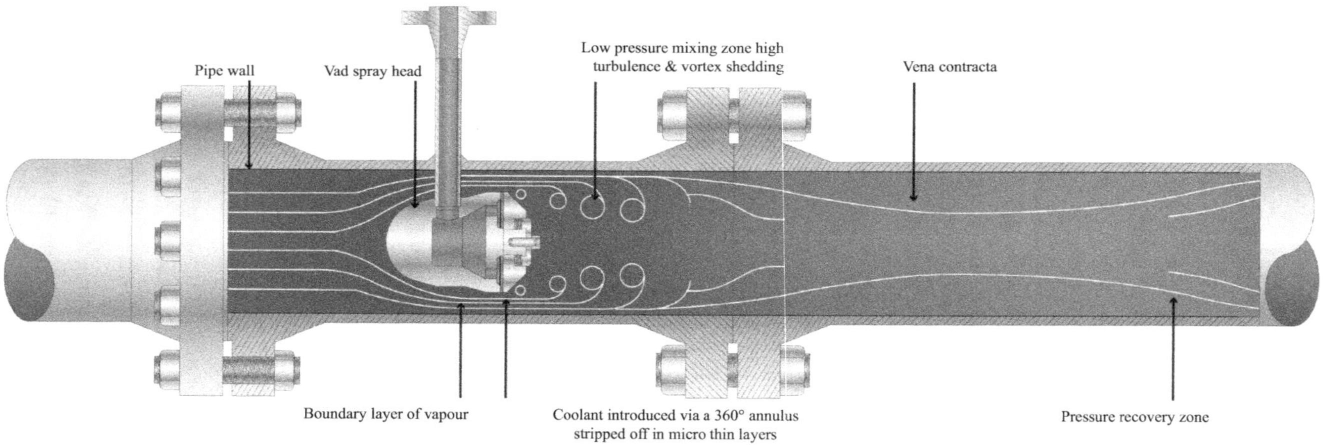

Figure 7.3 Principle of VAD. © SPX Flow, reproduced with permission

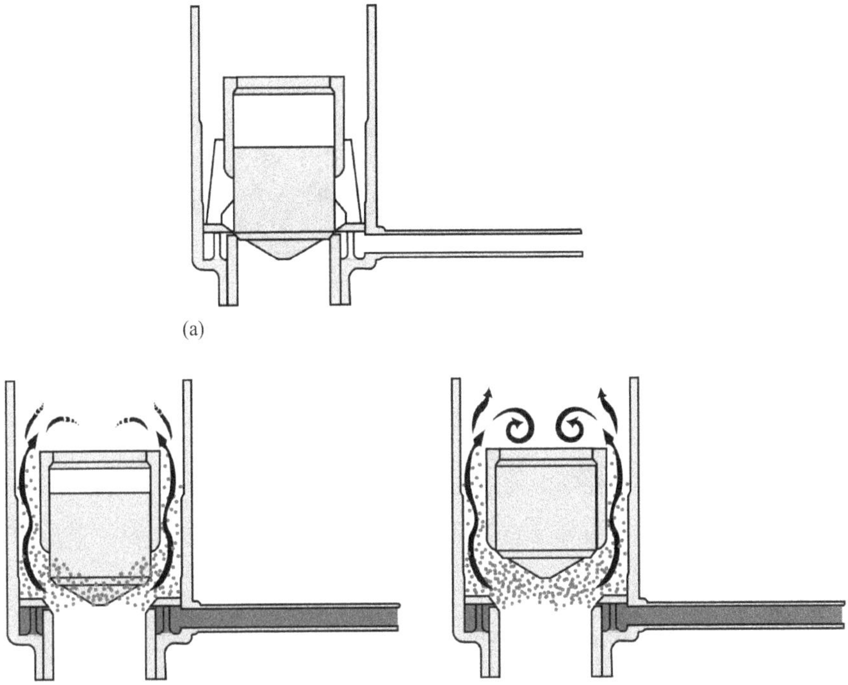

(a)

(b) (c)

Figure 7.4 (a) No load variable orifice attemperator; (b) lightly loaded variable orifice attemperator; (c) variable orifice desuperheater at full load. © SPX Flow, reproduced with permission

spray water pressure to ensure good atomisation. The ports are connected externally allowing a single control valve to be used for all the ports (see Figure 7.5).

7.2.6 Location of temperature sensors

Because the steam and water do not mix immediately at the plane of the nozzle or nozzles, great care must be taken to locate the temperature sensor far enough downstream of the attemperator for the measurement to accurately represent the actual temperature of the steam entering the next stage of tube banks. Direct impingement of spray water on the temperature sensor will result in the final steam temperature being higher than desired. Table 7.1 shows some typical values. An estimate of the minimum required distance may be derived from the lowest operating steam velocity and an estimated time for the droplets to evaporate. A typical time is 0.2 seconds.

7.2.7 Control systems for spray water attemperators

The simplest possible type of control would be based on measuring the temperature of the steam leaving the final superheater, and modulating the flow of cooling water to the spray attemperator so as to keep the temperature constant at all flow

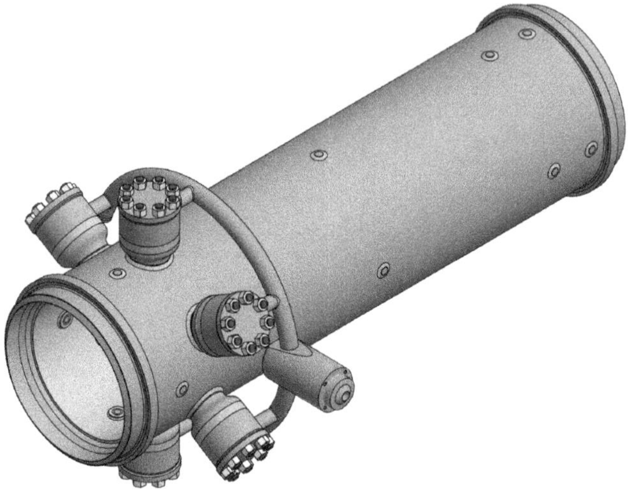

Figure 7.5 Radial discharge mechanically atomised attemperator. © SPX Flow, reproduced with permission

Table 7.1 Distances between spray and measuring element: based on information from SPX Flow and published by permission

Desuperheater type	Mechanically atomised	Multi nozzle	Variable annulus	Variable orifice	Radial discharge
Figure	7.1	7.2	7.3	7.4	7.5
Coolant turn down ratio	2:1	Up to 150:1	15:1	100:1	Dependant on number and size of nozzles selected
Control accuracy (°F)	±15	±10	±5	±5	±10
Control accuracy (°C)	±8	±5.5	±2.5	±2.5	±5.5
Typical distance to temperature sensor (ft)	30–50	16–40	20–33	12–20	26–40
Typical distance to temperature sensor (metres)	9–16	5–12	6–10	4–6	8–12

conditions. Unfortunately, because of the long time constants associated with the superheater, this form of control would produce excessive deviations in temperature, and a more complex arrangement is required.

Two time constants are associated with the superheater. One represents the time taken for changes in the firing rate to affect the steam temperature, the other is the time taken for the steam and water mixture leaving the attemperator to appear at the outlet of

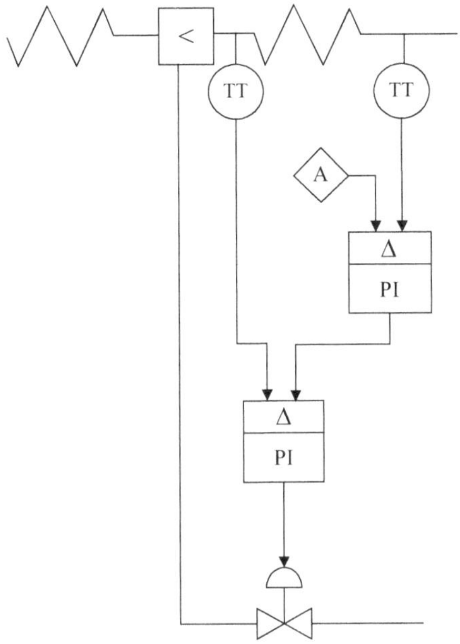

Figure 7.6 Steam temperature control with a single inter-stage spray attemperator

the final superheater. In terms of temperature control, it is the latter effect which predominates because, although changes in heat input will affect the temperature of the steam, a fast-responding temperature control loop will be able to compensate for the alterations and keep the temperature constant. It is the reaction time between a change occurring in the spray water flow and the effects being observed in the final temperature that determines the extent of the temperature variations that will occur.

Another problem with a simple system, as outlined earlier, is that it does not permit any monitoring and control of the temperature to occur *within* the steam circuit only at the exit from the boiler. These difficulties are addressed by the application of a cascade control system as shown in Figure 7.6. This shows a simple steam temperature control system based on the use of an inter-stage attemperator which is located in the steam circuit between the primary and secondary banks of superheater tubes. Since it is the temperature of the steam leaving the secondary superheater that is important, this parameter is measured and a corresponding signal fed to a two- or three-term controller (proportional-plus-integral or proportional-plus-integral-plus-derivative). In this controller, the measured value signal is compared with a fixed desired value signal and the controller's output forms the desired value input for a secondary controller. (Because the output from one controller 'cascades' into the input of another, this type of control system is commonly termed 'cascade control'.)

The secondary controller compares this desired value signal with a measurement representing the temperature of the steam immediately after the spray water attemperator. It is a matter of some debate as to whether a two-term or three-term controller should be used in this type of application. Because the steam temperature sensors used are subjected to the high pressures and temperatures of the superheater, they have to be enclosed in substantial steel pockets. Even with the best designs, pockets are usually slow responding, with the result that any high-speed fluctuations in the measured value signal will be smoothed out and the resultant signal will be fairly stable. The use of a derivative term is therefore easier than in, say, flow measurement applications where small-scale but sudden changes in flow can occur. When rapid input changes are differentiated, the controller output changes by a large amount, and for this reason tuning three-term flow controllers for optimum response can become difficult. This is not a problem with the temperature controllers described here, and the application of derivative action may be viable if it is felt that this could provide improved performance. As usual, it is important that the controller design should be such that the derivative term affects only the measured value signal (not the desired value or error signals), since differential response to operator-induced setpoint changes is always undesirable.

In Chapter 6, reference was made to the requirement for the spray water to enter the attemperator at a pressure which exceeds the steam pressure by a minimum value. It is worth remembering the point made there: that, where a variable speed feed pump is used, care must be taken to ensure that adequate water/steam differential pressure (DP) is available under all operational conditions.

7.2.7.1 Controller saturation effects

The type of control system described earlier is commonly encountered in a wide variety of applications, and it is subject to an effect which must be understood and adequately addressed by the design of the controllers used in the system. The effect is known as 'integral saturation' or 'reset windup', and it is a characteristic of integral action controllers whose output commands are fed into the inputs of cascade or secondary controllers. It sometimes confuses people when they are first introduced to this saturation effect in steam temperature control applications since the word 'saturation' is also applied to a thermodynamic property of steam. It is therefore important that the point is clearly understood that in this context the word 'saturation' refers to a controller output reaching a limiting value and then attempting to exceed that figure. While distributed control system (DCS) suppliers have standard methods to overcome this effect, steam temperature control has particular characteristics that require special consideration.

In a cascade design the primary controller produces a setpoint for the secondary (faster response) process – the attemperator outlet temperature in our case. The secondary controller moves the actuator to keep the secondary process close to its setpoint. If the actuator reaches its end limits, there will be a deviation from setpoint in the primary process (i.e. the superheater outlet temperature) and the primary controller will respond by adjusting the secondary setpoint.

Since the actuator is at its limits the deviation will not be restored, and while the primary controller's proportional component may not vary more than a few per cent, because the deviation continues to exist the integrator will 'windup' towards its limits. When the superheater outlet temperature returns to setpoint (perhaps by a boiler load change or after sootblowing), the actuator will not start to respond until the primary controller's output returns to the current value of the secondary process. This will only occur after the primary process has significantly overshot the setpoint.

A common solution is to provide directional blocking on the primary controller; thus, windup is prevented, but the controller is allowed to respond once the primary process recovers. However, this solution assumes the secondary process value has not changed during the saturation event. Such assumption is valid on loops where the secondary process value is only affected by the actuator position, but on steam temperature loops many factors can change the attemperator outlet temperature while the actuator is fully open or closed.

A better approach is to force the primary controller's output to track the secondary process value while the actuator is at limits *and* while the primary process is either above or below its setpoint (depending on whether the actuator is fully open or closed). Once the primary process returns to its setpoint, tracking is released and the actuator can immediately start to move. Some DCS suppliers provide this facility in their standard controllers; others require external logic to be configured.

7.2.7.2 Prevention of overcooling

In steam temperature control applications it is important to prevent the temperature being reduced too far. If the temperature at the inlet of the secondary superheater falls to a value approaching the saturation temperature, water droplets could form in the flow stream, raising the possibility of thermal shock to the pipework, and in addition the steam circuit could become partially plugged. The flow through the obstructed tubes will then be reduced and their surface temperature will rise, possibly causing premature tube failure.

7.2.7.3 Multistage attemperators

Some boilers have several banks of superheater tubes. In these cases, spray attemperators are normally provided between the major banks, as shown in Figure 7.7.

It will be seen that the control systems around each superheater comprise cascade loops that are quite similar to those discussed earlier. However, the set-value signal for the first stage of spraying is derived from the output of the controller regulating the final steam temperature. In fact, the signal may be characterised in some way to accurately represent the relationship between the temperature of the steam leaving the second stage of attemperation and that at the exit of the first stage. When the system is operating correctly, with the final slave controller maintaining its desired value and measured value signals at the same value, the effect is to maintain a constant temperature differential across the second attemperator. The temperature drop across the attemperator is a measure of the work being done by it, and by controlling this to a known value the cooling load through the entire string of tubes can be apportioned as required.

The function of the maximum selector unit (item 9) is to prevent chilling as described in Section 7.2.7.2. The steam pressure at the drum is measured and its value characterised (8) to produce a signal which represents the saturation temperature. Item 8 also incorporates a bias to represent a safe margin of operation and the resultant signal is fed to the maximum selector unit. If the output of the secondary

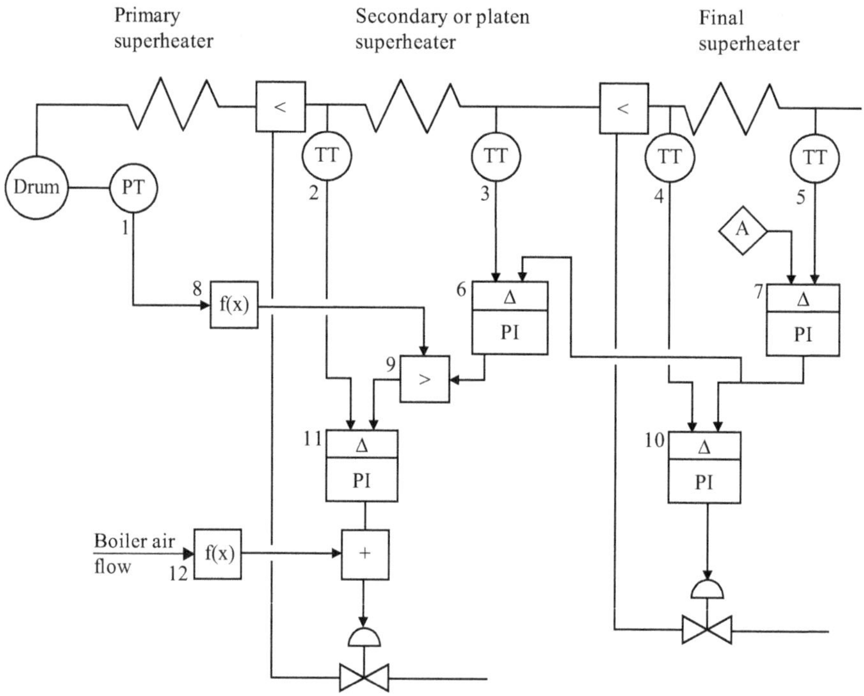

Figure 7.7 Steam temperature control with two-stage spray attemperation

superheater controller should fall to a value that is at or below the safety margin above the saturation temperature, it is ignored by the maximum selector unit, which clamps the desired value signal for the first-stage attemperator at this limit.

Another feature of the system shown in this diagram is the programming signal which presets the opening of the first-stage spray water valve according to a characteristic that the boiler designer has predicted. The temperature control systems then trim this opening to eliminate any residual error. These programming signals can overcome some of the boiler's time delays, producing a better and faster response to changes in load.

The cascade system is used to overcome the long time response of the superheater. It works but requires careful setting up of the track and reset functions to prevent the controllers saturating if the spray valve goes fully open or fully closed. It is also relatively difficult to commission. There are now alternative strategies that are widely used.

7.3 Advanced steam temperature control

There are many ways to improve the response of the steam temperature control. They range from term adaption of the P + I settings in a cascade loop, through a range of formalised algorithms to full optimisers. The Smith predictor is standard algorithm in many DCS systems and is designed to overcome pure dead time. In large boilers with a single stage of sprays the process response can be up to fifth order and so the effective dead time is very significant. In these cases, the Smith predictor can be used to advantage, but as it is a less robust design the dead time and delay model values must be adapted with boiler steam flow. Two more popular approaches are explained next. The ultimate is a multivariable optimiser, either as a stand-alone processor or, more commonly, integrated in the DCS. In both cases the IP is closely guarded by the supplier and setting up is beyond the scope of many control and instrumentation (C&I) engineers. All major DCS suppliers will have their own optimiser or other advanced control options for boiler controls. Optimisers typically modify control setpoints and so do not have a significant impact on the boiler vendor's standard control logic. However, the boiler control engineer has to take part in a design review of the integration of the optimiser with their standard logic.

7.3.1 Two-loop control

This is standard approach for Siemens on both once-through boilers and drum boilers. For once-through boilers it is used in combination with the enthalpy control system and together they can typically control to better than ± 3 °C.

It uses a model of the superheater response and the spray is adjusted to control the model. If the model is accurate the response is great. The design concept is shown in Figure 7.8.

The controller's operation can be understood by considering how the loop responds following a step change in setpoint. The loop has several interacting functions and the first description ignores the effect of blocks 6 and 7.

PT3 is the DIN symbol for three time delays in series, that is, it represents a third-order delay. In steady-state conditions, the output is the same as the input. The changes in output of each logic block can be seen in Table 7.2.

We start with the loop in equilibrium at time step 0, the setpoint and measured value both being 600 °C. The input to the final superheater is assumed as 520 °C, and the output of the PT3 is also 520 °C.

At time step 1 the operator changes the setpoint to 605 °C – this has an immediate increase on the controller (15) setpoint and the control valve starts to close in response to this. While it takes a finite time to increase the temperature of the steam, immediately following the attemperator, it is fast compared to the time taken to increase the final outlet temperature. For this explanation, I have assumed that it happens in the next time interval to step 2. Note the steps are instances in time when the temperature has moved by a significant amount. They are not equal steps. They are chosen to make the explanation easy to prepare and follow.

At step 2 there has been a step changed in attemperator outlet temperature but nothing else has changed.

At step 3 the change in inlet temperature is being sensed at the superheater outlet and the temperature has increased by 1 °C. which reduces the error by 1 °C. Coincidently the output of the PT3 has also increased by 1 °C so the controller setpoint (PT3 + error) remains the same at 525 °C.

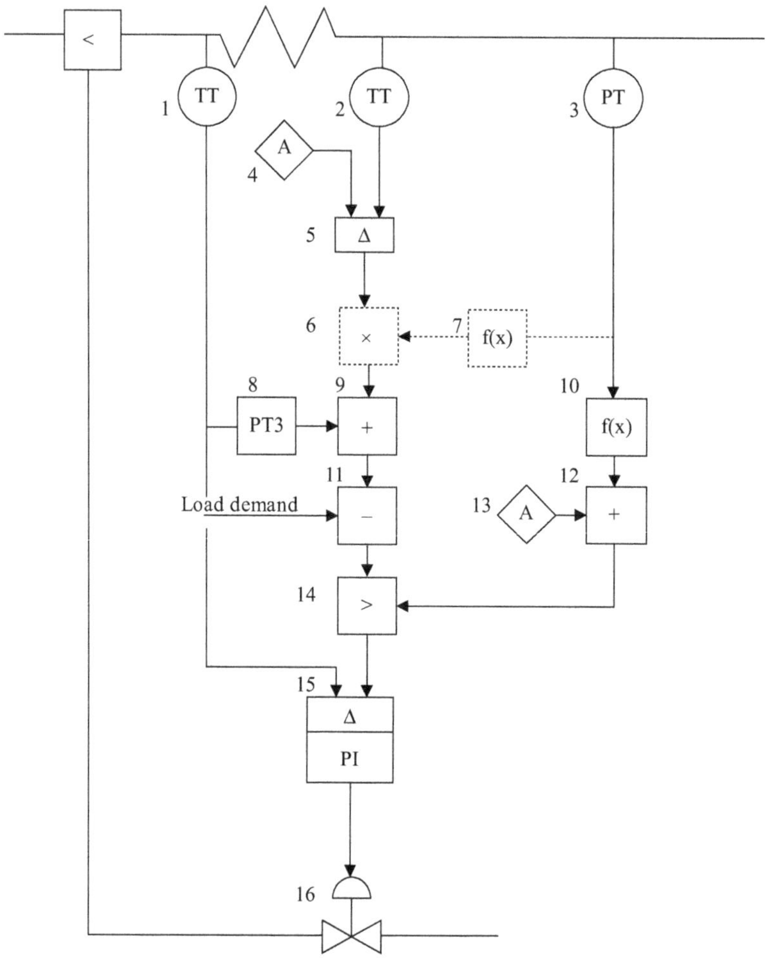

Figure 7.8 Steam temperature control with a two-loop control system. Based on Siemens' Benson Technology, published by permission

Steps 4–7 see the superheater outlet continue to rise in 1 °C increments, until the new setpoint of 605 °C has been achieved. The output of PT3 is rising at the same rate and so when added to the reducing error the controller setpoint remains unchanged at 525 °C.

The obvious good point is that the spray controller only moved once and having set up the correct spray no further movement was required. The question is how does the PT3 output coincidently move at exactly the required rate? The answer is that it is not a coincidence. The PT3 terms have been selected to model the superheater response, which will typically provide a good fit for boilers with

Table 7.2 Numerical explanation of two-loop control system

	Final SH			Final SH			Controller set point block 15
Time steps	**Set point block 4**	**Outlet block 2**	**Error block 5**	**Inlet block 1**	**PT3 block 8**	**Summer block 9 (8 + 5)**	
0	600	600	0	520	520	520	520
1	605	600	5	520	520	525	525
2	605	600	5	525	520	525	525
3	605	601	4	525	521	525	525
4	605	602	3	525	522	525	525
5	605	603	2	525	523	525	525
6	605	604	1	525	524	525	525
7	605	605	0	525	525	525	525
8	605	605	0	525	525	525	525

two stages of attemperators. Where the actual response is a third or fourth-order lag, a PT4 function can be applied to provide a better model fit.

The process response is comprised of a short transport delay which is proportional to the velocity of steam, and a series of process delays which occur along the length of the superheater tubing as the inertia of the metal masses absorb the increasing temperature. The process delay times decrease with increasing steam flow, approximately proportional to the inverse of the per cent boiler load.

The values to be used are determined during commissioning. A step change is introduced at, say, four different loads and the resultant changes in steam temperature analysed. The time T may be determined by measuring the slope of the response and referring to a look-up table or more normally by a system identification tool such as such as provided by MATLAB® and others.

At the start of this description we ignored some of these blocks which we will now consider. The premise assumed earlier was that to increase the superheater outlet by 5 °C you need to increase the superheater inlet by 5 °C. This is not true as the specific heat of the steam at the inlet and outlet is not the same. Blocks 6 and 7 are used to make this correction. For example on a drum boiler with steam at 540 °C, for a 5 °C change in outlet, the inlet only needs to change by 4.85 °C: a gain of 0.97 on block 7. The specific heat calculation is made at, say, three points over the load range. It can be a manual calculation looking up steam tables or derived from spreadsheet-based steam tables which are available from several internet sources.

The aforementioned logic can be applied to first- and second-stage sprays for both drum and once-through boilers.

7.3.2 State controller with observer

The cascade control system was developed as one way of overcoming the dead time and process delays of the superheater system. The system is effectively broken down into two parts, the attemperator and the superheater. Better control could be achieved by breaking the loop down into more components. The state controller

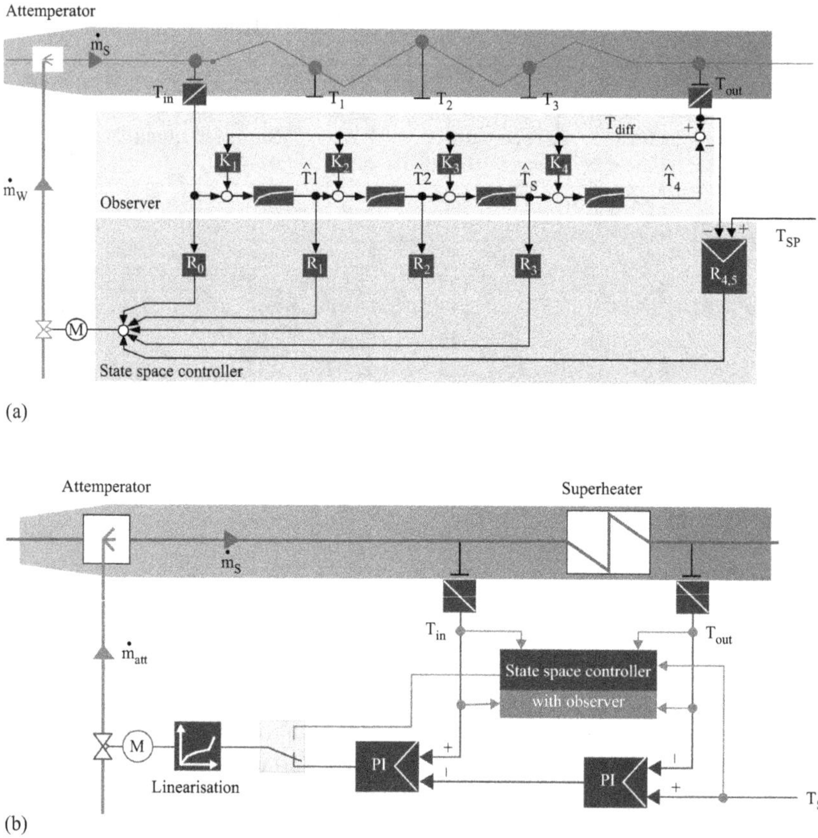

Figure 7.9 (a) State–space controller; (b) concept of SCO and cascade controller. *© ABB, reproduced with permission*

with observer does this. It does not use any additional real temperature measurements. A Luenberger observer estimates the intermediate states from measurements of the input and output of the real system (see Figure 7.9(a)).

The state controller with observer (SCO) is supplied as standard by IMI/Sulzer and ABB. While their performance is clearly better than the cascade system they are more difficult to commission. The ABB DCS controller module has the advantage that it has both a cascade controller and the SCO. This allows the loop to be initially commissioned as a conventional controller and then when more plant response data is available the ABB engineer can tune the state space controller. Open-loop response tests and identification analysis determine time constants and gains for different load points. Between the load points, the time constants will be interpolated.

Disturbances, such as changes in the heat transfer from slagging or soot blowing, are not modelled (although the SCO observer will identify them and respond). The SCO has a PID-controller in parallel with the observer and the state–space controller to correct any remaining deviation (see Figure 7.9(b)).

7.4 Controlling the temperature of reheated steam

In boilers with reheat stages, changes in firing inevitably affect the temperature of both the reheater and the superheater. If a single control mechanism were to be used for both temperatures the resulting interactions would make control system tuning difficult, if not impossible, to optimise. Such boilers therefore use two or more methods of control.

7.4.1 Reheat sprays

When hot sprays are used in the superheater they do not affect the efficiency of the boiler. The spray supply and feedwater inlet to the boiler are at the same temperature and are both converted to steam. If cold sprays are used (taken from upstream of the HP feed heaters) it has a small impact on efficiency because we make steam from the sprays starting from a lower temperature, although a smaller spray flow is required. Since in either case the efficiency effect is minimal, spray attemperators are the most common solution for superheater temperature control.

In the case of reheater spray water the source is usually an interstage tapping from the boiler feed pump, so it already has the low-temperature aspect of the reduced efficiency as with cold superheater spray. However, this spray water completely bypasses the HP cylinder of the turbine and so loses the opportunity to generate 30%–40% of the electricity that it would have done if it had entered the superheater system. Many clients will specify zero reheater spray flow during normal operation and impose severe contract penalties if spray is used.

For this reason, it is preferable for the reheat stages to be controlled by other methods under normal conditions, as described in the following sections.

7.4.2 Gas pass biasing dampers

In boilers with fixed burners, steam temperature control may be achieved by adjusting the opening of dampers that control the flow of the furnace gases across the various tube banks. In some cases, two separate sets of dampers are provided: one regulating the flow over the superheater banks, the other controlling the flow over the reheater banks. This is called parallel backend or split backend damper control. Between them, these two sets of dampers deal with the entire volume of combustion gases passing from the furnace to the chimney. If both were to be closed at the same time, the flow of these gases would be severely restricted, leading to the possibility of damage to the structure due to overpressurization. For this reason, the two sets are controlled in a so-called split-range fashion, with one set being allowed to close only when the other has substantially opened. See Figure 7.10 that shows a 1,000 MW two-pass boiler with a parallel backend. The furnace and pendent tubes above it are in the first (gas) pass while the reheater occupies one half of the second (gas) pass and the first-stage superheater sits above the economiser in the parallel half of the second pass.

These dampers provide the main form of control, but the response of the system is very slow, particularly with large boilers, where the temperature response to changes in heat input exhibits a second-order lag of two-minute duration or more.

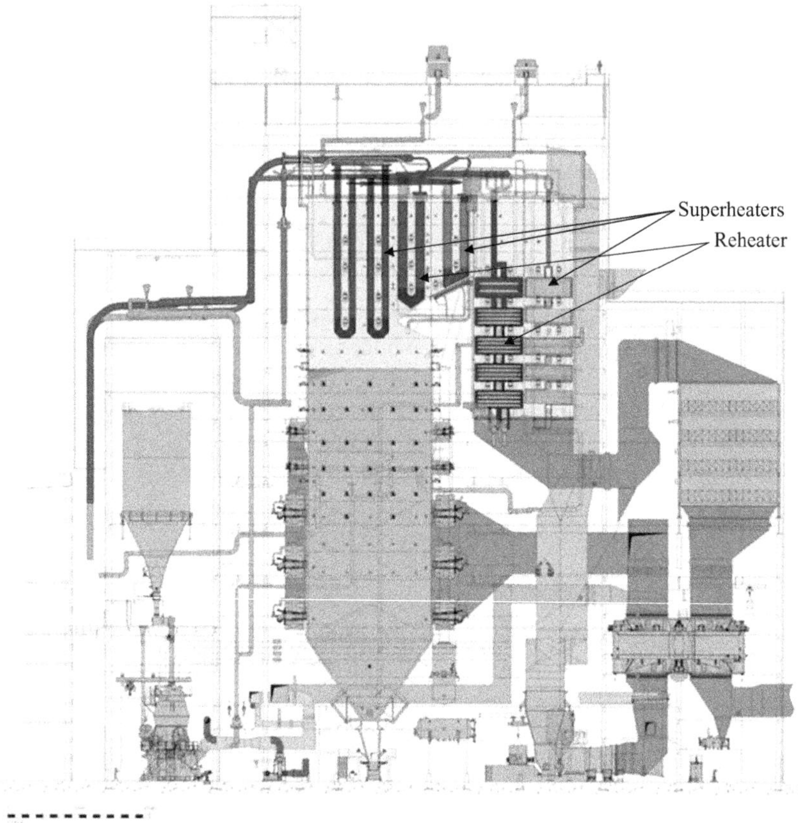

Figure 7.10 1,000 MW two-pass boiler with parallel backend. © Doosan Babcock, reproduced with permission

For this reason, and also to provide a means of reducing the temperature of the reheat steam in the event of a failure in the damper systems, spray attemperation is provided for emergency cooling. Reheater sprays are often seen to come into play briefly during load increases or sootblowing.

The spray attemperator is shut unless the temperature at the reheater outlet reaches a predetermined high limit, for example, 5–10° above the base reheater outlet steam temperature setpoint. When this limit is exceeded, the spray valve is opened. In this condition, the amount of water that is injected is typically controlled to bring the exit temperature back into the region where gas-apportioning or burner tilting can once again be effective. Cold reheat steam saturation protection is provided to prevent overspraying.

If a turbine trip occurs the reheat flow will collapse. In this situation, the reheat sprays must be shut immediately in order to prevent serious damage being caused by the admission of cold spray water to the turbine.

7.4.3 Temperature control with tilting burners

As explained in Chapter 3, the burning fuel in a corner-fired boiler forms a large swirling fireball which can be moved to a higher or lower level in the furnace by tilting the burners upwards or downwards with respect to a mid-position. The repositioning of the fireball changes the pattern of heat transfer to the various banks of superheater tubes and this provides an efficient method of controlling the steam temperature since it enables the use of spray water to be reserved for fine-tuning purposes and for emergencies. In addition, the tilting process provides a method of controlling furnace exit temperatures.

With such boilers, the steam temperature control systems become significantly different from those of boilers with fixed burners. The boiler designer is able to define the optimum angular position of the burners for all loads, and the control engineer can then use a function generator to set the angle of tilt over the load range to match this characteristic. A temperature controller trims the degree of tilt so that the correct steam temperature is attained.

While burner tilts affect both reheat and superheat steam temperature, they are typically used to control reheat temperature as tilting burner boilers do not normally have gas pass dampers, but do have superheater sprays.

7.4.4 Flue gas recycling

Where boilers are designed for burning oil, or oil and coal in combination, they are frequently provided with gas recirculation systems, where the hot gases exiting the later stages of the boiler are recirculated to the bottom part of the furnace, close to the burners. This procedure increases the mass flow of gas over the tube banks, and therefore increases the heat transfer to them. Because the gas exiting the furnace is at a low pressure, fans have to be provided to ensure that the gas flows in the correct direction. Controlling the flow of recycled gases provides a method of regulating the temperature of the reheated steam. The superheater temperature is also affected but corrected by the spray system. Interlocks are provided to protect the fan against high-temperature gases flowing in a reverse direction from the burner area if the fan is stopped or if it trips. The high temperature and high dust burden gives maintenance problems for the fan and the flow measurement. The dust burden is significantly reduced when oil or gas firing.

The control logic cascades a temperature controller onto the recirculation flow controller.

An alternative cold gas recycling configuration is to provide a return duct from after the ID fan back to the furnace. This then has two dampers: a stack damper and a recycle (or furnace) damper (see Figure 7.11).

The stack damper cannot be allowed to close so will have a mechanical stop in the 50%–70% range. It is most closed, to say 60%, at low loads but must start to open above this to allow the flue gases to pass to the stack. Conversely the furnace inlet damper is fully open at low loads but might close in to say 50% at full load when the reheater is most efficient.

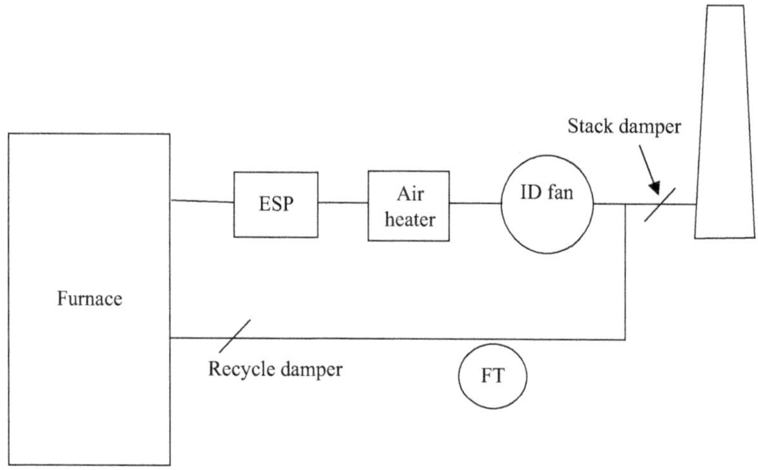

Figure 7.11 Gas recycling arrangement

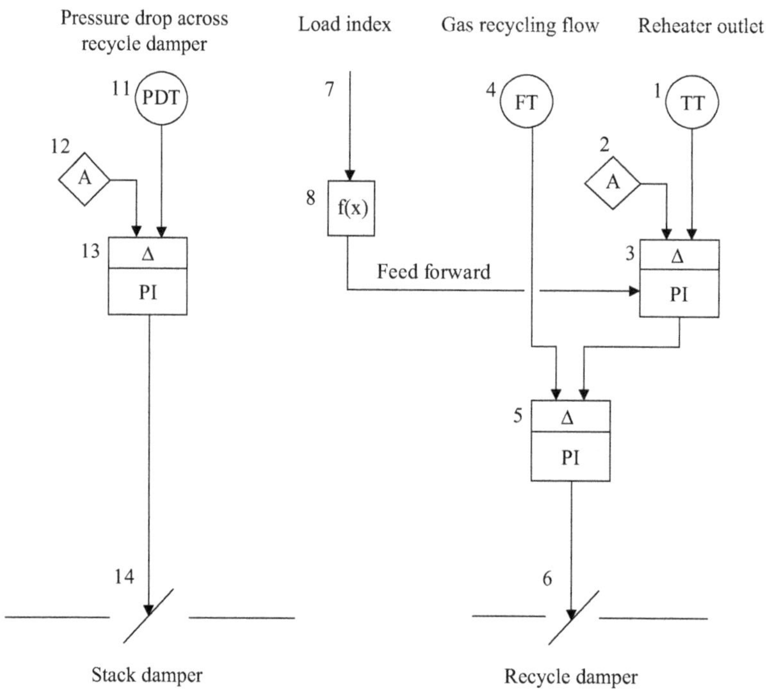

Figure 7.12 Flue gas recycling logic

Simplified logic is shown in Figure 7.12. The reheater temperature (1) is typically measured by redundant thermocouples. The average is compared in controller (3) with the setpoint (2). The output of this controller is shown as the setpoint for the flow controller (5).

In fact, it only acts as a trim on the load-dependent gas recycling flow setpoint (8). The flow is measured (4) and corrected for changes in temperature. To help this loop function a separate control loop is used to control the pressure drop across this damper. The DP is measured (11) and compared in a controller (13) with the required DP (12) and the controller output regulates the back pressure by moving the stack damper.

7.4.5 Hot air injection

This works in the same way as gas recycling but uses combustion air instead of flue gas to increase the mass flow of gases over the super heater and reheater banks, but in this case extra air is introduced. Any effect on the superheater is trimmed via the spray loop. Typically, the reheater underperforms at low loads which requires a lot of extra air to correct it, with associated increase in O_2 but performs better at high loads and maybe only 2% BMCR additional air is required for control. This approach is only suited to base load stations. A similar effect can be achieved by increasing the O_2 setpoint.

7.5 Temperature measurement

Temperature measurement of water and oil is well understood and comprises a thermocouple in a thermowell, usually with a head-mounted transmitter connected to a DCS. An electronic cold junction compensation is included in the circuitry of the instrument. The same applies to these in the boiler industry. The aim of this section is to make the reader aware that some boiler related measurements are more complicated and to indicate typical solutions that need to be agreed between the boiler supplier and end user. Vibration is one issue that is common to most boiler measurements. For this reason, many suppliers prefer to use thermocouples instead of resistance temperature detectors (which are less robust) for all measurements except those embedded in motor bearings and windings.

7.5.1 Measurement of air and flue gas temperature

The basic air/gas temperature measurement design follows the above selection principles. Problems encountered are related to the size of the duct and the low specific heat of the media. The ducts can be large enough to put a bus in, so multiple measurements may be considered.

The C&I engineer needs to consider whether these need to be averaged or the median used. Secondary air in a windbox is expected to be at a uniform temperature, so you should use the median temperature, while there will be a temperature profile across the air heater discharge and so an average is more appropriate.

Any changes in air or flue gas temperature take some time to be seen by the thermocouple because of the thermal inertia of a thermowell. One approach is to use an open-ended carrier tube instead of a thermowell. This just takes the weight of the thermocouple, if horizontal, or stops it moving if vertical. The thermocouple protrudes past the end of the tube and so is exposed to the flue gas and quickly responds to changes in temperature. Not all Engineer Procure Construct (EPC) organisations will accept this design as there is no proper seal between the flue gas and the outside world.

7.5.2 Measurement of steam temperature

Steam measurements use the same approach of thermowell, thermocouple and transmitter. All thermowells used on main steam pipework are welded in place. Some suppliers offer a quick response thermowell with a more complicated welding and installation procedure. The material and weld procedures must be specified by the boiler piping engineer. A Wake frequency analysis is carried out for all thermowells used in high-velocity systems to ensure that they can withstand the vibrations caused as the thermowell sheds vortices. If necessary the thermowell can be manufactured with helix strakes on its outside to reduce the vortices.

Head-mounted transmitters may be used on pipework outside of the boiler 'penthouse' (above furnace) region but cannot be used in the penthouse, having a surrounding temperature above 350 °C. The thermocouples are wired to a transmitter in a local junction box. It makes sense to use the same type of transmitter for all the temperature measurements in a particular control loop.

It is necessary to record some steam temperatures with high-accuracy instruments during the boiler performance test. So that this can be done, without disrupting the control measurements, test thermowells are provided. These sometimes use a larger diameter. If used in the penthouse, then either the boiler must be shut down and the boiler allowed to cool so test thermocouples can be fitted or permanent test thermocouples must be fitted during construction. These spares can be fitted in a conventional thermowell and then cabled to the outside of the penthouse so they are available for the test engineer, or larger extensions can be fitted to the thermowells and these pass through the penthouse walls. In theory, these thermowells allow the thermocouple to be replaced without shutting down the boiler.

7.5.3 Measurement of metal temperature

The rate of firing is controlled during start-up to limit the rate change of temperature of thick-walled components such as the drum and superheater headers. The drum is monitored to ensure it does not distort and the headers to avoid thermal stress. On a once-through boiler the storage vessel metal temperature is also measured. Sometimes thermocouples are used to measure the outer wall temperature, and sometimes the wall is drilled and a mid-wall temperature is also measured. An alternative approach is chordal drilling to allow the thermocouple to be fitted (see Figure 7.13). Arguably this is the best sensor but has the disadvantage that the thermocouple is very fragile and must be installed in the factory.

In a once-through boiler the tubes run is a spiral and it is assumed that they all pick up the same amount of heat. To check that there are no major anomalies, some, say one in three or all the tube temperatures, are measured as they leave the spiral. Figure 7.14 is the transition between spiral and vertical furnace wall tubes and shows the task that the C&I engineer has in finding a suitable location and technique to make these measurements.

Metal surface temperatures can be made using a standard type K thermocouple. There are two problems to consider. The first is ensuring an accurate measurement, and the second is to protect the thermocouple from excessive temperatures.

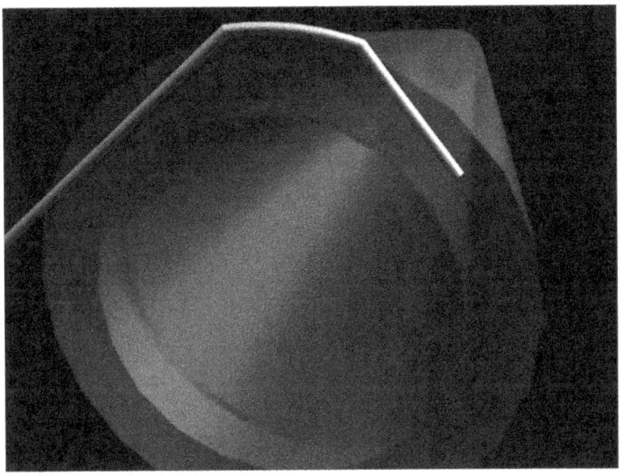

Figure 7.13 Chordal thermocouple. © WIKA, reproduced with permission

Figure 7.14 Transition between spiral and vertical wall. © Doosan Babcock Ltd. reproduced with permission

If we consider skin temperature measurements for the spiral tubes, the drum and the storage vessel, and headers in the boiler penthouse then the thermocouple needs only standard protection, but care must be taken to ensure good contact between the thermocouple and the metal. One way is to weld a threaded connection into which is screwed a spring-loaded thermowell attachment. The mechanical engineer must be aware of this to help decide when to weld on the attachment, to agree that the material is appropriate and to decide if the weld needs stress relieving.

The superheater pendant tube temperature can be estimated by comparison with the same tube temperature in the penthouse and by a knowledge of the furnace temperature from furnace pyrometers.

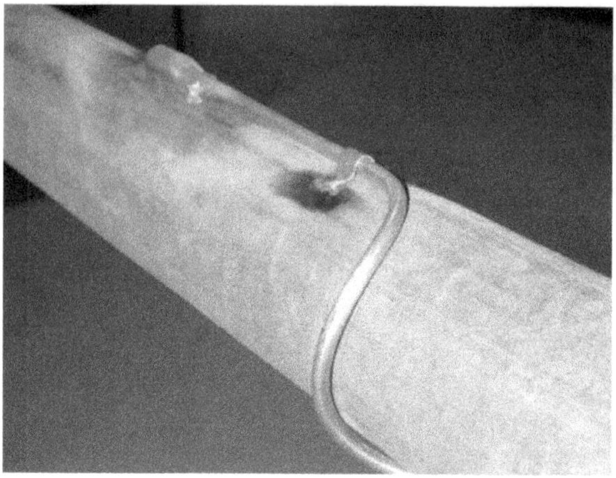

Figure 7.15 V-Pad® thermocouple. © WIKA, reproduced with permission

However, if actual metal temperatures are required then several techniques might be considered. One requires building up a weld deposition with the same material as the tube, drilling it and inserting the thermowell so that its tip is touching the original tube. The thermocouple is then peened in place. Another is the V-Pad®, as shown in Figure 7.15.

Both types are fitted in the factory and again guidance must be sought on material and weld procedures. Some suppliers recommend that the thermocouple is protected by a carrier tube that is welded to the superheater tube and cooled by the steam at 600 °C, to protect the thermocouple from overheating in the furnace. Others as with the V-Pad® claim that clamping the thermocouple to the tube is sufficient to protect the thermocouple from the furnace gases which are more than 1,000 °C hotter.

Metal temperatures are not used in the conventional control schemes, but may be used for alarms, or for design checks, to resolve issues or for stress calculations. As a once-through supercritical boiler may have between 200 and 1,500 metal temperature thermocouples, cables, transmitters and DCS analogue inputs can be saved by collecting this data via a multiplexing or remote I/O unit. One note of caution is that those used to calculate stress and remnant life may need to be monitored in real time.

The following standards define thermocouples IEC 60584-1: Thermocouples: basic and tolerance values of the thermoelectric voltages IEC 60584-3: Thermocouples: Thermocouple cables and compensating cables.

ASTM E230: Standard specification and temperature-electromotive force (EMF) tables for standardised thermocouples.

7.6 Summary

This ends our system-by-system survey of boiler and HRSG C&I systems, and we return in Chapter 11 to look in more detail how the dynamics of these systems are taken into account by the boiler and turbine master controls to match the unit load demand, particularly for units that are required to change load frequently in response to the grid's requirement for power.

Chapter 8

Control equipment practice

David Lindsley[1] and John Grist[2]

On an operational plant, the control systems that have so far been examined may be implemented in any of a variety of ways, ranging from pneumatics to advanced computer-based systems, but in all cases, it should be possible to identify the various loops within the relevant configuration. These days, most control functions are implemented by means of a computer-based system, so we shall now briefly look at a typical configuration. After that, we shall examine some of the other hardware used in the systems and then consider the environmental factors that influence the selection of control and instrumentation equipment.

8.1 A typical DCS configuration

DCS stands for 'distributed control system'. The term 'distributed' means that several processors are operating together. This is usually achieved by dedicating tasks to different machines. It does not necessarily mean that the separate computers are physically located in different areas of the plant.

Figure 8.1 shows physically how a generic DCS system may be arranged. Figure 8.2(a) and (b) shows alternative presentation emphasising the communication links and includes a demilitarised zone (DMZ). This is a small network with incoming and outgoing firewalls as a 'neutral zone' between the DCS and the outside public network.

The following notes relate to individual parts of that system. In practice, each manufacturer will usually offer some variant of the system shown in this diagram, and the relevant description should be consulted, but the comments made here are general ones which may help to identify points which should be considered and discussed when a new or refurbished system is being considered.

8.1.1 The central system cabinets

The cabinets house the control processors (CPs) or controllers that execute the control functions. These cubicles also contain the attendant interface and

[1]Retired
[2]Consulting Engineer

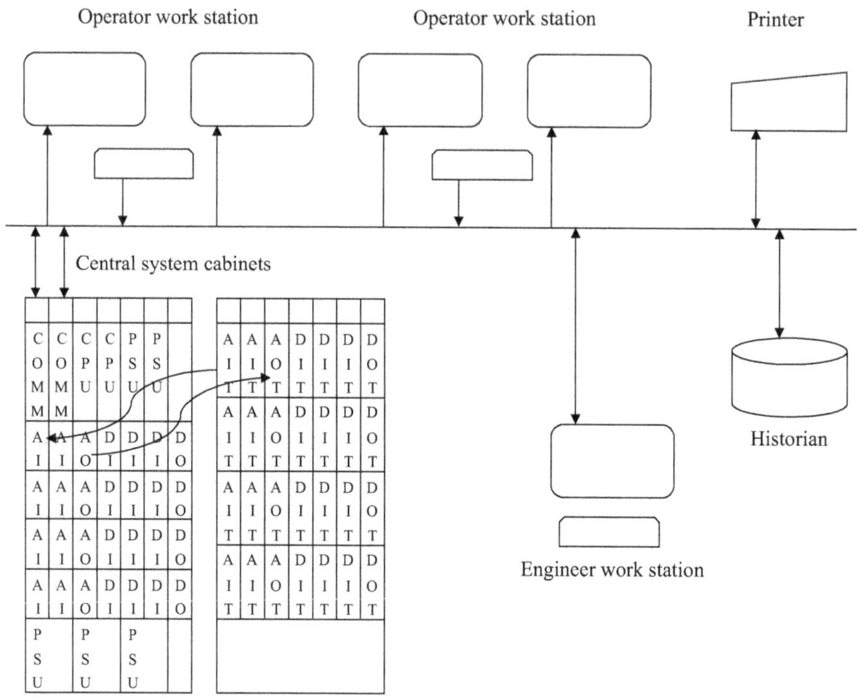

Figure 8.1 Generic DCS hardware configuration

input/output (I/O) cards and the necessary power supply units (PSUs). The latter will usually be duplicated or triplicated, with automatic changeover from one to another in the event of the first failing. (This automatic changeover is often referred to as 'diode auctioneering' because silicon diodes are used to feed power from each unit onto a common bus-main. In the event of the operational PSU failing, its diode prevents a power reversal while the back-up power unit takes over.) At this time, it is important that the system should raise an alarm to warn that a PSU failure has occurred. Otherwise the DCS will continue to operate with a diminished power supply reserve and any further failure could have serious consequences.

Clearly, the DCS cannot operate continuously from batteries alone. A reliable and stable source of power will therefore need to be available (usually 110 or 240 V AC). If the DCS includes internal back-up batteries it will continue to operate if the AC feed is lost, but such batteries are normally sized to retain essential data in the memory and to provide a limited amount of functionality. They may also allow limited control to be performed, but all this will function for only a short period (typically 30 minutes), and it is therefore usual to provide an external uninterruptable power supply (UPS) system which can allow the plant to be operated for a longer time. The duration of this period warrants very careful consideration. Long periods require large and expensive batteries and charger systems, and this expense can rarely be justified (especially since such a major power loss will probably have disabled all

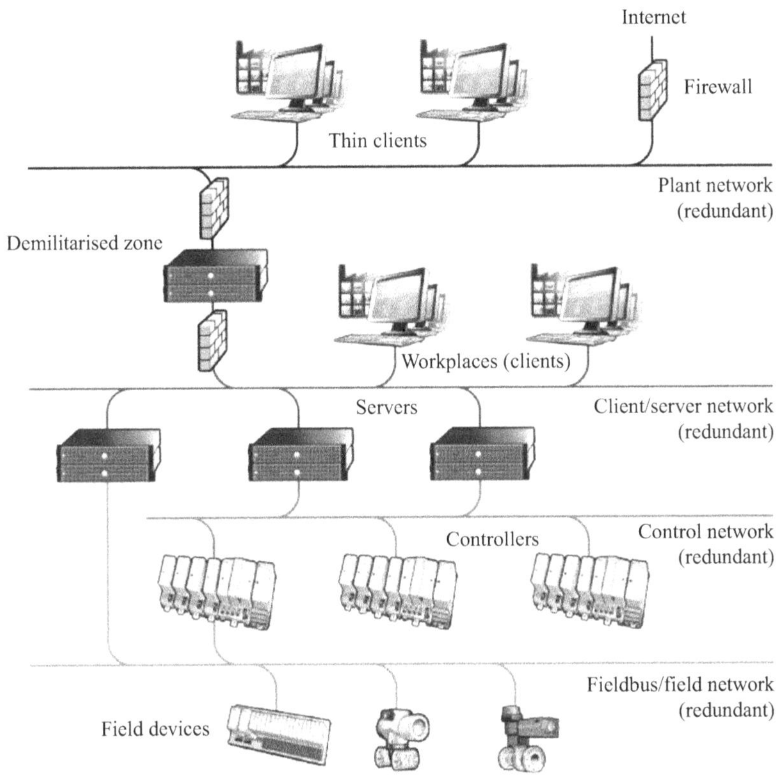

Figure 8.2 (a) Generic DCS communication configuration. © AAB, reproduced with permission

pumps, motors, etc.). Instead, it is common to provide a battery capacity that will allow the plant to be safely shut down in the event of power failure. The determination of the time required for such an operation is a matter of discussion with the process design engineers and the plant management.

In addition to supplying the computer system, the power supply system will usually also have to provide DC supplies for 4–20 mA transmitters and for limit-switch contacts. (The voltage connected to a contact and thence to the DCS input channel is often referred to as the 'wetting voltage'.) Transmitters operating on the 4–20 mA range which are powered from the DCS are sometimes called 'passive'. In comparison, those that operate from local power supplies are called 'active'.

The I/O cards consist of analogue and digital input (DI) and output channels. Analogue inputs (AIs) convert the incoming 4–20 mA signals to a form which can be read by the system. The printed-circuit cards for AIs may or may not provide 'galvanic isolation'. With a galvanically isolated device the signal circuit is electrically isolated from others, from the system earth and from the power supply common rail. Galvanic isolation simplifies circuit design since it prevents inadvertent short-circuiting, but consideration should be given to the possible build-up

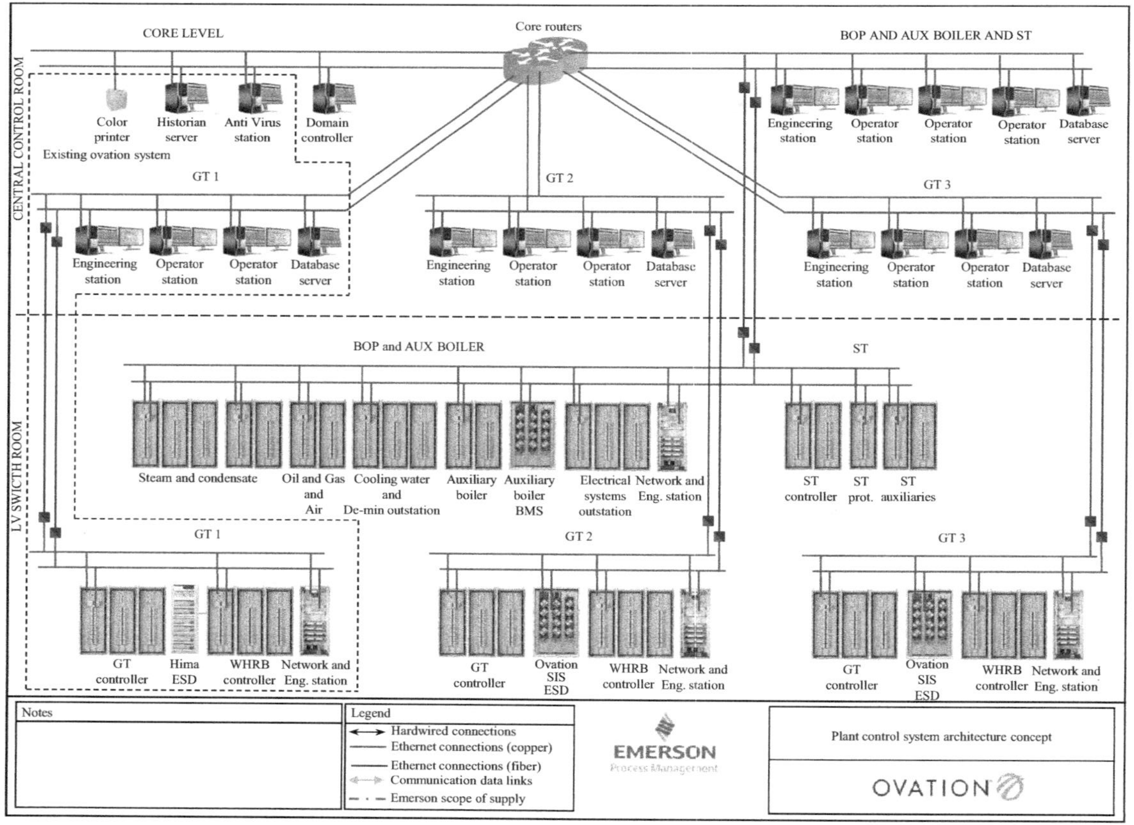

Figure 8.2 (b) Alternative generic DCS communication configuration. © Emerson, reproduced with permission

of static charges on completely ungrounded circuits, which could cause damage to input devices (which are usually rated for not more than a few tens or hundreds of volts). This is normally an important consideration only in areas of very low humidity or where there is a strong presence of charged particles.

The commissioning process, and the task of identifying and correcting faults, are operations which are considerably assisted by the provision of light-emitting diode (LED) status indicators on the digital output (DO) cards and comprehensive software diagnostics. Some systems provide switches on the DI cards, which can be of assistance with commissioning and fault-finding. However, inadvertent or deliberate maloperation of such switches can have serious consequences since the DCS is then provided with incorrect plant status information and it may take inappropriate action. (The use of logic probes, which inject signals into a system to check its operation, is also to be deprecated, for similar reasons.)

Analogue and digital I/O channels are normally grouped into 8, 16, 32, or 64 channels per printed circuit card. Also, 8 or 16 AI channels are commonly accommodated on a card, but analogue output (AO) channels consist of current generators and so occupy more space and are more expensive than AI channels, which are based on small operational amplifier devices. DI channels are very simple and cheap and may be grouped into 16 or even 32 inputs to a single card. DO channels driving lower power devices are also simple and cheap, and may also comprise 16 or 32 inputs to a single card, but DOs for higher-power devices (such as solenoid valves) usually require the provision of relays. These may be included on the card or they may be separate.

When considering the provision of spare I/O channels, careful thought must be given to the grouping of channels. If a system has 256 AI channels available, of which only 230 are actively used, it may be said to have 11% spare capacity in this area. However, the grouping of functional areas into cards will inevitably result in the occurrence of more spare channels in one area than in another. It is possible, therefore, to have the required amount of spare I/O capacity available in terms of the overall system, but to be unable to modify or extend a particular part of the system safely, because no spare channels have been provided in the required area.

Spare capacity should be provided both in the form of 'populated' channels (i.e. spare inputs and outputs on individual cards) and 'unpopulated' space (i.e. spaces for additional cards). To avoid a spaghetti-like tangle of cross-connections, the spare spaces should be sensibly distributed through the system.

8.1.2 Termination and marshalling

It is important to understand that the grouping of inputs and outputs on the I/O cards does not always correspond with the grouping of signals into multipair cables, which is dictated by the physical arrangement of equipment on the plant. While it is sensible to avoid mixing different control systems (e.g. feed water control and combustion control) onto a single card, the signals associated with a single system will not necessarily all be carried in the same cable. The result is that a certain degree of cross-connection or 'marshalling' is always required.

Well-designed systems will provide adequate facilities for neatly marshalling the signal connections, but this inevitably requires that the identification of signal connections and their location in the cable system is known at an early stage of the contract. The later this problem is resolved, the more complex and untidy the system will become. Complexity and untidiness can be dangerous because it can lead to mistakes occurring during commissioning or afterwards.

One approach uses ribbon cable connections between the I/O cards and the terminals in the marshalling cubicles. This approach is shown by the double ended connector in Figure 8.1. This has the advantage that if the configuration is known they can be prewired and shipped to site before the computer cubicles, allowing the plant cabling to be started earlier.

Unless care is take the marshalling cubicle becomes a 'bird's nest' of tangled wires, one solution is to use a software package to organise the cabling for you. Some of the features of smart plant instrumentation (SPI) are described in Section 8.9.

8.1.3 Operator workstations

The operator workstations consist of screens on which plant information can be observed, plus keyboards, trackballs or 'mouse' devices allowing the operator to send commands to the system. They also comprise printers for operational records, logging of events (such as start-up of a pump), or alarms.

The main operating screens are typically PC based in a single or dual-stacked arrangement and may include large-screen displays used for process and alarm monitoring. See Figure 8.3(a) and (b) for arrangements for both PC-based and large

Figure 8.3 (a) Typical control room. © ABB, reproduced with permission

Figure 8.3 (b) Typical control room showing operator displays. © Emerson, reproduced with permission

screen displays. The selection of the type of screen depends on the operational requirements, but will ultimately be determined by the available budget. Critical ergonomic factors affect the optimum design of the workstations, and great care must be exercised to ensure that the plant can be operated safely under all conceivable modes of failure, and that no computer-assisted errors can occur due to the operator being confused by the information presented to him or her.

An important consideration is the screen update time. This is the time between the occurrence of an event and its appearance on the screen. As system loading is increased, this time can become extended, but the operator will need to be made aware of each event as soon as possible after it occurs, so that corrective action can be taken. An update time of 1 s is barely adequate to deal with fast-moving events, but it can be quite difficult to achieve.

In considering the operator displays associated with a typical boiler control system attention should be given to the vast amount of information that must be provided. The logic diagrams in this book are necessarily simplified and exclude the many interlocks and other functions that are required in reality. When a practical plant is considered it soon becomes apparent that accommodating the amount of information and control facilities can lead to very cluttered display screens. This has led to the development design principles that follow a hierarchical structure, minimise distracting detail from graphics, emphasise reliance on well-designed actionable alarms and provide 'vectoring' from a selected alarm directly to the associated control screen. Some of these principles are discussed further in Chapter 12.

Let us take mill group control as an example to see the importance of having unambiguous graphics and the advantages of a hierarchical 'overview' design. Clearly, the mill group graphics are 'carbon copies' of each other, varying only with respect to the tag numbers of each item and the dynamic information relating to each area of the plant. It is therefore reasonable to display only one group at a time on the screen, allowing it to be started, adjusted or stopped as required. However, to avoid making any mistakes, the operator should be very clearly and unambiguously informed of which group is displayed at any time. Also, a master display should enable the operator to view the status of the entire set of mills feeding the boiler. This overview allows relative mill performance to be continually compared and biasing or other adjustments to be applied from the higher level.

The development of these operator displays is therefore unusually demanding, and if insufficient time or money is allocated to the performance of this task, the results can be at best unwieldy and at worst dangerous.

For Engineer Procure Construct (EPC) contracts the DCS vendor will typically develop the graphical displays with review and approval by the end client. Where a system is being upgraded the owner will typically assign operators to assist in development and review of the displays. This ensures that colour codes and alarm handling follow the station standards. However as discussed in Chapter 12, operators involved in screen development should be given training in modern design principles and the standards that are to be followed.

8.2 Interconnections between the systems

The considerations applying to field cabling are dealt with in Section 8.8. However, special thought needs to be given to the data highway. This is a high-speed link over which a great deal of information is transmitted. The cable employed for this purpose is very specialised, and great care has to be taken in its installation. Physical damage, severe bending or incorrect termination can cause maloperation. If a fibre-optic cable is used, the considerations that apply to this type of cabling must be meticulously followed.

The integrity of the data highway is crucial to the safety of the plant and therefore it is usually duplicated in a 'double-ring' arrangement. This allows for two faults (one in each cable), and in some cases a third simultaneous fault to exist and still maintain normal operation.

8.3 Equipment selection and environment

Although modern gas-fired plant naturally tends to be clean in comparison with its coal-fired equivalents, any power station environment still presents a severe test for electronic systems. The control system designer has to deal with the problems of operating low-voltage, potentially interference-prone, electronic equipment in close proximity to electrical plant operating at 11 kV and above, with all its attendant switchgear and transformers. The situation is exacerbated when considerations of safe operation in hazardous environments are brought into the picture. It becomes even worse when considering the dust, dirt and vibration that are significant factors in practical power plant environments. Naturally, the latter problems (dust and dirt) become particularly acute in coal-fired plant.

The success of a control system depends on the designer understanding and addressing these factors. To assist in this process the following text provides an outline of good equipment design and installation practices. Because the subject covers so many different disciplines, the chapter is divided into three sections:

- Mechanical factors: the ground rules for providing good facilities for control and instrumentation equipment.
- Electromagnetic compatibility: guidelines for minimising the risk of mal-operation caused by interference.
- Physical environmental considerations: dealing with dust, dirt, vibration and hazardous atmospheres.

These matters must be understood and judiciously applied when an installation is being planned, but doing this involves considerable interplay with the civil and mechanical engineering disciplines, and appropriate action must therefore be taken at a very early stage in the design and construction phases of the plant. In a new plant, given diligence and understanding on the part of all the disciplines involved, one can hope to achieve this goal. But in the case of a refurbishment project the task becomes much more difficult, because here one is dealing with a plant whose construction is already complete. In this case, the control system designer must work with what already exists. In the end, it may come to a matter of fighting a ditch-by-ditch battle, eventually retreating to the last principle – the one that must never be sacrificed – which is to obtain an installation that is safe to operate and maintain.

8.4 Mechanical factors and ergonomics

In this section, we shall consider the mechanical installation of electronic control equipment. This is necessarily a summary, and like most aspects of technology it is affected by changing requirements and technologies. Because requirements, technologies and the availability of materials are always changing, some form of guidance on up-to-date practice should be sought and it is tempting to think that the selected system vendor will be able to provide this.

Reputable vendors should be pleased to provide guidance on installation practices to be employed with their equipment and systems. This is partly because by providing such information they demonstrate that they are experienced in power station work and are able and willing to help with such matters. It is also in their own interest to do everything possible to ensure that their systems will not be exposed to mechanical or electromagnetic conditions that could jeopardise their performance.

However, it will be unwise to wait until a specific vendor has been selected because in most cases on 'greenfield' sites this action occurs when the basic concrete and steel construction is almost complete, by which time it will be too late to make any changes. It is useful to obtain guidance on current practice from a range of system vendors.

8.4.1 Site considerations

The electronic assemblies that comprise a control system will generally be located in three areas:

- The field: where transmitters, sensors, detectors and actuators are sited. Some I/O cabinets may also be field-located.
- The equipment room(s): accommodating the control cubicles, processors, I/O facilities and power supplies.
- The control room: housing the operator facilities (screens and keyboards), plus the system printers, etc.

These areas represent very different environments for the equipment they contain, ranging from the severe conditions of dust, dirt, humidity, heat, vibration and hazardous areas that are to be found on the plant, to the comparatively quiet and clean conditions that should be found in the control room.

8.4.1.1 Field equipment

Every control system depends for its operation on accurate information on the plant being controlled (which is the duty of the process transmitters) and the ability to apply the resulting commands to the plant (which is done by actuators). These electromechanical transducers are vital to the proper operation of the system. They must operate efficiently and reliably. Unfortunately, items of equipment in these critical areas are subject to particularly severe difficulties. For a start, the design of electromechanical transducers requires a blend of good electronic engineering and mechanical engineering. A thorough understanding of metallurgy and engineering chemistry is often also required. Designing equipment where several disciplines are involved is much more difficult than working in only one discipline. If this hurdle is overcome successfully a good device will result, but it will then be installed on the plant where it will be exposed to the severe environments that often exist there. The equipment's operation will then depend heavily on the application of the best possible installation and maintenance practices.

A successful control system requires detailed definition of each component part, and in the case of the transmitters this is achieved by meticulous specification of the transmitter itself, and by careful definition of how it will be installed.

Figures 8.4 and 8.5 show one step in the latter process, a so-called hook-up diagram for pressure transmitters. The hook-ups define how the transmitter is piped up to the process. Two types of hook-up diagram are illustrated here, one showing all the mechanical items that will be needed to complete the assembly (such as elbows, tees and unions), the other outlining in schematic form how the system operates. Normally, only one of these types will be used for a given contract. Such diagrams define the connections and as such are an essential prerequisite for installation of an instrument. The detailed version of the hook-up is useful for costing/estimating purposes, although the same information may be provided in a simpler form.

A few points about this diagram warrant further discussion. The tapping point isolating valve is usually provided by whoever installs the main high-pressure plant pipework. It is connected to the transmitter subsystem by a small-bore line, known as the 'impulse' pipe, and the selection of the correct type and size of pipe will have a considerable bearing on the accuracy, reliability and maintainability of the installation. The standards that are applied change from time to time and vary between countries and users. It was usual for HP pressure and differential pressure transmitters to be connected to the process via 15 mm OD stainless steel impulse pipes and 20 mm nominal-size valves, while drum-level transmitters were connected via 16 mm OD pipes and 32 mm nominal-size valves. However, for supercritical boilers the 16 mm OD pipe does not have an adequate pressure rating but 14 mm does in both 1.5 and 2 mm wall thickness. The 2 mm thickness is recommended as it is easier to butt weld. Care must be taken if the pipework is bent during installation as tight bends reduce the wall thickness and pressure rating. Good practice is to have the rating checked by a piping engineer.

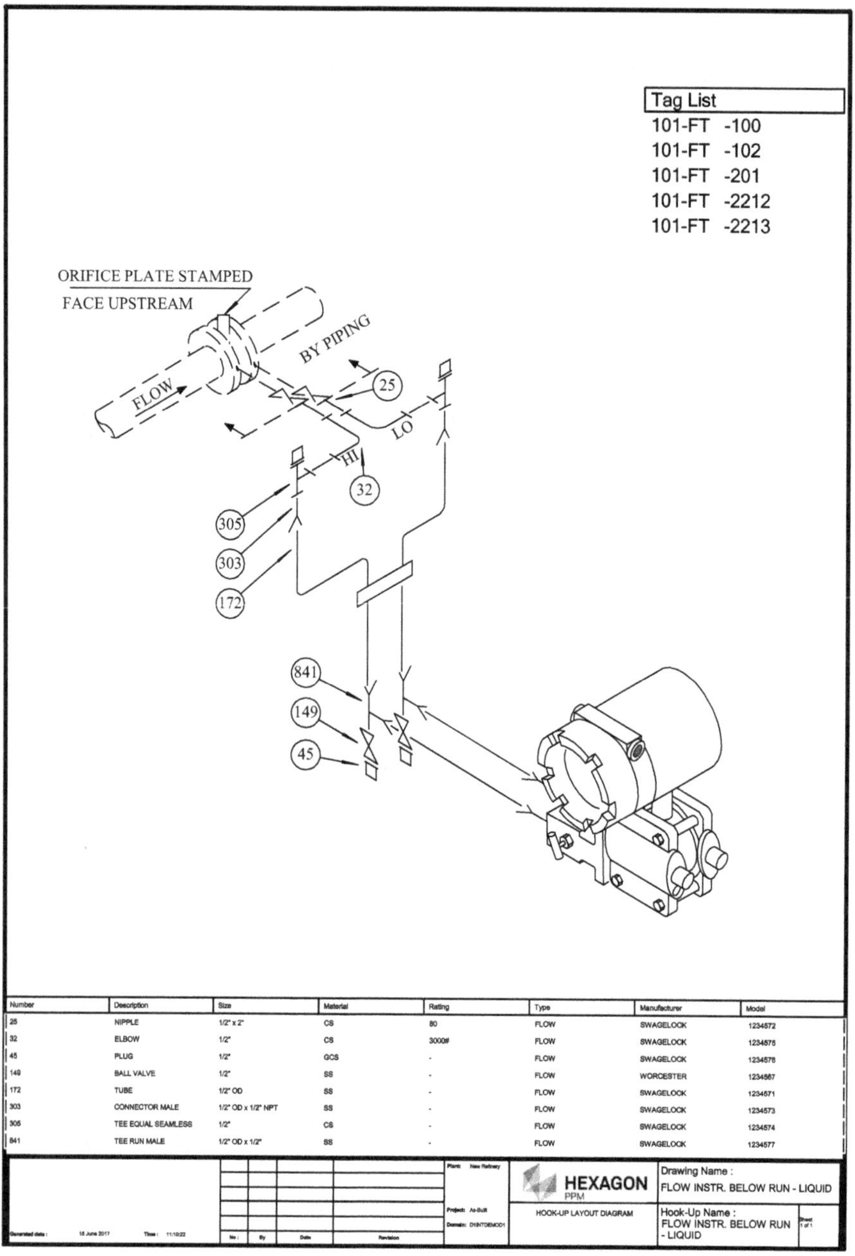

Number	Description	Size	Material	Rating	Type	Manufacturer	Model
25	NIPPLE	1/2" x 2"	CS	80	FLOW	SWAGELOCK	1234572
32	ELBOW	1/2"	CS	3000#	FLOW	SWAGELOCK	1234575
45	PLUG	1/2"	GCS	-	FLOW	SWAGELOCK	1234576
149	BALL VALVE	1/2"	SS	-	FLOW	WORCESTER	1234567
172	TUBE	1/2" OD	SS	-	FLOW	SWAGELOCK	1234571
303	CONNECTOR MALE	1/2" OD x 1/2" NPT	SS	-	FLOW	SWAGELOCK	1234573
305	TEE EQUAL SEAMLESS	1/2"	CS	-	FLOW	SWAGELOCK	1234574
841	TEE RUN MALE	1/2" OD x 1/2"	SS	-	FLOW	SWAGELOCK	1234577

Generated date : 18 June 2017 Time : 11:10:22

No :	By	Date	Revision

Plant: New Refinery
Project: As-Built
Domain: D18NTDEMOD01

HEXAGON PPM
HOOK-UP LAYOUT DIAGRAM

Drawing Name : FLOW INSTR. BELOW RUN - LIQUID
Hook-Up Name : FLOW INSTR. BELOW RUN - LIQUID
Sheet 1 of 1

Figure 8.4 DP transmitter hook up. © Hexagon Technology, reproduced with permission

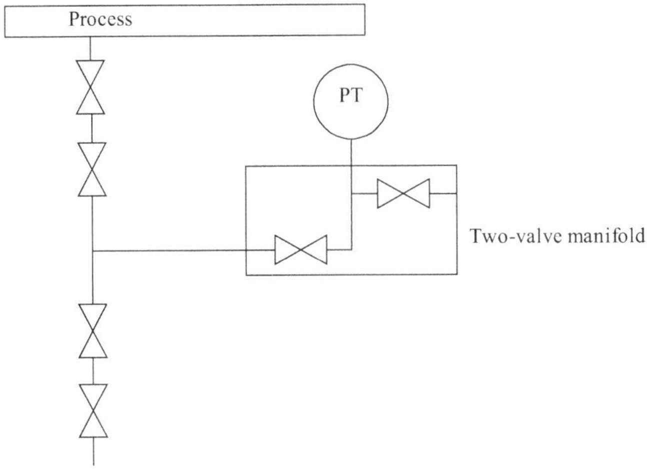

Figure 8.5 Pressure transmitter hook-up

Commercial pressures have nowadays led to a situation where some purchasers merely stipulate that the whole system should operate satisfactorily (sometimes within defined parameters) for a given time (say, 20 years). Little or nothing is said about matters such as maintainability or ease of access. Using the aforementioned connection sizes will lead to an installation that is workable, reliable and easy to maintain.

The sizes of the connection pipes and valves form only one part of the picture. Other matters concern the lengths, lagging and slope (rise or fall) of the connecting pipes (which will allow them to be vented or drained), etc. Transmitter location is also important. In general, pressure and DP transmitters are situated below the tapping point for liquids (to avoid air entrapment) and above the tapping point for gases (to minimise dust collection).

The two-valve instrument manifold shown in Figure 8.5 is a standard subassembly which may either be integral with the transmitter or provided as a separate item as shown. It allows the transmitter to be connected, calibrated and vented before removal.

The blowdown valve assembly enables the pipework to be flushed through to remove entrained gases, deposits, etc., to a suitable drain or vessel. In this example it comprises two valves, a 'master' and 'martyr' – an arrangement that enhances long-term maintainability in a high-pressure application. If a single blowdown valve were to be provided in such an application, the differential pressures and velocities to which it would be subjected each time it is opened or closed would quickly erode the internals. The 'master and martyr' assembly operates as follows to avoid this problem. Prior to initial commissioning of the system, both valves are closed. When pressure has been applied, the master valve is opened first, followed by the martyr. When any debris or undesired gas or vapour has been ejected, the martyr is shut off first, followed by the master. When the system is to be shut off again after use, the master is opened while the martyr remains closed, and then the martyr is opened. By this means the onerous duty of opening or closing of the pressurised

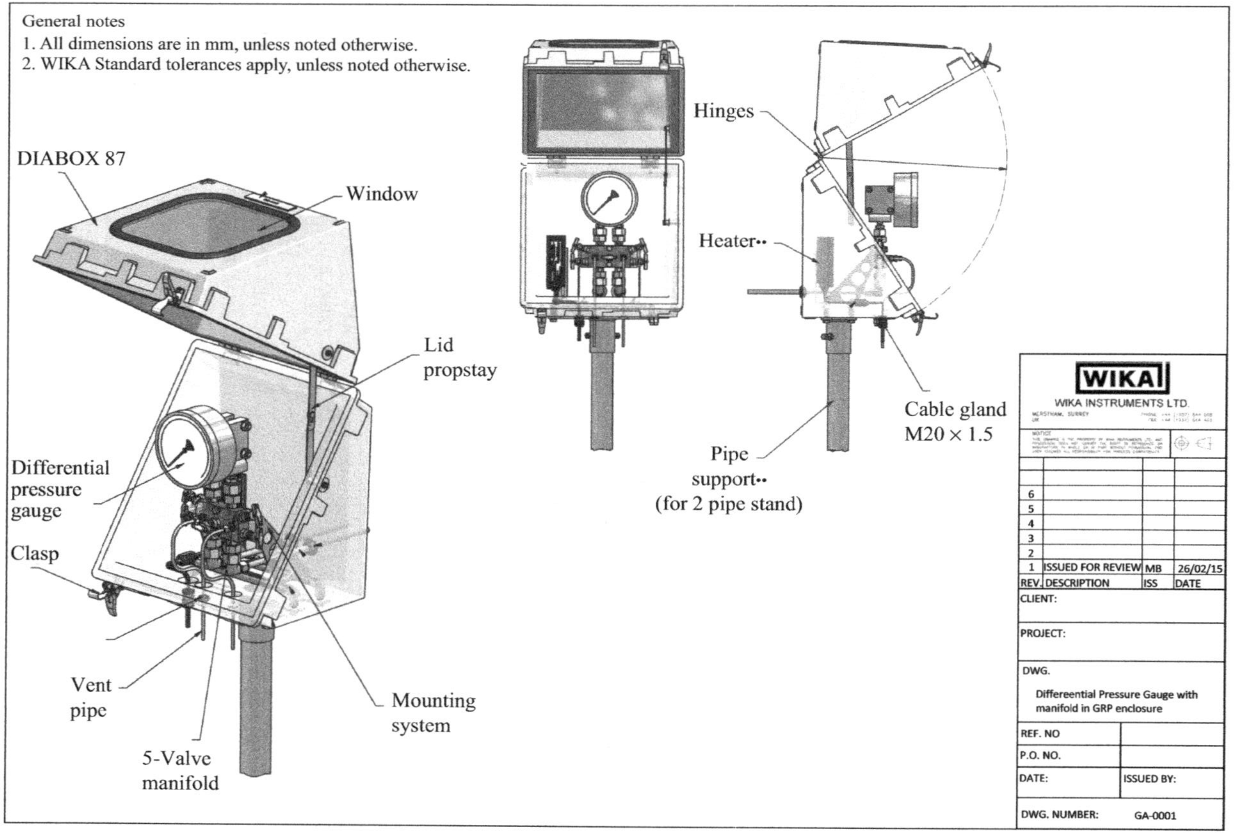

Figure 8.6 Typical assembly of pressure gauge in a fibre glass enclosure. © WIKA, reproduced with permission

system to the atmosphere is always handled by the martyr, with the master merely opening or closing without changing the flow. When the martyr eventually succumbs to these harsh conditions of use it can be quickly and easily replaced while the master is closed without having to isolate the transmitter at the tapping point (a process that may necessitate shutting down the plant). Since the master valve is never subjected to harsh operational conditions it should survive for an indefinite period.

As a rule of thumb single blowdown valves are used on low-pressure systems below say 40 barg while master martyr valves are used on both superheat and reheat systems. The blowdown must be drained to a safe location. Note some EPC contractors consider that the blowdown valves are a safety hazard and do not provide them.

Detail similar to that provided in the hook-up Figure 8.5 must be provided for each instrument. Each of these has its own peculiarities, and neglect of a simple requirement can render a vital measurement inoperative, inaccurate or unreliable. Again, reputable manufacturers will be able to provide detailed guidance for each device. But beware of those who claim that their instruments are so simple that no such guidance is needed!

Figure 8.6 shows pictorially all of an installation's components, including a heater. At this point some typical pressure gauge requirements are noted:

The gauge glass should be non-splintering laminated safety glass.

High-pressure gauges are fitted with a blowout disc so that if the bourdon tube ruptures the disc blows out before the glass face. The blow out disc shall operate at a maximum of a half of the glass burst pressure (EN837).

Standard accuracies in accordance with EN837 are ±1% over the full-scale deflection (FSD). Many EPC will specify a better accuracy, and ±½%FSD can be achieved.

8.4.1.2 Actuators

In the chapters of this book dealing with various control loops, reference has been made to the controlling devices (such as valves and dampers) which translate the control system's demands into changes of flow and pressure. The modulation of these devices is the duty of an actuator, and in the next few paragraphs we shall briefly survey some of the actuator types to be found in power plant.

Pneumatic actuators are well established and cost-effective. In addition to being reliable, accurate and capable of fast response, they are simple to use and maintain. They are therefore found in a great many installations, and where they are used the control system's commands have to be processed from electronic form (generally 4–20 mA) to pneumatic form (usually 0.2 to 1 barg) and then used to drive the valve via a positioner.

It is now normal to use a smart positioner that accepts a 4–20 mA signal makes an electronic position correction and outputs directly to the control valve. While there are many suppliers offering this the ABB TZID shown and the Siemens Sipart (among others) have the advantage that it can be configured to fail in a fixed position on loss of air supply and/or on loss of the 4–20 mA signal (Figure 8.7).

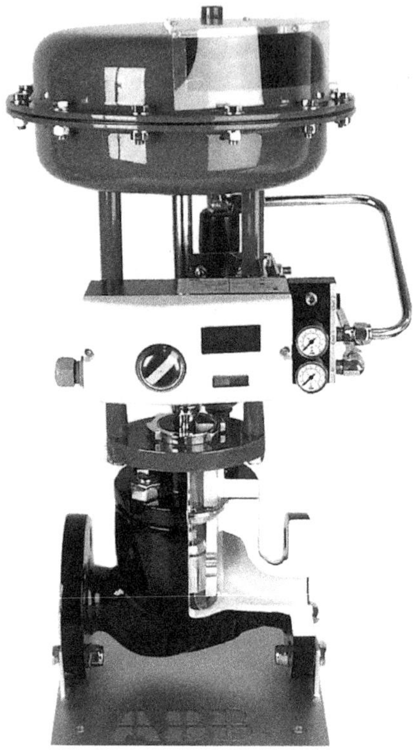

Figure 8.7 *ABB TZID electro pneumatic positioner. © ABB, reproduced with permission*

In many cases, the pressure available from the I/P converter will not exert sufficient force on the diaphragm to move it against the reactive force applied to the valve stem by the process fluid. Unless an extraordinarily large diaphragm is used, the 1 barg output signal produced by commonly used I/P converters is unlikely to generate stem forces greater than 20 kN. Also, I/P converters are designed to feed into small volumes so that, even if they could generate enough pressure to move the valve through its full travel, they will be unable to do so quickly since it will take time for them to build up sufficient pressure in the large volume above the diaphragm.

Two solutions are available to overcome this problem: boosters and positioners. A booster is a pneumatic relay that converts small pressures to large ones. A positioner is essentially a power controller whose function is to translate the pneumatic control signals into mechanical movement of the valve stem. A positioner applies a high-pressure air signal to one side or other of the diaphragm, adjusting the relative pressures until the stem position corresponds with the demand signal. The mechanical design of the positioner may incorporate a cam which can enable the demand/position relationship to be shaped to a linear, square-law or other non-linear characteristic.

Although the use of a positioner is often essential, it is not always necessary. A positioner contains a feedback mechanism and difficulties can arise because of the inclusion of an additional integral term into the overall control system. This was particularly important when the control parameters of the positioner were not adjustable. However modern positioners have adjustable tuning parameters, so you should expect to always use a positioner. The positioner ensures consistent response even with different actuator loadings.

They come with a self-tuning option that is useful for a quick set-up. However, the self-tuning facility should not be used for difficult loops. The positioning loop should be tuned with both large (10–15%) steps and small (2%) steps to ensure fast response with minimal position overshoot.

A point which can also raise difficulties is where mechanical stops are introduced to limit the range of stem movement. In such cases it can be extremely difficult to adjust the way in which the positioner operates at the extreme limits of the valve travel. When the valve reaches the limit of movement set by the mechanical stop this will not necessarily correspond with the command signal. The positioner will therefore attempt to eliminate the apparent error by applying additional pressure to the diaphragm. If the command signal holds the valve against the mechanical stop for an extended period the volume of air on one side or other of the diaphragm will build up to the full pressure of the air supply. Once this has happened any subsequent command to move the valve off the end stop will be followed by a delay as the air pressure within the diaphragm vents off. As this venting can occur only through the pipes and valves of the system, and as these are often constricted, the vent period can easily become extended. In time-critical applications the results can be unacceptable or even dangerous.

Where such mechanical stops are fitted it is also important that the positioner is set-up to correspond with the actual range of free movement of the valve. If one stop is set at a position that is, say, 10% from the fully closed position of the valve and the other is set at 90%, the positioner should be set-up so that the stem is moved to the 10% position when the command signal is at the minimum of its range (e.g. 4 mA), and to the 90% position when it is at the maximum (e.g. 20 mA). If this is not done and the positioner is ranged over the theoretical stroke of the valve stem (not between the stops), saturation will again occur as the positioner attempts to move the valve past one or other of the stops. However, even with the correct settings, it will be almost impossible to avoid the problem of saturation since, even if the difference between the position of the mechanical stop and the corresponding signal is microscopically small, over an extended time the integral action effect of the positioner will cause the air pressure in the diaphragm to build up (or vent), with all the implications already described.

In such cases (and, in fact, wherever a fast-responding loop already exists around the valve), it may be advantageous to abandon the positioner. This aim will be easiest to achieve if sufficient force can be exerted by the signal air pressure alone but, if this is done, the length of pneumatic pipework between the I/P converter and the diaphragm should also be minimised in order to reduce the delays that can otherwise be caused.

If the pressure available from the I/P converter is inadequate, or if it is impossible to obtain a short pipe run between the converter and the diaphragm, consideration should be given to using booster relays to provide adequate air pressure to the diaphragm without the use of a positioner.

8.5 Electric actuators

Although pneumatic actuators are inexpensive, reliable and fast-operating, their use necessitates the provision of compressed air supplies. The air must be clean and dry, entailing the use of filters and driers. It is therefore attractive to consider devices that do not require such expensive ancillary plant. In addition, the compressibility of air makes it difficult to provide a 'dead-beat' response when dealing with large masses.

With the evolution of reliable solid-state position controllers for electric motors, the scope has opened up for avoiding the use of pneumatic operators by the use of electric actuators. These are self-contained and only require an ordinary source of electric power. They can provide dead-beat response and also have the advantage of providing inherent 'fail-fix' operation since on loss of power they lock in position. On the other hand, making an electric actuator fail to the open or closed position on loss of power is not so simple.

When specifying an electric actuator, it is important to state the required failure mode as well as the operating speed (time to travel from fully closed to fully open).

On/off and regulating electric actuators are made by several companies. Those most specified by end users are Auma, Rotork and Limitorque. These can be fitted with local hand wheels, limit switches and accessed via hard-wiring, Profibus or Modbus. However, for modulating control, much more frequent operations are required and end users in the United States, the United Kingdom and Europe often specify the Beck and ABB Contrac range of actuators. These are designed for 100% duty without overheating. The Auma and Sipos range have a lower duty cycle, but are effective if the control loop is properly tuned and are also frequently used in Europe, as well as Australia and Asia for modulating control.

Early electric actuators included solid-state relays to switch the motor on and off. These were not reliable. Once this problem was resolved the motor itself become the weak point in the chain and as the control electronics improved the motor over heated due to the high number of starts per hour.

Many regulating actuators will offer 1200 starts per hour, which is ideal for intermittent operation. Most DCS specifications ask for AO to be capable of

Figure 8.8 Contrac electric actuator. © ABB, reproduced with permission

being updated four times a second. If the actuator was reversed on each update it would be started 4 × 60 × 60 or 14,400 times an hour. In practice this does not happen as the controller tuning provides some deadband for noise suppression, but control actuator specifications should ask for as high a start rate as available for high-activity applications – around 1,800–3,600 starts per hour if available.

The ABB Contrac electric modulating actuator is specified as standard by major utilities in Europe for key loops such as spray valves used for steam temperature control. Their key specification includes stall-proof motors capable of 100% duty cycle (Figure 8.8).

Similarly, the Beck actuator is widely specified in the United States and elsewhere especially for applications requiring unlimited modulation (100% duty), precise control and high levels of reliability. The Beck design incorporates a single-purpose motor that cannot burnout or overheat. The motor design also enables the actuator to start and stop instantly, thus making it a popular choice for control loops requiring continuous, yet precise, modulation (Figure 8.9).

8.6 Hydraulic actuators

Hydraulic actuators offer another way of dispensing with air compressors and their ancillary equipment. This type of actuator is powerful, fast and accurate, and can be provided in fail-open, fail-closed or fail-fix configurations. Originally, instead of being used to avoid the air compressors, they were used only for applications

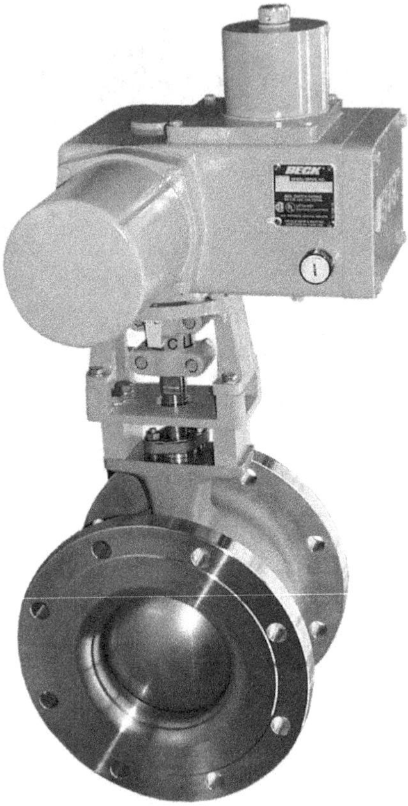

Figure 8.9 Beck electric actuator on quarter turn valve. © Beck, reproduced with permission

requiring large forces such as feed valves and actuators on forced draught (FD) and induced draught (ID) fans. They also have the advantage that 'within reason' their torque can be increased just by increasing the pump pressure up to 207 barg. The thrust and speed can both be adjusted on site. On one power station only two actuator sizes were required to cover every boiler actuator requirement.

The hydraulic components are standard and may be readily sourced and serviced in Third World countries. See Figures 8.10 and 8.11 showing hydraulic actuators.

The points to consider all centre on the nature of the hydraulic medium employed. Is it flammable or corrosive, what provision is made to guard against leakage, etc.? Also, if for reasons of economy a centralised hydraulic reservoir is shared between several actuators. The centralised hydraulic reservoir is not usually used as a failure can jeopardise the whole boiler. Early hydraulic fluids were flammable and so common systems with potential leaks went out of favour to be replaced by stand-alone actuators with their own pump, a much smaller

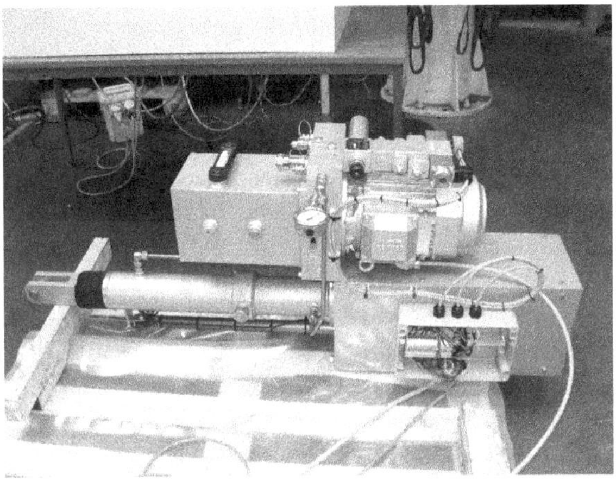

Figure 8.10 Typical hydraulic actuator as used on a coal-fired power station. © Advanced Actuators Ltd., reproduced with permission

Figure 8.11 Large hydraulic actuator used to modulate 10 m × 10 m damper. Output torque is 1.2 million Nm. © Advanced Actuators Ltd., reproduced with permission

reservoir and electronic controller. A more recent concern is, 'Is the fluid carcinogenic?'

One potential disadvantage is that the self-contained units do not store hydraulic energy. While they can be driven by springs to an end position the operator cannot move them if the power fails. The solutions to this are to provide a lever-operated pump.

8.7 Cabling

8.7.1 Summary

The cables linking transmitters and actuators to the control system will be installed in areas where they may be exposed to impact from passing vehicles or falling objects. In addition, they may be subject to movement of the structure. For these reasons, cables should be adequately protected and well supported. It is common practice in some countries to use steel wire-armoured cable to provide protection, but if adequate mechanical support and protection is provided by other means (such as cable trays) there is an argument in favour of using cheaper, unarmoured cable, even in the most severe plant environments.

Often, a boiler is actually suspended from a steel frame to allow it to expand and contract as it heats. The movement between objects on the boiler front and a fixed reference point can be quite considerable and, unless this effect is considered, the cable can be damaged (Figure 8.12).

8.7.2 Fire-resistant, fire-retardant and low-smoke cables

Because of the need to understand the difference between fire-resistant, fire-retardant and low-smoke cables, we included the following based on an explanation provided by Gavin Clements of FS Cables. The terminology often gets very

Figure 8.12 Fire integrity cables. © FS Cables, reproduced with permission

muddled, and other names such as circuit integrity and fire performance cables have been introduced that may help.

8.7.2.1 Fire resistant

A cable that is fire resistant will contain fire barriers to prevent the cable short circuiting and failing during a fire situation. They are also called circuit integrity cables. They are designed to continue to work even if subject to a flame. One obvious use is in fire detection and alarm systems. Originally this would have been a mineral insulated cable with a copper tube sheath (which may or may not have had a plastic sheath over the top) This is still the best product that you can use but has the disadvantage that installing it and terminating it is quite complex and specialised.

'Softskin' fire-resistant cables are common now days. These have a low-smoke halogen-free (LSHF) outer sheath and various combinations of aluminium tapes, mica glass tapes and rubber insulation materials to provide the fire-resistant barriers. They are installed with conventional wiring methods and much more flexible. They are more suitable for the larger electrical loads found with emergency lighting, smoke extraction fans and even main FD and ID fans! They are specified to maintain their integrity in a fire for a prescribed period of time. British tests are most stringent requiring three hours of continuous operation in a fire once subjected to a fire they must be replaced.

8.7.2.2 Fire-retardant, flame-retardant or fire performance cables

These terms are used to describe a cables resistance to propagating fire. They can be manufactured from many different materials including poly vinyl chloride (PVC), LSHF plastics, rubber or fluoropolymers to name a few.

A good level of flame retardance will make a cable safer in as much as it will help limit fire spread but on its own it does not offer any form of circuit integrity performance. Fire performance is a term generally used to describe circuit integrity cables (Figure 8.13).

Figure 8.13 Fire performance test. © FS Cables, reproduced with permission

8.7.2.3 Low-smoke halogen free

A more common requirement is to ask for LSHF or PVC-free cables. The reason for this is that if there is a fire PVC cables may give off chlorine-filled smoke. This is life threatening, and if the chlorine combines with water from a sprinkler system, hydrochloric acid is produced. If the moisture lands on circuit boards, etc., it will corrode them creating long-term failures; also it is a conductive acid so energised equipment may short circuit on contact.

A wrong assumption people often make is that LSHF materials are flame retardant; whilst many are, an equal proportion are not. It is not a mandatory requirement.

8.8 Electromagnetic compatibility

The high voltages, heavy currents and large magnetic fields associated with power station equipment give rise to the risk of interference with electronic systems. It is an important requirement that the system designer recognises this fact and pays careful attention to dealing with the risk. Guidance on design and installation practice is available from several sources [1], and it is nowadays generally a mandatory requirement that systems comply with electromagnetic compatibility (EMC) rules defined by the country in which the plant is to be operated [2].

In general, a system designed to be immune to interference should employ optocoupled inputs and outputs for digital signals, and its AIs should include low-pass filters to provide a high level of attenuation to frequencies above say 20 Hz. Such filters provide discrimination against 50 or 60 Hz pickup from mains-operated devices and against any high-frequency disturbances that may be generated by switchgear, variable-speed devices and the like. If the roll-off characteristics of these filters are adjustable, the input channels of the system can be made to recognise legitimate variations in the measured signals (e.g. dealing with rapid pressure changes or slow-changing temperatures) while effectively ignoring interference-induced signals (such as 50 Hz pickup).

8.8.1 Earth connections

Good earthing practice demands the use of a star-point connection for all screens. Figure 8.14(a) shows how current flowing through a common impedance affects other circuits connected to it. In this example, the voltage appearing at the input of a device (with reference to earth) is given by:

$$V_i = (i_, \times z) + (i_3 \times z_3).$$

It must be recognised that a major fault in an electrical machine is a transient phenomenon which can result in a very large current flowing to earth in a very short time. In addition, lightning strikes on structural steelwork, cables and machine frames can cause currents of hundreds of kiloamps to flow to ground. Because of the high-frequency nature of the current in all such cases, the complex impedance of the earth connection becomes dominant (i.e. its resistance and its inductance).

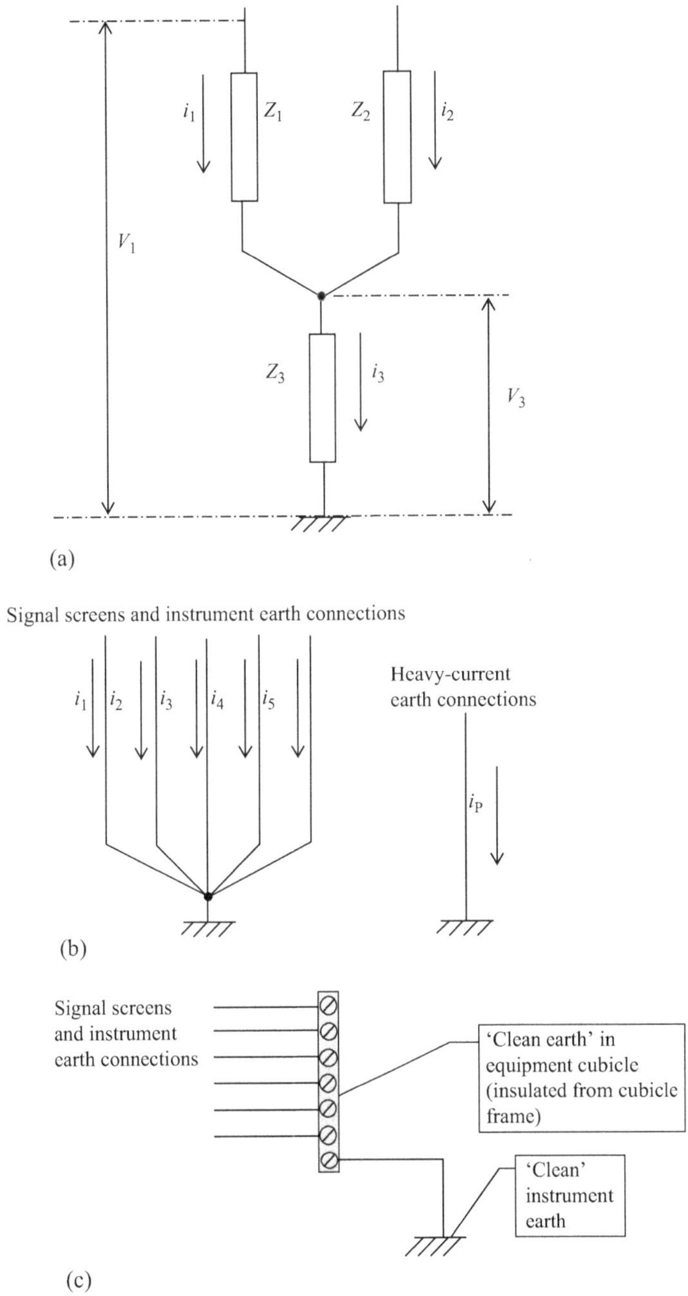

Figure 8.14 Good earthing practice: (a) current flow through common impedance; (b) segregation of instrument and electrical (safety) earth connections; (c) a practical arrangement for an instrument earth connection

If the device represented by z is an AI channel of the plant DCS, it would normally be handling 4–20 mA signals and its input impedance would be say 250 Ω. The voltage across this input channel at full-scale would therefore be expected to be no more than 5 V (0.02 A × 250 V).

Now, assume that the common earth impedance is IO Q and that a transient fault current (i_3) of 100 A flows through it. (In practice, fault levels can be much higher even than this.) The voltage occurring at this input channel of the DCS under these conditions would therefore be:

$$V_3 = (0.02 \times 250) + (100 \times 10) = 1005\ \text{V}.$$

This example shows that input circuits of the DCS can be subjected to voltages several hundred or thousand times higher than expected due to such fault currents.

Figure 8.14(b) shows how this effect can be minimised by segregating signal earth connections from the earth connections of machines and their switchgear. The currents flowing to the instrument earth will all be of the order of milliamps, and if the connection to earth has a low impedance the maximum voltage appearing at the common point will be no more than a few hundred millivolts.

Figure 8.14(c) shows how such segregation can be achieved in practice. A low-impedance busbar, insulated from the metalwork of the cable itself, is provided in each equipment cubicle and connected to a ‘clean earth’ point by a low-impedance cable. All instrument earth connections (including cable screens) within the cubicle are connected to this busbar. The metalwork of the cubicle is connected to a separate safety earth point (often through the mounting bolts).

8.8.2 Cables: armouring screening and glands

Reference has been made earlier to the use of steel wire armouring to provide mechanical protection for interconnecting cable, and it is sometimes argued that armouring provides immunity to interference. This is not totally correct. Although steel wire armouring does provide some degree of protection against magnetic fields, its performance as an electrostatic screen is poor. For this reason, it is essential to use cable with a braided (or foil) screen, with or without overall steel wire armour, for all signal connections. In the United States it is common to use conduit, instead of armour, to provide the mechanical protection. While in Europe armoured cable is common. Sometimes unarmoured, but screened cable is run in a cable tray. Then it is good practice to cover the top tray to provide additional mechanical protection.

Figure 8.15 shows how a screened cable is used to connect between the component parts of a system. It also shows how the various conductors should be connected to earth.

Because it is very difficult to spot missed, duplicated or badly made earth connections once a cable installation has been completed, it is vital that work is properly supervised and very carefully checked during the installation of a system. In this respect, useful assistance is provided in standards such as BS 6739:2009 ‘Code of Practice for instrumentation in process control systems: installation, design and practice’.

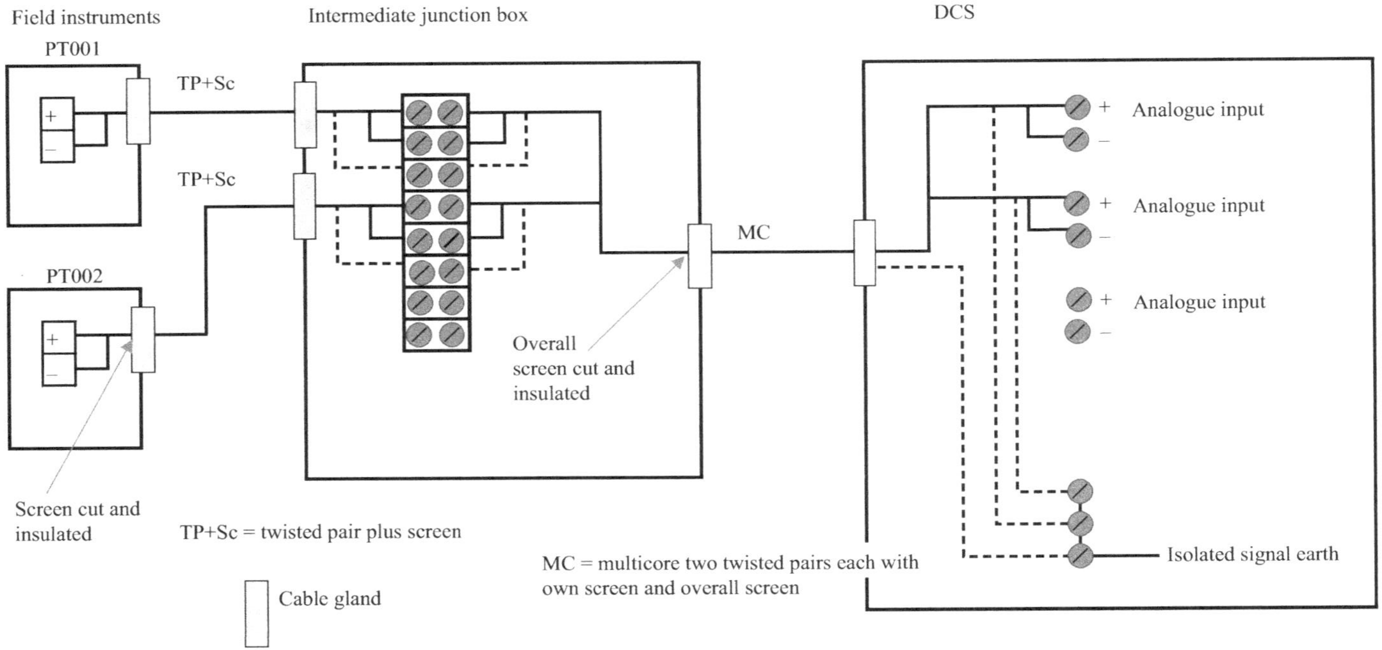

Figure 8.15 Interconnections using screen cable

8.9 Database tools

As each project is engineered you start to see many similarities between projects and recognise that some short cuts can be taken. So often a process and instrumentation diagram (P&ID) may be reused, but in the detailed engineering instrument suppliers and ranges are changed. Changes to a PID following a hazards and operability (HAZOP) may require changes to the instrument schedule, the I/O schedule, the cable schedule and the preparation of a new hook-up and installation material take off (MTO). The updating of these takes many hours. The actual time can be reduced to the time needed to pick a range and agree a location if a 'smart' engineering tool is used. Many companies have used an Access database to manipulate information and even linked it to AutoCAD attributes to enable standard template drawings to be replicated for each instrument. A complete suite of such tools can be purchased. One of the most common was called INtools, now renamed as Smart Plant Instrumentation. It is part of a suite of tools provided by Hexagon PPM. The tools most used by control and instrumentation (C&I) engineers are SmartPlant Instrumentation and SmartPlant Electrical. These work best when linked to SmartPlant PID, SmartPlant 3D and SmartPlant Foundation. The latter is an overall database that allows information to be exchanged between the other packages via controlled workflows. If the complete suite is used and a control valve is added in the PID, the work flow will automatically suggest adding it to the 3D model and the valve list. The valve will be shown the correct scale and any insulation included, Allowing clash detection to be run (see Figure 8.16).

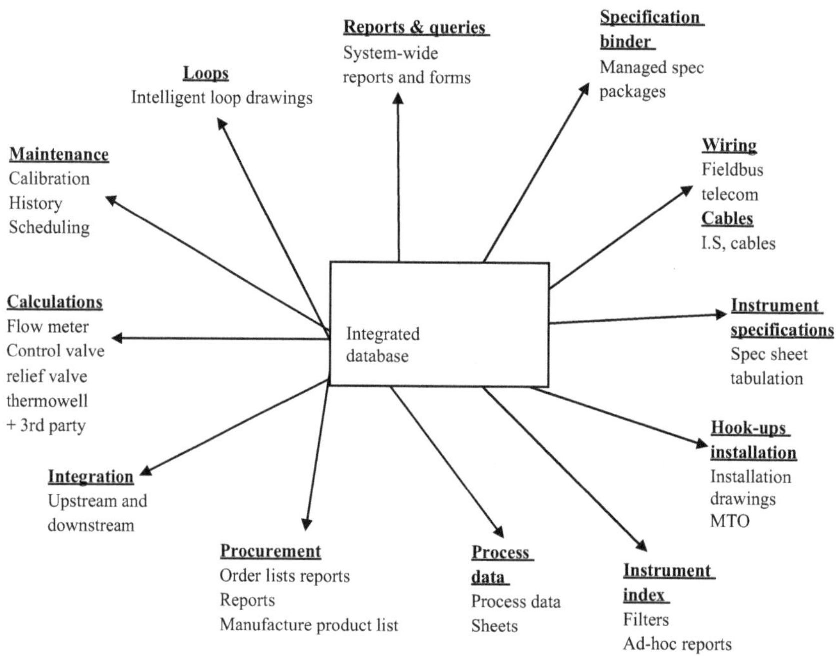

Figure 8.16 SPI database based on information provided by Hexagon PPM

The package requires careful setting up but once configured enables cabling between transmitter and local junction box (JB) to be predefined in SPI and all field cables from the transmitter via local junction boxes and marshalling cubicles to the DCS to be shown (see Figure 8.17).

SPI in a relational database with predefined and user-defined engineering interfaces starts with the instrument database. This is a list of all instruments with the usual attributes such as tag number, range, and supplier and some SPI-specific attributes such as 'type'. The type CV, PG, etc., is associated with a predefined wiring diagram, hook-up and specification; these can be out-of-the-box specifications similar to the International Society of Automation and generic hook-ups, or you can develop project-specific documents. Several variations of each type can be developed so, for example, valves can have electric or pneumatic actuators. Some instrument suppliers such as Endress and Hauser and valve suppliers such as Emerson have developed SPI interfaces so that their standard specifications can be accessed and used without retyping (see Figure 8.18). The instrument index can be created by hand imported from Excel or Access or created automatically if linked to SmartPlant PID, where the instrumentation attributes in the PID database are downloaded to SPI.

In real life, engineers try to limit the instrument index to say 20 columns so it can be printed on A3 and possibly even completed by the end of the project. With SPI the project instrument index can have say 40 columns, even though the formal list still only shows 20.

This enables multiple schedules to be prepared, for example, a list of instruments with locations and hook-up details for the installation team. This is a schedule of pressure transmitters with ranges and materials for the procurement department. While the data manipulation to produce multiple schedules is possible with an Access database, SPI has many advantages including: an Engineering Data Editor that enables the schedules and data sheets to be developed using engineering terms rather than SQL. SPI usually uses Oracle as a database platform as this has proved very robust and stable database. It provides the ability to regenerate all the schedules hook-ups, wiring diagrams and specifications (if required) whenever instruments are added or changed.

SmartPlant Electrical can be used to prepare single-line diagrams and specify motors. For the C&I engineer, it can be used with SP3D and a C&I cable list to route all C&I cables and C&I cable tray. Then it can check the C&I tray sizing and calculate total cable quantities. The C&I cable list is created in SPI.

SmartPlant PID produces normal PIDs but in addition to the pictorial relationship is a database containing line sizes, design temperatures and pressure. When linked to SPI this extra data is published to the instrument index, so valves know if they have flanges or butt wed ends, the pressure transmitters know the process design conditions and flow elements know the line size.

SmartPlant 3D produces drawings that can be viewed as plans, elevation or in 3D. The plant can be built up in layers so that during a design review the stairs and galleries can be highlighted. Or just the ducts shown so that the C&I engineer can check straight length requirements for flow devices. There is a walk-through

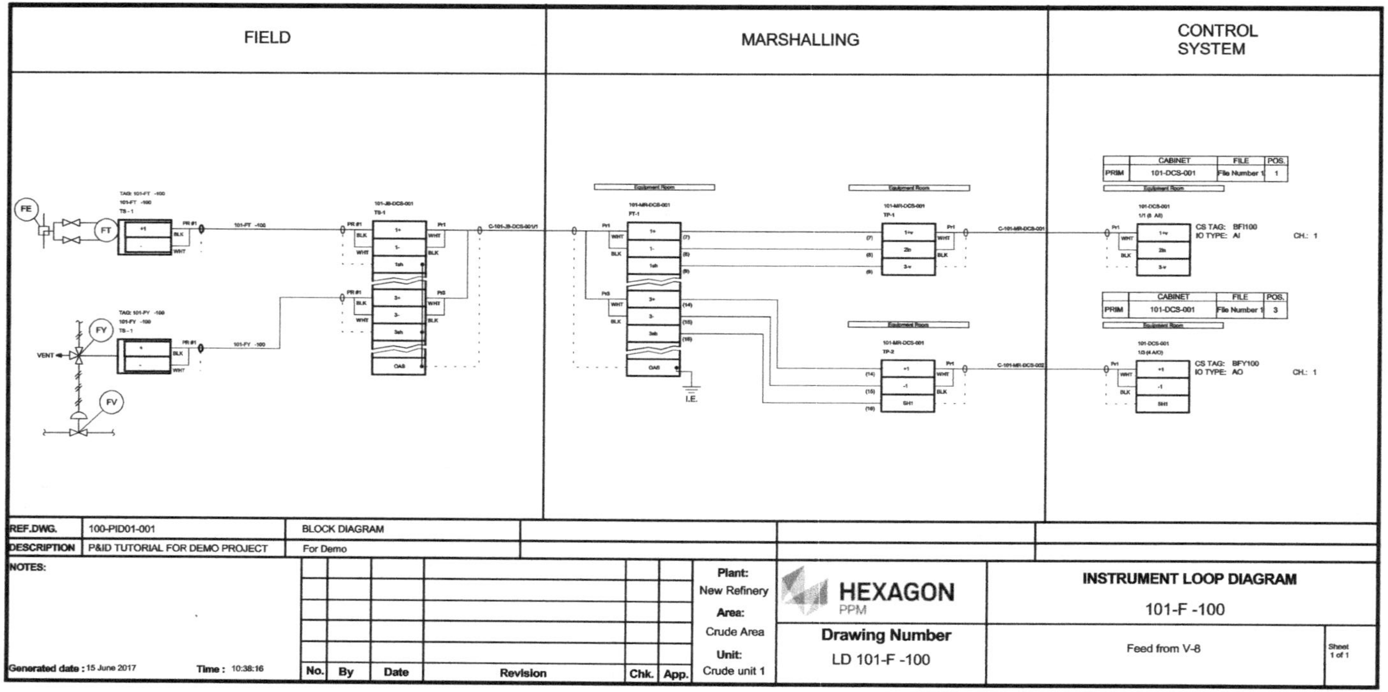

Figure 8.17 SPI loop drawing. © Hexagon PPM

GENERAL	1	Tag Number		101-FT -2212	
	2	Service		Feed to B-101 Pass A	
	3	Location	P&ID No.		100-PID01-001
	4	Function		D/P FLOW INDICATING TRANSMITTER	
	5	Mounting		2" PIPE	
	6	Area Classification		CLASS 1, DIVISION 2, GROUPS C&D	
	7				
	8				
PROCESS CONDITIONS	9	Fluid		Naphtha	
	10	Pressure Max	Oper.	14 bar-g	13 bar-g
	11	Temperature Max.	Oper.	150 °C	150 °C
	12	Oper. Spec. Gravity	Oper. Viscosity	0.891	1 cP
	13				
	14				
UNIT	15	Instrument Range		25-625 mmH2O 4°C	
	16	Calibration Range		0 - 24 Am³/h	
	17	Elevation	Suppression	500% of span	500% of span
	18	Element Type		Piezo Resistive	
	19	Element Material		316ss	
	20	Body Material	Body Rating	316 ss	3000 psi
	21	Process Flanges Material		Hastelloy C	
	22	Wetted O-Rings Material		Viton	
	23	Fill Fluid		Silicon Oil	
	24	Bolts	Housing	Cadmium Plated Carbon steel	Aluminum
	25	Paint		Epoxy	
	26	Process Connection	Electrical Connection	1/2" NPTF	1/2" BSP
	27				
DIAPHRAGM SEAL	28	Process Connection		N/A	
	29	Rating		N/A	
	30	Diaphragm Material		N/A	
	31	Upper Housing Material		N/A	
	32	Lower Housing Material		N/A	
	33	Fill Fluid		N/A	
	34	Capillary Material		N/A	
	35	Capillary Type	Capillary Length	N/A	N/A
	36	Flushing Connection		N/A	
	37				
	38				
OPTIONS	39	Integral Meter		Yes	
	40	Integral Meter Scale		Yes	
	41	Hydrostatic Testing		No	
	42	Cleaning		No	
	43	Calibration		Yes	
	44	Certification		No	
	45				
	46				
	47				
	48				
	49				
PURCHASE	50	Manufacturer		ROSEMOUNT	
	51	Model		1151DP4E22S2B1M2	
	52	Purchase Order Number		E-D-456-97-1	
	53	Price	Item Number	1095 $	11
	54	Serial Number			

Notes: See notes

No.	By	Date	Revision
0	ms	21/12/1998	For bids

INSTRUMENT SPECIFICATION

Diff. Pressure Instr. (flow)

HEXAGON PPM

Sheet 1 of 1

Code: 56 | Doc. No.: 101-21-134435 | Rev.: 0

Figure 8.18 SPI data sheet. © Hexagon PPM

feature, so you can guide an 'icon' through the plant to check head room, make him rotate and see the plant from his viewpoint. Cable tray can be added and then by clicking on an instrument and the associated marshalling cubicle the cable length is automatically calculated for SPI. SmartPlant Electrical allows this to be calculated on a batch basis provided the start and end locations have been preloaded.

As you can imagine there is a significant capital investment and learning curve for the first project. The learning curve can be eased by recruiting trained operators and buying training as part of the Hexagon support package. The commercial costs could be reduced by starting with just one or two packages. One approach is to start with the instrument index, its associated schedules and the instrument specifications on the first project. Then on the next project add loop drawings and hook-ups and termination drawings.

8.10 Reliability of systems

Because of the large numbers of electronic components that are manufactured, and because component manufacturers keep good records of failure rates, etc., it is fairly easy to obtain statistical information on reliability that will provide a good indication of the predicted reliability for any given system. In practical terms, what really matters is the length of time for which a system will be capable of remaining in operation over the course of a year or over its operational lifetime. This is governed by both the reliability of the equipment and the speed with which repairs can be effected. For example, it would be theoretically possible to construct a very reliable system by arranging for all functions to be performed by a few very large-scale integrated (VLSI) circuits connected together without the use of plug-and-socket connectors. Because VLSI devices are inherently reliable and because connectors are a source of failure such a system would offer a very high level of reliability. Unfortunately, it would be very difficult to repair if it did fail.

The reliability of any electronic system can be predicted with a high level of confidence by referring to statistical data produced by manufacturers, independent test laboratories or bodies such as the defence or nuclear authorities.[1] Such data can be used to calculate the predicted failure rate, or mean time between failure (MTBF), of the system, and by using ultra-reliable components and eliminating all less reliable devices, it should be possible to achieve MTBF rates of perhaps one failure in a million hours of operation (i.e. one fault in just a over a century of operation!). However, if a failure did occur in such a system, locating its source and repairing the fault would be extremely time-consuming. Here, another statistical calculation is used: the mean time to repair (MTTR). This figure is based on factors such as the diagnostic tools available to locate the source of a fault, the availability of spare parts, the work involved in removing the faulty component and then replacing it.

[1] For example the Systems Reliability Service Data Bank of the United Kingdom Atomic Energy Authority, AEA Technology, Thompson House, Birchwood Technology Centre, Risley, Warrington, Cheshire WA36AT, UK.

A useful way of looking at the practical aspects of reliability is to combine the two factors. This leads to another statistic, the system 'availability', which is a combination of the MTBF and MTTR:

$$\text{Availability}(\%) = (\text{MTBF} \times 100)/(\text{MTBF} + \text{MTTR}).$$

Using this formula shows that the availability of a system with a MTBF of 80,000 hours and an eight-hour MTTR is 99.99%. Achieving the mean time to repair of eight hours is reasonable. This is the time from the fault occurring, through the process of locating maintenance staff to carry out a repair, through the fault-finding process, to locating a replacement component, to installing it and restarting the system. If the diagnostic tools are very powerful, enabling the location of a fault to be quickly and easily pin-pointed, and if spare printed-circuit cards are mounted nearby in the system cabinets, already powered (and therefore warmed up), then it may be possible to reduce the MTTR and if this is cut to say four hours, the same system will now offer an availability of 99.995%.

When evaluating the likely reliability of a system, all three of the these factors should be examined together because it may be that a high level of availability is based on a less reliable configuration but an impossibly short MTTR.

At first glance an availability of 99.98% may appear to be very good, but if this is based on a four-hour MTTR it implies that the MTBF is 20,000 hours. This means that the system is likely to suffer failures on about nine occasions over an operational life span of 20 years. A system with the same availability, but with a more realistic 8-h MTTR would have a 40,000-hour MTBF, meaning that over the same lifetime the system could be expected to fail on about four occasions.

It must be remembered that availability, MTBF and MTTR are all statistical predictions. Nothing in them will guarantee that a system will operate without fault for a defined time. (In fact, the system may still go wrong on day 1, though the likelihood then is that it should not go wrong again for a very long time, although that may seem to be poor consolation at the time.)

It must also be appreciated that it is not realistically possible for these statistics to be confirmed by measurement during the test phase. At best, a so-called reliability run may extend for a few weeks, but this represents only a few hundred hours of operation, which is a small fraction of a typical MTBF prediction (which is usually in the order of tens of thousands of hours). A reliability run will only show up problems where the reliability is seriously deficient.

When considering Safety Integrity Level-rated systems, or where availability is paramount, a more rigorous approach including a deterministic assessment of components plus failure mode and effect analysis and random access memory assessments, followed by reliability-centred maintenance will increase the level of confidence.

To realistically evaluate a supplier's predictions, the best that can be done is to obtain the data on which the calculations have been based and compare one system with another, while at the same time asking whether any assumptions that have been made are reasonable. Supplier's predictions are regarded as optimistic and hence the need for a third part evaluation. Beyond that, the designer should look at what is likely to happen when the chips are down.

While it is normal to have an availability guarantee for a power station, it is often impossible to hit the target if every failure is considered. For example, if one furnace pressure transmitter has failed is the power station considered to have failed? It will of course continue to produce electricity with the furnace pressure controlled by the two other functioning transmitters. When evaluating control system availability, failures that do not completely stop the system can be given a lower weighting factor. Hence if all five displays on a boiler turbine DCS control system fail, it is counted as down time. But if only one fails it counted as down time ×0.2.

Since this was first published new standards and techniques have become available. As the technology has developed it has become more specialised and is no longer the sole realm of the C&I engineer. It is therefore not appropriate to develop this further in C&I book. However, the following are recommended for further reading,

BS 5760-0:2014	Reliability of systems, equipment and components – Part 0: Guide to reliability and maintainability.
BS EN 60300-3-11:2009	Dependability management – Part 3–11: Application guide reliability centred maintenance.
BS EN 60812:2006	Analysis techniques for system reliability – procedure for failure mode and effects analysis (FMEA).
BS EN 61025:2007	Fault tree analysis (FTA).
BS EN 61078:2016	Reliability block diagrams.
BS EN 62308:2006	Equipment reliability – reliability assessment methods
BS EN 62740:2015	Root cause analysis (RCA).
IEC TR 62380:2004-08	Technical report – reliability data handbook – universal model for reliability prediction of electronics components, PCBs and equipment.

8.10.1 Analysing the effects of failure

In the course of designing a control loop careful thought must be applied to the effects of failure of any component. If any risk can be posed by such a failure, precautions must be taken to limit its effects. Such considerations must be applied to transmitters, process switches and actuators, as well as to the DCS itself. It will usually be necessary to have the design confirmed by some form of risk assessment procedure such as a HAZOP [3].

The HAZOP procedure has traditionally been applied by considering the results of failure of each and every item on the plant. One of the approaches that is adopted is for a team from each discipline to look at each item and ask a series of questions such as for a valve: what happens if it opens, shuts or locks in position? Otherwise, the questions may be aimed at assessing the effects of more or less pressure or temperature on the device in question. The HAZOP procedure is very specialised, and the audit of the plant is usually conducted during the design stages of the project by a team of process engineers, control engineers and others, the whole being coordinated by a specialist organisation.

The emergence of programmable systems has raised several questions as to the validity of this type of study. For example, a traditional HAZOP may lead to the conclusion that if one valve fails to open the situation may be dealt with by the opening of another valve or by the tripping of a pump (either action being initiated by the human operator or by a safety interlock system). However, with a programmable system it may be necessary to consider the possibility of a failure in the DCS causing multiple failures to occur at the same instant, while at the same time any corrective action that the operator may wish to take, and the protective systems themselves, is disabled or seriously impaired. Such questions have recently been addressed, and the matter must be considered in the light of the new guidelines [4].

The following provides an overview of some of the safety-related matters that will need to be considered during the design procedure.

8.10.1.1 Failure modes in electronic systems

Oversimplistic approaches are sometimes adopted towards the analysis of failure, and the field of industrial control systems design is littered with ill-considered ideas. One of the most notable of these is that 'an electronic signal will always fall to zero under a fault condition'. Worse still is the theory that this is the reason for the use of the 'live zero' (e.g. 4 mA) in signals such as the well-established 4–20 mA range.

Though an electronic signal may well fall to zero under certain conditions (e.g. breakage of the connection between the transmitter and the receiver), it is as likely to rise to 20 mA (or above), to lock at an intermediate value (even though the measured parameter has changed), or to slowly drift away from the correct value. If the signal source is provided with self-diagnostic facilities (as described in the following section), the output can be configured to rapidly change to a high or low value, but this is the only condition where such an action can be assured.

As for the reason behind the use of a live zero, this has little or nothing to do with failure. The real reason is that the live zero provides a minimum current for the signal source, enabling the device to be powered from the receiver. This is known as 'two-wire transmission' since it eliminates the need to provide separate conductors to power the transmitter.

Given that a process transmitter is capable of detecting problems within itself and warning of this, so that the DCS can take the appropriate action, what can be done about the signals transmitted by the DCS to actuators? A modern DCS will incorporate powerful diagnostics facilities and watchdogs that will warn of incipient or actual problems. But the action that can be taken in response to these warnings is more difficult to determine. If the entire DCS is failing, then it may be necessary to shut down the plant. On the other hand, if only one output channel has failed it may be possible to override the commands or operate emergency devices that bypass the fault.

8.10.1.2 Transmitters

Many modern transmitters are able to run self-diagnostics routines that provide a high level of confidence that a failure will be detected. In general, these routines are arranged to drive the transmitted signal to one extreme or other (usually outside the normal operating range of the instrument). The direction in which the signal is driven must be defined when the instrument is first specified.

Thus a 4–20 mA transmitter can be specified to drive the signal to under 4 mA or above 20 mA when a fault is detected. The DCS must then be configured to raise an alarm and take the necessary action to protect the plant when such an event is detected. Depending on the hazards which may arise, the required action may be to 'freeze' the relevant system output command at the value it had held prior to the fault being detected, or it may be to open or close an isolating valve, or it may be to operate a separate emergency shutdown system.

The fault detection process will drive the signal to the selected point virtually instantaneously, and if the DCS is configured to respond quickly to such an event the necessary action will usually be taken before any hazard can arise.

If the nature of the process or its instrumentation is such that it is not possible to provide adequate protection by the means outlined earlier, some other form of protection will be needed. This may take the form of a measurement and control system that monitors the process by an entirely separate set of instruments and takes action if a dangerous discrepancy arises between the two sets of measurements or it may be a sophisticated two-out-of-three voting system. Various guidelines are available for making such decisions [3,4], and these should be consulted to justify the cost of providing adequate levels of supervision and back-up. It is unlikely that any authority would be able to supervise each step in the design, construction and installation of a power station DCS. But once the plant has commenced operation, if an incident does occur and results in damage to the plant and/or injury to personnel, the entire design and construction process will be very closely examined. The consequences will be severe for all involved if at that stage it is not possible to demonstrate compliance with the various guidelines and standards.

8.11 Summary

Armed with an understanding of the plant and of its control and instrumentation systems and equipment, we can now move on to look at how the operational requirements are defined on a project, and some aspects of how equipment on a plant is identified.

References

[1] Bull J H. 'Code of practice for the avoidance of electrical interference in electronic instrumentation systems'. ERA 75-31, ERA Technology Ltd., Leatherhead, Surrey, UK.

[2] 'Electromagnetic compatibility for industrial process measurement and control equipment'. IEC Publication 801-1. International Electro-technical Commission, Geneva, 1987.

[3] 'A guide to hazard and operability studies'. Chemical Industry Safety and Health Council of the Chemical Industries Association, London, 1992.

[4] BS EN 61882:2016 Hazard and operability studies (HAZOP studies) – application guide.

Chapter 9

Requirements definition and equipment nomenclature

David Lindsley[1] *and Don Parker*[2]

9.1 Overview

The provision of a control and instrumentation (C&I) system for a power station, whether it be for a greenfield project or as a retrofit at an existing station, is a complex matter which requires careful and comprehensive administration. The task is demanding: the design of the equipment must be correct, systems must be designed on time, equipment has to be carefully specified and purchased, and everything has to be delivered to site, installed and commissioned to a tight programme which interweaves the C&I system with the many other activities that will inevitably be taking place on site at the same time. When complete, the system should be fully supported by comprehensive documentation which enables maintenance staff and users to deal with it.

The actual process of designing the C&I systems forms only one part of the many activities that go together in the task of engineering a complete contract. Although some of the other operations may seem mundane and trivial, they are really anything but that. They are as essential to the contract as the technical design work.

9.2 Defining the requirements

The outcome of a major engineering project such as the design, installation and commissioning of the C&I systems of a power station will to a large extent depend on specifying the requirements at the outset so that what is provided fits within the budget and is fit for the intended purpose. Equally exact documentation is required at each stage as the system metamorphoses from the original concept to the final, functional form.

This procedure requires a considerable amount of definition, and the following outline lists the documents that might be required over the lifetime of a typical project, listed in the order in which they may be expected to be generated. This does

[1]Retired
[2]Provecta Process Automation LLC, USA

not pretend to be an absolute definition that must be rigorously followed on every installation. It is a practical system that has produced good results when followed on several projects. Other documentation systems provided by the plant owner as design inputs or offered by a system vendor as part of the design process may be perfectly acceptable, provided that the same degree of definition is achieved at each stage.

9.2.1 The Project Plan

The coordination of a control system project will require comprehensive project management documentation. A key component in that documentation is the Project Plan. While it does not define the actual operational requirements of the control system that is being installed, this document defines the management and control mechanisms that will enable the project to proceed and be completed successfully. The Project Plan will set out the scope of work; a baseline schedule of main activities and critical milestones; responsibilities of the involved parties and arrangements for management of the design process (including change management), quality, schedule, cost, risk and safety.

9.2.2 System descriptions

System descriptions (SDs) are useful control system design input documents that describe the main components, operation and control of particular physical systems in the power plant. For example, an SD may describe the secondary air system, including design parameters and operation of the forced draught (FD) fans, air heater, secondary air dampers and related instruments and flow elements. For a new power plant, the SDs will be produced in the early project design phase. Where a control system is being replaced on existing plant, the SDs will generally be found with the original design, operating or maintenance documents.

Each SD will set out the design parameters for main drives, such as motor power and pump pressures and flows, as well as the operational intent and process limits. The document provides guidance for the design of aspects of the control system including operating logic interlocks (the requirement for certain conditions such as flows, temperatures or valve positions to be met before proceeding), drive protection, alarming, modulating loops and start and stop sequence operation. The SD will make reference to associated piping and instrumentation diagrams (P&IDs) other relevant design and operating documents.

While the SDs are important for equipment design, operation and maintenance, they are not intended to comprehensively define the requirements of the control system. The SD should be treated as a reference document for operating requirements to assist in the preparation of a specification, and later in developing the detailed design.

9.2.3 The Functional Design Specification

A system can be defined in several different ways, but an essential requirement is the definition of what the system should do *in relation to the process* it is monitoring and/or controlling. This requirement is met by the Functional Design Specification (FDS).

This is a process-related definition of the functions that the system will be required to perform. It does not provide detailed descriptions of the system hardware and software, such as response times, power supplies, environment, etc., except where these are critical to meeting the functional objectives of the installation. Because it defines the requirements, the FDS should be one of the documents against which the vendor is invited to submit a commercial bid for a project.

A typical FDS will describe the plant as a whole, and then discuss the control loops with the required accuracy, response times and dynamic range of each loop.

On occasion, the FDS has been considered as the *only* definition document required for selection and implementation of a replacement distributed control system (DCS). In this case, plant owners or construction firms prepare an FDS with a little more detail than described earlier by including items such as general requirements for cyber security, corporate interface, an input and output (I/O) database, actuator list and a set of broad performance clauses for vendor quotation. An invitation for expressions of interest (EOI) may be issued with the expanded FDS, and the vendors are required to investigate details of the plant and operations, making comprehensive submissions with their quotations. The detail in the submitted proposals must be sufficient for a vendor to be selected.

One difficulty with this approach arises if the vendor's detailed technical proposal is not then bound with the FDS to form a working contract. However other potential pitfalls also exist, not the least being the existence of 'unknowns', which the client has not identified, and the vendor had no cause to include in the quotation. Any such lack of clarity can lead to disagreements over such matters as interface terminal points, cabling material, extent of training and documentation, level of assistance by the plant owner during commissioning and the reliability levels of supplied hardware.

The approach can also lead to higher-priced quotations. Any supplier must mitigate the risk of being exposed to unknowns with a price 'buffer' (i.e. an amount added to the price as a contingency value based on what the unknown factors might cost to accommodate). The increased effort in preparing a tender from an incomplete technical definition may also increase the price from all tenderers if they include a large 'bid cost recovery' amount lumped with the engineering component, which will raise the overall project cost.

It is therefore very advantageous to prepare a more detailed component of the specification, issued, for example, after EOIs are received, or concurrently with the FDS as a single, comprehensive requirements definition. This detailed component is referred to as the Technical Specification.

9.2.4 The Technical Specification

While the functional specification describes *what* the functions of the control system are, the Technical Specification (TS) defines in some detail *how* the functions are to be achieved and the levels of performance for those functions. The TS is best prepared by the plant owner or a specialist representative to provide sufficient detail in the technical requirements for vendors to prepare a fully scoped offer.

The TS will describe the extent of the equipment supply, the physical environment for that equipment, the design requirements for both the physical equipment and application software and, finally, the performance requirements of the system. It should include the following definitions:

1. Equipment environment:
   - Locations of control and field termination cubicles; cable access; location of connected systems
   - The environmental conditions in which the equipment will be expected to operate (defined in terms of temperature, humidity, vibration, shock, etc., as well as dust levels to be encountered, hazardous area requirements, etc.)
   - Power supplies available (including voltage and frequency excursions)
   - Access coordination for installation (requirements for managing de-commissioning of existing equipment, listing of concurrent projects that could interfere with installation unless all parties agree to coordinate activities)
   - Off-site location of project configuration files and historical logs during commissioning, in case of fire or other major incidents
2. Equipment design and capability:
   - Instrument redundancy principles for safety instrumented systems and critical modulating control loops
   - Special instrument requirements: hazardous areas, intrinsically safe instruments, etc.
   - IT technologies: thin client and remote terminal facilities, operating systems, hardware (printers, monitors, servers, workstations, network equipment); networking requirements (field I/O communications, zone segregation, corporate-level access)
   - Cybersecurity: access levels, password management, system patch and malware protection management, network interface protection
   - Historian: storage, logs, trending and analysis capability, remote access
   - Alarm system capability: facilities for setting priorities, filtering, shelving, etc.
   - Engineering facilities: number of workstations and configuration software licenses
   - Plant process performance calculations: unit heat rate, boiler heat loss, heat exchanger efficiency, etc.
   - Control room arrangement: desks, human–machine interface (HMI) screens, lighting
3. Control system design requirements:
   - Design quality management plan and process
   - Design documentation: design procedures; functional design descriptions; wiring and control drawings; settings lists and digital data provisions
   - Functional Safety design requirements: design management plan; design process, redundancy, fault tolerance, etc.
   - Change management: documentation; approvals, retesting

- Level of sequence automation: ranging from basic equipment operation up to single push-button unit start-up
- Modulating control design outlines or enhancement requirements: inclusion of advanced control strategies; integration of field-located controllers, etc.
- Contingency functions: runbacks, load holds, rundowns, additional plant trips, etc.
- Automatic generation control (AGC) and ancillary services capability

4. Control system performance and testing requirements:
   - Equipment performance: control loop cycle times; graphics update period, alarm event time resolution; processor start-up and switch-over
   - Controlled plant performance: sequence operations, ramp rates, stability, load range, frequency response, AGC operation, contingency event operations
   - Alarm system management performance: alarm frequency, handling of bursts and alarm toggling, filtering of downstream initiations, etc.
   - Specific tests to be conducted, for example, hot and warm unit start-ups; load ramps between certain ranges at set ramp rates; major auxiliary trip and runback; simulated frequency steps
5. Operation and maintenance aspects:
   - System maintenance documentation
   - Operator and technician training requirements

We shall now consider several aspects of these requirements in more detail.

9.2.5 Design of functional safety systems and related boiler control logic

Boiler control systems, particularly protection (tripping) logic and associated instrumentation, must be designed in accordance with standards and guidelines to ensure the risks to plant and personnel from any potentially dangerous event are minimised. For new plant, the boiler supplier will be responsible for provision of the instrumentation, interlocking and protection logic design to be given to the DCS supplier for implementation. For retrofits, the plant owner will generally need to set out the design criteria and process to be followed. This may involve a review of the requirements of updated codes such as the National Fire Prevention Association document NFP-85 and the introduction of Functional Safety design and review processes to assess the need for additional instrumentation and tripping functions.

This is a large topic requiring careful design management. A new chapter has been prepared in this second edition to describe the design process in detail.

9.2.6 Testing requirements

The Technical Specification will also lay down the requirements for testing, such as:

- Factory-acceptance tests: Where the system is set up in the supplier's premises, connected to a simulator, or to switches and signal sources for inputs and to meters and other indicators to show outputs, and then put through a series of routines to show that it performs as required. The factory acceptance test (FAT) will also define the environmental tests to which the equipment must be subjected – heat soak temperature and duration. In cases where these tests are done as part of manufacturing, the supplier's documentation may suffice.

- Site-acceptance tests: Performed after the system has been installed and commissioned on site, when it is subjected to changes in desired value settings, simulated and actual equipment trips, etc., to prove that it reacts correctly and in good time to such events. Typically set out are the load ranges, ramp rates, equipment to be tripped and unit start-up types to be tested. An ANSI/ISA standard [1] provides a framework for specifying suitable site tests on power plant control systems. Once all I/O is connected and the plant is in service, several system tests should be conducted, including per cent processor loading, graphics response times and proportion of spare I/O (if required by the TS).
- Reliability run: Where the equipment is left in full control of the process, to demonstrate that is capable of operating correctly for a defined period, with no malfunctions. The document should state the duration of the reliability run, the conditions under which this test will be expected to operate and what should happen if a failure occurs (e.g. start again and repeat the test).

Any commercial requirements relating to guarantees, performance bonds, etc., should be defined separately, although these will interrelate with the TS and should therefore be referred to.

Note that the various acceptance test procedures will at this stage be defined only in general terms. A full definition is provided by test specifications, also known as inspection and test plans (ITPs), prepared as detailed components of the Project Quality Plan that the vendor should provide and adhere to. These are provided for each phase of testing, both for preliminary checks by the vendor and for acceptance by the client. Depending upon contract arrangements, the final acceptance ITPs may be prepared by either the DCS vendor or the client.

9.2.7 Making provision for site tests

It is usual to retain a sum (typically 5% of the overall contract value) which is paid only when the vendor has proved that the system is capable of performing satisfactorily. However, it should be recognised that a control system supplier can demonstrate that his equipment and systems are capable of functioning as required only if the plant is made available for testing the system's performance on the operating plant. It is unreasonable to retain what may in fact be a substantial sum of money without giving the supplier a reasonable opportunity to prove that his equipment is as accurate, fast-responding, reliable and correctly designed and tuned as required. Yet, it is quite common for the plant owner to procrastinate over performing such tests. The reason for this is that a power plant represents a major investment, and starting the recovery of that investment must naturally commence as quickly as possible. As soon as the plant has been completed to the point where it can start earning its keep, strong commercial pressure comes into play, often requiring the plant to operate at maximum output. The pressure will be to start earning revenue as quickly as possible and for as long as possible to maximise earnings. Reducing the output of the plant for the purpose of carrying out tests is therefore unpalatable.

It is in everybody's interests that this dilemma is recognised and a suitable form of words developed to cover the situation, yet this is not often done. One solution is to relate the financial retention to the test programme, defining when the tests will be performed and adding words to the effect that if the tests are not carried out within a defined time after commissioning has been completed, performance retentions will be released provided that the delay is for commercial reasons outside the control of the C&I supplier.

To further facilitate completion of tests, the specification should set out requirements for a level of attendance flexibility by the vendor to enable the tests to have a minimal impact on production. This may include the requirement for extended overnight tests and have provision to return to site for completion of tests after a period of interruption.

9.2.8 Vendor documentation systems

Each control system vendor will propose his own system for managing the design processes and the vast quantity of associated data. It is important that the information in these design systems is transferrable to the plant owner after completion in an accessible and maintainable format. For example, a spreadsheet-based listing of I/O and terminations, delivered in paper format at completion, may provide complete information but will require extensive work by the owner before it is a useful maintenance tool.

Some of the capability of modern documentation systems include:

- Automated loop-diagram generation from a master I/O database.
- Management of all drive protection and interlock signals in a database, generating both design lists for review and check documents for field tests.
- 'Self-documenting' DCS control logic drawings. The acceptability of these systems to act as both detailed configuration documents and functional-level design outlines for ongoing training and maintenance is highly dependent on both the client's expectations and the vendor's system outputs.
- Management of the DCS I/O, control block tags and HMI graphics in a single database.
- Alarm management facilities including embedded operator guidance, alarm settings and test descriptions accessible from the DCS.

Standards exist to classify documentation. IEC 61355-1:2008 (and its British equivalent EN 61355-1:2008) 'Classification and designation of documents for plants, systems and equipment' particularly applies to power and process plant systems. Part 1 sets out the application rules and classification tables to identify different document functions, with categories such as management, contractual, technical, functional, location and listings. It also provides letter codes to identify different document forms. For example, C denotes a chart, D is for a diagram,

M for a map, P for a plan, S for a sketch and T for a table. L denotes a drawing and X is for a document in textual form.

9.3 The KKS equipment identification system

Each item of equipment on a power plant site has to be identified by a method which enables it to be uniquely defined, specified, purchased, installed, commissioned, used and maintained, and this requires some logical way of numbering equipment.

Although several systems of nomenclature can be identified, two methods are in widespread use: one American the other European. The latter is known as the KKS (Kraftwerk Kennzeichen system translated as power station designation system), which was developed by a consortium of large manufacturers and users under the auspices of the German VGB Technical Committee on technical classification systems [2].

A further development from the KKS designation system, known as RDS-PP® 'Reference Designation System for Power Plant', was first published as a DIN standard in 2007 and introduced as an ISO/IEC international standard in 2015 [3] with application guidelines provided by VGB [4]. The RDS nomenclature is based on the principles of the ISO/IEC standards and incorporates most of the KKS formats, with around 90% of the KKS code letters being utilised in the new system key. RDS-PP is intended to progressively replace KKS, although at the time of publication of the second edition of this book, KKS remains the most common designation method. The following description is therefore limited to KKS.

KKS nomenclature is extremely comprehensive, and once it has been understood the system provides a very useful method of identifying any piece of equipment and its operational role in a plant. The system has its weaknesses, but it is so widely used, both in Europe and wherever European influences are felt, that it is very important for the control system designer to have at least a rudimentary understanding of its operation.

KKS defines everything on a plant, from the smallest electronic component to the largest turbo-alternator and even covers the buildings that contain it all.

The system also uses alphabetic characters which do not readily relate to English-language names of the equipment or function. For example, valves are prefixed AA, and the designation letter C is used to identify the purpose of an instrument. This can add to the confusion since many engineers relate C to a controller. Further confusion can be caused because, for example, KKS uses the letters DP to indicate a pressure instrument ('P') used in conjunction with a control system or DCS (designated 'D'), when DP is usually taken as referring to differential pressure.

To help overcome this confusion, many users have customised the KKS code to match their requirements. So, while DP does mean a pressure transmitter connected to the DCS, some will use CP for all pressure instruments since 'C' is also a valid designation for a control system. The numeric part of the tag can also be allocated in groups to differentiate transmitters (e.g. 001–099), switches (100–199), gauges (500–599) and calibration and commissioning devices. Using this convention example, a device tag ending with CP001 is a transmitter and with CP501 is a pressure gauge.

Those who are used to the American system of nomenclature may find it confusing that the KKS codes for all equipment are determined by the functional area of the plant *to which the equipment relates*. For example, in a steam temperature loop where the measured variable is the temperature of the steam at the final superheater outlet, the transmitter will have a KKS code beginning with LBA, but the spray water control valve, as part of the feed-water system, will be given a code beginning with LAA. This is very different from the American system, which allocates a consistent area code *to the whole loop,* and this can be confusing. It is particularly important that this distinction should be clearly understood by plant personnel since with the KKS nomenclature the tag number allocated to the valve merely indicates that it is handling water, and gives no indication that the valve is controlling the steam temperature. Critics of the KKS system use this as evidence against the system, stating that when the completed installation is in place, the nature of the medium being handled by the valve is fairly obvious (because of the size, nature or colour-coding of the pipework), whereas the effect of the valve is not. With the older system, the fact that it is a spray water valve is apparent from the tag number.

The differences are illustrated in Figure 9.1, which shows how the two systems are applied to the same type of control loop. It will be appreciated that, on the actual plant, the tag number allocated to this valve will merely identify it as being part of the feedwater system, and it will not be apparent that its controlling effect is on the steam temperature.

9.3.1 The importance of agreement and coordination

The KKS system allows different users to adopt different approaches, stressing that all parties involved on the project should agree on the selected path to be followed. Although this does show a degree of flexibility, on a given project it becomes confusing to those who were not party to the decisions that were originally made. If only for this reason, it is important that an agreement on the numbering methodology is reached as soon as work commences on a project. It is equally important that a single individual is appointed to control and coordinate the allocation of numbers. This will avoid duplication and different symbols being applied to instruments that serve the same sort of function.

It is also important to understand that the KKS code for an item is almost invariably dictated by the functional area of the plant on which it is used (an exception is the DCS itself since this generally serves almost all areas of the plant). Therefore, before a number can be applied to, say, a temperature transmitter, it is necessary to know the KKS code for the pipe to which it is fitted. This dependence renews the pressure for the plant piping and instrumentation diagrams (P&IDs) to be defined at an early stage in a project and for them to be as accurate and complete as possible before work begins on specifying the instrumentation and systems. This point cannot be overemphasised. Problems will inevitably arise as soon as any attempt is made to allocate KKS numbers to instrumentation before the P&ID stage has been completed.

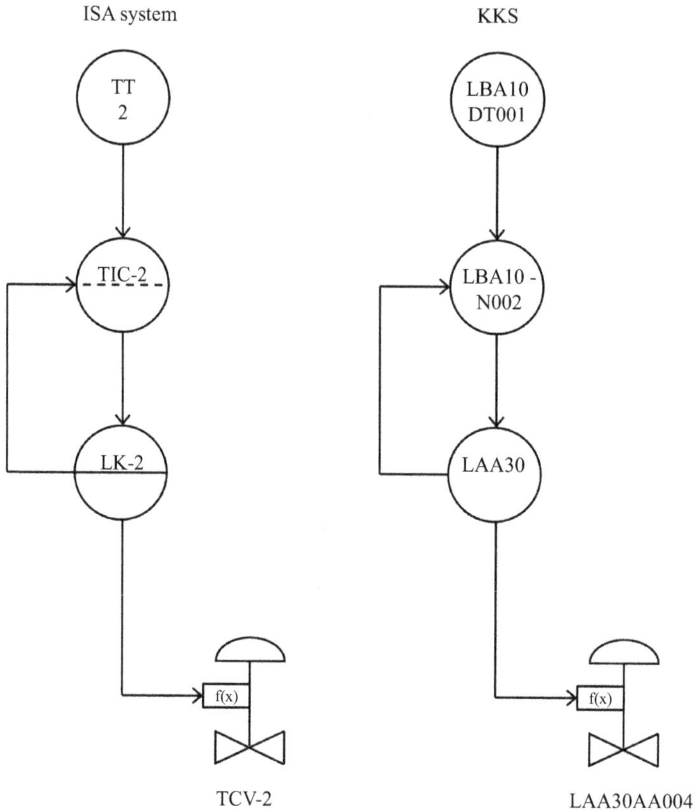

Figure 9.1 Comparison of KKS and ISA numbering systems applied to a steam temperature control loop

It must be recognised that the KKS system offers a comprehensive method of housekeeping which can greatly enhance the task of allocating nomenclature to all equipment on a plant. This is a routine, but surprisingly costly, facet of the administration of a project. Properly used, KKS can yield considerable economies and efficiency: poorly administrated it becomes a nightmare.

9.3.2 Review of KKS

The following is an outline of how KKS identifies the components that go to make up the C& I systems (and *only* the C& I systems) of a power plant. The structure of this book does not allow for a comprehensive guide to be provided, but the following brief explanation should at least enable the reader to grasp the principles. However, it is emphasised that KKS has a complex set of capabilities which have been customised by many users with their own numeric codes and display shortcuts. For each project there will be a project-specific coding manual that must always be followed in preference to this explanation.

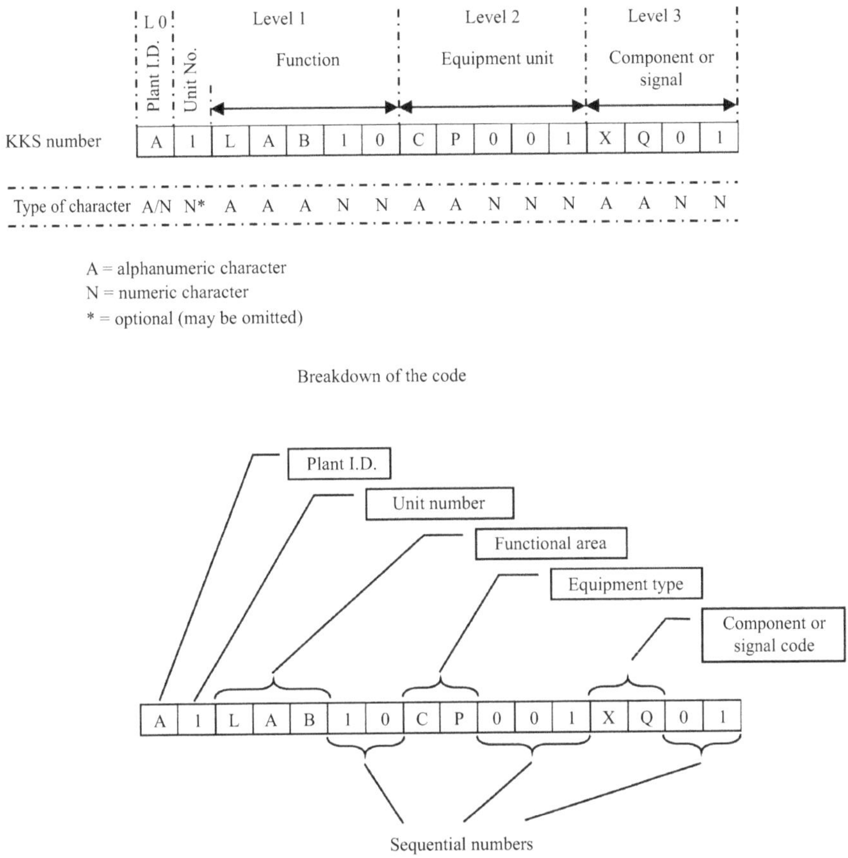

Figure 9.2 Overall KKS numbering philosophy

9.3.2.1 The levels of coding

The number allocated by the KKS system to a piece of equipment is broken down into a number of sections or 'levels' (see Figures 9.2 and 9.3) and within each of these levels is a field or set of fields, each field being occupied by a letter or number. Each letter or number is allocated a field identification: G for the plant, F for the function, A for the equipment and B for the component.

- The Level 0 code is a single digit used where there are, for example, two power plants on the same site. These are usually designated as 'A' and 'B', but the system allows numeric characters to be used if desired. If only one plant exists on the site the first character is omitted altogether.
- The first digit of the Level 1 code identifies the boiler/turbine unit on which the relevant piece of equipment is fitted. This is always a numeric character and is usually 1 for boiler 1, etc. Where the equipment is common to all areas of the plant (e.g. a cooling tower) the number 0 is allocated to it.

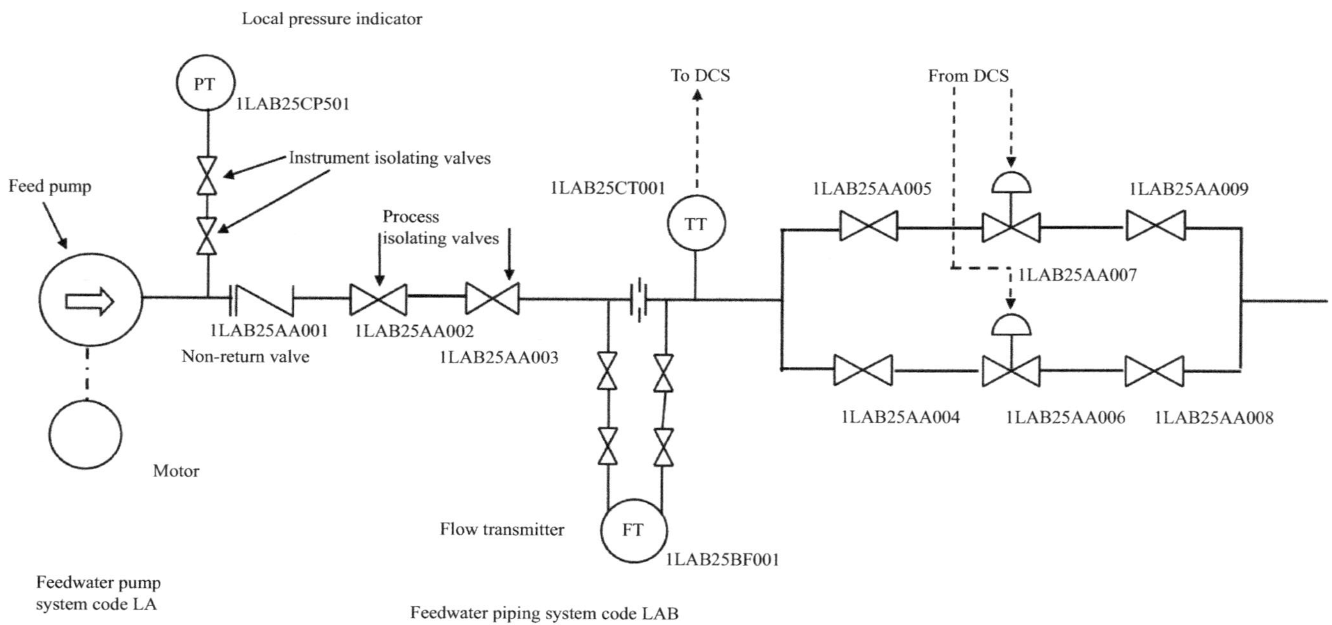

Figure 9.3 KKS coding applied to a feedwater pump system

- The remaining five characters of part of the Level 1 code define the function of the system and the sub-system (e.g. a feedwater pump system, a particular pump and all equipment associated with it). This group of digits (the first three alphabetic, the next two numeric) is combined with the unit-identifying prefix, into what is called the 'function code'.
- The Level 2 code has five characters that define the particular piece of equipment (such as a pressure gauge, CP) and its unique sequential number (e.g. 501).
- The Level 3 (component or signal) code has five characters that identifies the component itself, and in the case of a device that generates some form of electrical signal, also defines the nature of that signal. The Level 3 component is not usually displayed on P&IDs. This level is useful for identifying signals states, such as normally open and closed switch contacts, in loop and logic diagrams. It could also be used to describe instrument manifold valves supplied with a pressure gauge or to differentiate between a thermowell, thermocouple and head-mounted temperature transmitter.

It is important to understand that if there are, for example, 200 pressure gauges on a plant, the sequential numbers for these do not start at 001 and continue through to 200. The numbers relate to the functional area of the plant on which they are used. Therefore, they start afresh each time the preceding code changes. For example, the first pressure gauge on the HP steam piping system of a plant (where the piping system is numbered LBA10) will be allocated a sequential number LBA10CP001, the next will be LBA10CP002, etc. The numbering starts again on another system: the first pressure gauge on the hot reheat steam piping system (numbered LBB10) being allocated the sequential number LBB10CP001, the next being LBB10CP002, etc.

The structure of this coding system is illustrated in Figure 9.2. In this example, the designation A1LAB10CP001 may describe the first pressure transmitter on the feedwater piping system LAB10 located on Unit 1 in power station A. The signal code XQ01 refers to the first analogue output of the transmitter.

9.3.2.2 Explanation of the main coding principles

There are several hundred categories within the KKS codes but the following is a brief summary which will provide some assistance in dealing with boiler C&I systems. Where it is felt that the wording is ambiguous, some examples have been provided. However, it should be remembered that these interpretations are not universally implemented, and the importance of project-specific agreement and coordination (as mentioned in Section 9.3.1) cannot be too strongly emphasised.

Level 1: function
B = power transmission and auxiliary power supply
C, D = instrumentation and control equipment
E = conventional fuels supply and residue disposal
H = conventional (i.e. nonnuclear) heat generation
L = steam, water and gas cycles
M = main machine sets

P = cooling-water systems
Q = auxiliary systems (e.g. air compressors)
R = gas generation and treatment
S = ancillary systems (e.g. heating and ventilation)

It will be found that may of the control functions, signals, instruments and equipment associated with boiler controls begin with C, D, E and L. Each code will have many subdivisions. For example, L (steam, water and gas cycles) has over 50 subdivisions. These include:

LA = feedwater system
LAB = feedwater piping system
LB = steam system
LC = condensate system
LCH = HP heater drainage system

Level 2: equipment unit
A = mechanical equipment (e.g. valves, dampers, fans, including actuators)
B = mechanical equipment (e.g. storage tanks)
C = direct measuring circuits (e.g. local indicators)
D = closed-loop control circuits
E = analogue and digital signal conditioning
F = indirect measuring circuits (e.g. sensors feeding remote indicators)
G = electrical equipment (e.g. cubicles, junction boxes, generators, inverters, batteries, lightning-protection system)

Some codes often found on P&IDs include:

AA = valve
C = field instrument (e.g. CP, CT, CL, etc.)
D = control-loop related instrument (e.g. DF, DT, etc.)

9.3.3 An example of how the codes are used

Figure 9.3 shows how the KKS codes may be applied to a part of the feedwater pumping system on a plant. In this example, the pump relates to boiler 1 and the pressure indicator at its discharge is therefore numbered 1LAB25CP501, the C being the Level 2 code for a direct-measuring device indicating pressure (P). The feedwater-flow transmitter 1LAB25DF001 forms part of the three-element feedwater control system and its Level 2 code therefore uses D for a closed-loop control circuit and F for flow. As suggested earlier, others may replace D with C to avoid confusion with DP instruments, giving the designation 1LAB25CF001.

When it comes to transmitters whose signals do not form part of a closed-loop control system, different interpretations are applied by various users to the Level 2 codes. In some cases, the classification letter D (or C) is applied to any transmitter which feeds the DCS, irrespective of whether or not the signal is used in a closed-loop control system. Other users extend the meaning of F (indirect measuring circuits) to include this type of measurement. In Figure 9.3, the former of these interpretations is used, and the temperature transmitter that is used to provide a signal for the operator display is therefore allocated the letter D as the first character of its Level 2 code (the second character is T, for temperature).

This point illustrates the importance of coordination and agreement between all parties.

9.4 Summary

In this chapter we have considered some of the documentation systems that may be used to define and manage the wide range of requirements to be met by a C&I system. In the next two chapters we will delve into some of the processes used for designing modern boiler protection systems and the structures of advanced modulating and sequence controls employed to enhance the operational flexibility of power plant.

References

[1] ANSI/ISA-TR77.30.01-2015, 'Power plant control system dynamic performance test methods and procedures' (USA, International Society of Automation, 2018).

[2] DIN 40719 Part 2 (IEC 750) 'Circuit diagrams, circuit representations, definitions, classification (diagrams, charts, tables, item designations) alphabetical classification examples', 1978.

[3] ISO/TS 81346-10:2015 (En) 'Industrial systems, installations and equipment and industrial products—structuring principles and reference designation—part 10: power plants', 2015.

[4] VGB-S-821-00-2016-06 'Reference designation system for power plants RDS-PP® – letter code for power plant systems'. ISBN: 978-3-86875-946-4 *VGB Power Tech Services* 2016.

Chapter 10

Functional safety and associated reviews for boiler protection and control systems

John Grist[1]

10.1 Introduction

The operation of boilers can pose risks to people and equipment. Instances have occurred where furnaces have exploded due to an accumulation of unburnt fuel and uncontrolled ignition, or imploded from incorrect loading of the induced draught (ID) fans or from furnace gas pressure excursions following a fuel trip. Rupture of boiler pressure parts cause plant damage and serious harm to personnel, and fuel equipment such as oil pumps and pulverisers can catch fire and/or explode.

For many years, the design of burner management systems, field equipment and operational practices for the safe operation of boilers has been guided by the National Fire Prevention Association (NFPA) 85 and EN 12952. Similarly, the protection of turbines from water induction is guided by the ASME code TDP-1. The more recent introduction of the life cycle approach and risk management process associated with the application of functional safety standards for boiler plant and their protection systems has found many points of intersection with these standards. It is therefore beneficial to apply design processes that achieve functional safety for boiler control and protection systems within the NFPA 85/EN 12952 framework. Note that functional safety design does not provide prescriptive guidance into the extent of these codes. Similarly, the NFPA 85/EN 12952 codes do not describe how to estimate control equipment failure rates, instrument testing frequencies and the reliability requirements. The two go hand-in-hand for securing boiler operation and protection. The International Electro-Technical Commission (IEC) 61511 does not attempt to replace the good engineering practices recommended by NFPA 85.

This chapter gives an outline of the functional safety processes and associated boiler-related reviews. Real examples from boiler reviews are included. It is an introduction to the processes. It is not intended as a formal guide. In all cases guidance should be sought from a competent engineer.

[1]Consulting Engineer

10.2 Background

When the first edition of this book was prepared the buzzwords referring to compliance requirements for boiler protection system design were health and safety but since then a new term has arrived in the industry – 'functional safety' and with it, specialists to interpret standards, design and implement these systems. IEC standards 61508 and 61511 on functional safety are becoming more prevalent in the UK-engineered designs of protection systems for new coal fired power stations. IEC 61508 deals with functional safety for those designing electrical electronic or programmable, equipment for protection systems and IEC 61511 with safety instrumented systems (SIS) in the process industry, for those specifying a protection system. In the UK, the Health and Safety Executive considers IEC 61511 as representing best practice. If the standard is followed a sound basis for the specification, design, implementation, operation and maintenance of boiler protection systems can be demonstrated. The importance of this is that following best practice will provide an audit trail for a competent design in the case of an investigation to an incident and will show that risks were assessed and managed.

IEC 61511 and the boiler industry expect the creation a series of documents and reviews that together demonstrate the process that has been followed to develop a safe operating plant. They will include basis of design, basis of safety, HAZard IDentification (HAZID) review, process and instrumentation diagram (P&ID) review, HAZard and OPerability study (HAZOP), determination of Safety Integrity Level (SIL), demonstration of SIL, factory acceptance tests (FAT), training of operators and maintenance staff, site acceptance tests (SAT) and ongoing reviews. The following descriptions give a control and instrumentation (C&I) engineer's perspective and do not detail the complete process as seen by safety, process, systems and civil engineers. It includes accepted practice in the power industry as well as those activities prescribed by IEC 61511.

10.3 Basis of design: basis of safety

These basis documents are sometimes combined. From a boiler supplier's C&I perspective the basis of design will

- Identify scope: Is the DCS in our scope? Who is installing the system? Are the sootblower controls to be programmable-logic controller (PLC) based or in the distributed control system (DCS)? The scope definition should also include relevant design codes.
- Restate any process safety and functional safety codes such as NFPA 85, ASME TDP-1 EN 61511. ASME1 EN 12952.
- Identify any environmental risks to the site, such as being in an area subject to lightning storms, tsunamis or earthquakes.
- Identify any known hazards, for example, if hydrogen or ammonia is to be used.
- State the processes that will be followed, such as HAZID, HAZOP, LOPA, hazardous area classification and other supplier or client-specific meetings or documents.

- Identify who will be responsible for these processes, defined in terms of competency and independence.
- Describe any preset requirements, for example, the biomass or ammonia unloading area maybe defined as hazardous areas.
- Discuss the use of fire and explosion protection systems.

10.4 Process HAZID study

IEC 61511 does not prescribe the hazard study to be used. Typically, it may be a combination of HAZID, HAZOP or failure mode effect analysis (FMEA).

The HAZID (hazard identification) identifies potential hazards at an early stage of the plant design process and allows alternative systems and equipment to be considered. For example, oil may be selected as a support fuel rather than gas to reduce the risk of an explosion. Similarly, urea rather than ammonia may be used in a selective catalytic reactor (SCR). Flood risks, earthquake zones, severe weather conditions and the proximity of people to plant should be considered and build these into his design. The HAZID is usually conducted before the boiler C&I engineer is involved.

If one participates in a HAZID study, note that the purpose is to identify hazards, but not necessarily fix them, just list known safeguards or safeguards that are typically applied in the industry, or are technically feasible, whether or not they are in the design for this specific plant. The study can also recommend safeguards for future consideration.

10.5 P&ID review meeting

The P&ID review is the most constructive review insofar as it identifies changes that may be made at a stage with minimal cost impact. It comes before the HAZOP and may not necessarily involve the client. A good review meeting is chaired by a person at the chief engineer level or well-respected project engineer not directly involved in the project. The rule is: 'Don't redesign it. If it works leave it.' There may be up to 20 attendees, hence the need for a senior chairperson. Technical authorities from each discipline will review the design for fitness for purpose, and compliance with the relevant codes and in-house practices. Lead engineers will review against the client's requirements. Anyone can raise safety concerns. The meeting is at a high level, and so checking the correct tag numbers, off-page connectors, etc., is not done here.

For a conventional boiler, these reviews take two full days, covering the steam and water P&IDs one day and the air and gas and combustion another day. Any novel design, such as carbon capture, would need extra time. The basic boiler type, number of fans, number of burners, turn down, fuels, etc., are agreed before this stage and recorded in the Basis of Design document. The P&IDs will therefore include the fundamental piping, ducting, drives, valves, dampers and instrumentation. The meetings are an opportunity for engineers to consider high-level questions such as those relating to maintenance requirements, duct crossovers for availability, start-up systems and fuel system changeover. Typical boiler-related examples of these follow:

- Maintenance: Do the fans need to be maintained without shutting down the boiler? If so are there sufficient isolation dampers? Do the dampers provide sufficiently good isolation to allow engineers to enter the duct work?
- Cross-overs: Is the boiler expected to operate with FD fan A and ID fan B? If so, are there sufficient ducting crossovers to allow this? Is the duct size shown on the P&IDs? (It is needed to show that the cross over area is big enough).
- Start-up: Is there a pinch valve or alternative arrangement to ensure adequate spray water pressure? How is the feedwater flow to be controlled? Often feed valves and feed pumps are not part of the Boiler Island scope so information may need to be sought from others.
- Is the system expected to recover from a 'black start'? If so, does it need steam and where does that come from?
- Fuel changeover: If switching between, say, light and heavy fuel oil, or between biomass and coal is it necessary to remove or shut down the burners?
- Commissioning: Is a flue gas analyser grid required at the economiser outlet? Who is providing calibrated flow elements for the performance tests?
- Are the failure modes of each control valve defined?

Then at the next level each P&ID is reviewed to ensure code compliance, for example,

At least three furnace pressure transmitters if NFPA 85 applies.
Pressure gauges to be fitted to pressure vessels such as air receivers and the boiler drum.

One problem with in-house design reviews is that they are often limited to the boiler P&IDs only, and the attendees have no knowledge of what lies beyond their plant scope terminal points, such as whether flow elements are available in the feedwater supply upstream of the boiler. Similarly, details of what is downstream of the ID fan, if a FGD fan is being supplied by others may be unavailable at this stage. These unknowns must be recorded and clarified later, ideally before the HAZOP.

10.6 Hazardous area classification

This stage looks only at the precautions to be taken to prevent a spark, or similar heat source, causing an explosion. In the general process industry hazardous area classification is considering the risk of a fault producing a spark or an instrument overheating and these igniting an explosive atmosphere. However, in a boiler a flame is always present and the likelihood of explosion is much greater. Hence some interpretations of the standards suggest that alternative methods should be used. For example, prevent flexible burner-hoses leaking by ensuring that they do not rub on the gallery and change them at every major outage.

The various areas of the boiler are considered and zoned according to the risk of an explosion. The lower the zone number the greater the risk. In Europe three zones are considered for gas and liquids:

- Zone 0: The substance likely to explode is present continuously or for long periods (more than 1000 hours a year). For example, inside a fuel oil tank.
- Zone 1: An area where explosive gases are likely to be present in normal operation (between 10 and 1000 hours a year).
- Zone 2: The substance is likely to explode is not normally present and if so only for a short time (less than 10 hours a year). Usually caused by a leak. For example, in the vicinity of a control valve in a gas line.

Note the hours per year are a guide only and not officially adopted.

Another classification but with zones numbered 20, 21 and 22 apply to explosive dust. Biomass transfer stations and inside coal silos are often classified as zone 20.

In the United States similar rulings apply but using class 1, 2 or 3. Classification can be determined by any competent engineer. The electrical and instrument engineer should be involved as the zoning has an impact on the equipment they provide. Once the area has been classified the electrical and instrumentation engineers choose the appropriate instrumentation lights and motors that have been certified for use in that hazardous zone. Protection can be achieved by a range of methods including encapsulation, pressurisation, increased safety, making them explosion proof or using intrinsically safe equipment.

10.7 HAZOP

The most important functional safety meeting is the HAZOP. It usually involves the client and an external chair.

A HAZOP is an industry-standard study carried out by a multidisciplinary team, who apply standard guidewords to identify deviations from the design intent of each system making up the overall plant. The team attempt to identify the causes and consequences of these deviations and the protective systems installed to minimise the consequences and thus to make recommendations which lead to risk reduction.

The HAZOP processes are defined in EN 61882:2016 Hazard and operability studies (HAZOP studies) Application Guide. Major boiler makers and Engineer Procure Construct (EPC) contractors may have their own in-house standards although these should in general follow the guidance provided by EN61882.

Guidance for the reader based on lessons learnt

- Make sure the chairman takes a copy, lunch time and evening, and backs it up to ensure continuity if the PC is damaged.
- Everybody from project manager to graduates will typically want to attend these. The EN standard recommends a team of seven so non-contributing attendees need to be controlled, maybe sitting at another table.

- The best chairman will draw out operational concerns that might need the client to answer as well as the expected hazards for the supplier to consider.
- An experienced chairman will explain the procedure at the start of the meeting so these are not repeated here.

Some interesting boiler-specific observations are the following.

In HAZOPing a steam inerting line, deciding that even if steam passed unexpectedly to the mill there was no process risk, but almost missing the fact that during maintenance there can be workers in the mill! They need to be protected by physical isolation such as a spectacle plate or removing a section of pipe before entering the mill. There is nothing new in this HAZOP finding. Health and Safety Executive (HSE) ask for the best isolation and spectacle plates are a standard maintenance feature. The warning is that person entry was not considered in the first set of questions.

Sometimes several HAZOPs are held. This may be for one or more of the following reasons.

- An early HAZOP is required to allow equipment to be purchased and engineering to continue. A follow-up HAZOP is held when details of specific mill or fan designs are available. Typically, they will have their own P&IDs.
- The follow-up HAZOP looks in great detail at sub-processes, such as a SCR or a hammer mill. Each of these will require specialists that were not needed in the earlier review.
- The HAZOP described earlier is for the boiler only; a similarly structured HAZOP will be needed for the turbine and balance of plant.

10.8 Determination of safety integrity level (SIL)

If a protection function or safety instrumented function (SIF) is used to achieve a required risk reduction, it should be assigned a SIL. This is an integer value (in the range 1–4, where 4 is the highest) denoting the level of integrity (i.e. reliability) the function must possess in order to achieve the required risk reduction. For a boiler SIL 3 is the highest level that might be discussed. Most SIFs are SIL 1 or SIL 2.

The target SIL for each SIF is determined by assessing the amount of risk reduction required to achieve a tolerable risk level for the hazard being considered. Tolerable risk levels need to be defined by the plant owner. They are typically defined in terms of a maximum tolerable frequency for an incident arising from a specific hazard that leads to a specific amount of harm. For example, a company may state that any specific upset condition leading to a single fatality on-site is tolerable no more than once in 10,000 years. Alternative tolerable risks may be defined for environmental risk and loss of assets.

In order to calculate the amount of risk reduction a SIF must provide: the analyst estimates the frequency of harm in the absence of the SIF. The most common method is to use layer of protection analysis (LOPA). This has replaced risk graphs as the method of choice for SIL determination although for more complex systems a fully quantitative method such as fault tree analysis (FTA) may be required. FTA will also be used if a risk reduction of greater than 10,000 is required from a combination of instrumentation and controls. Most of the following descriptions of LOPA have been taken (with permission) from a training presentation prepared by Dr Peter Clarke of xSeriCon.

Firstly, the frequency of one or more initiating events is estimated. A typical initiating event is failure of a control loop, trip of a fan or loss of utility supply.

If there are any factors that determine whether the hazard is present all the time, these factors are identified next. They are known as 'enabling factors'. For example, if the incident under consideration can occur only during start-up, the enabling factor is the fraction of total plant lifetime that the plant is under start-up.

Next, the failure probability of any applicable layers of protection is estimated.

A layer of protection is an engineering measure to prevent the initiating event from leading to the unwanted consequence. Typical examples are

- Alarms with operator response
- Control loops
- Check (non-return) valves
- Safety instrumented functions (apart from the SIF under consideration)
- Pressure relief devices
- Bund walls

Layers of protection must be specific, auditable, independent and dependable (SAID).

Each project needs specific rules which must be defined in an approved LOPA procedure. These might include:

- Corporately agreed tolerable risk criteria.
- Agreed risk reduction data to be used.
- 'How many layers of protection are allowed?' (For example, usually only one alarm per scenario.)

Then conditional modifiers (CMs) are identified. CMs are factors that affect whether the harm actually occurs and have an associated probability.

Examples are:

- Occupancy (people hurt only if present)
- Probability of ignition of flammable release
- Wind direction (toxic release harmless if blown away from people)

It is important to include CMs in a LOPA as they often make a significant difference to the final result. They lead to a lot of discussion, and the LOPA team must document all assumptions and conclusions carefully.

The expected frequency of harm in the absence of the SIF is found by multiplying each of these factors. If it is greater than the tolerable frequency then extra protection usually in the form of a SIF is added to reduce the risk to an acceptable level. The SIF is assigned a 'risk reduction factor' (RRF) which is the ratio of the expected frequency to the tolerable frequency. Finally, the target SIL is derived from the target RRF by reference to tables in IEC 61511.

For example, if a particular flange in a steam line leaked once every 10 years what is the risk of an operator being killed?

The initiating frequency is 0.1 per year.

A layer of protection is provided by the insulation. The effectiveness of the layer of protection will depend on size of leak and steam pressure. For this example, it is assumed to prevent death in 50% of the cases.

The CMs are identified as

1. An operator is not normally in that area but patrols it once a shift. Present 45 minutes in 24 hours (=0.031).
2. It may be possible to avoid walking into the steam if the operator hears the steam leak or sees condensate or steam leaking from the lagging. 50:50 chance of avoiding it.
3. Will the leaking steam be fatal? A higher risk exists in an enclosed space and a lower risk outdoors. For this example, it is assumed death only occurs one in 10 times.

So overall the expected frequency of fatality from this incident is: $1 \times 10^{-1} \times 0.5 \times 0.031 \times 0.5 \times 0.1 = 0.0000775 = 7.75 \times 10^{-5}$.

It is necessary to formally record all the assumptions and calculations either in a standard format using a spread sheet or using a computer package. There are several available, but often an independent chair will use his company's standard.

Having decided on the safety integrity requirements for each identified SIF, a safety requirement specification is prepared and a safety instrumented system (SIS) purchased against that specification.

10.9 Project liaison meetings

Having a HAZOPed set of P&ID and knowing the required SIL the project team can now purchase the DCS and SIS. In practice they are likely to have issued preliminary specifications in advance of this stage and may, by now, have selected a supplier.

This section looks at meetings that might be held between the boiler supplier, the turbine supplier, the balance of plant (BOP) supplier, the end user, the EPC contractor and the DCS supplier.

Each of these parties will have their own views on the best way to run a project, and this meeting is to ensure that only one way is agreed for the following.

Logic symbols	Usually already specified as the International Society of Automation, Scientific Apparatus Manufacturer Association (SAMA) or Deutsches Institut für Normung (DIN, German Institute of Standardisation) in the contract but client and DCS supplier may be familiar with multiple standards and accept alternatives up to the factory acceptance test. After which the DCS documentation becomes the standard.
Supporting descriptions	Are descriptions of logic diagrams required? If so are they to help the DCS engineer or the operator?
P&ID overview	Unlikely to be a formal review at this stage but an opportunity to ask for details of relevant systems not in your scope.

(Continues)

(Continued)

Integrated boiler turbine master concepts	All parties will have a view on this, but may be reluctant to take the overall responsibility. This should rest with the EPC contractor.
Tag numbering system	Basic requirements will have been specified, but coordinator must be agreed, and if KKS or RDS-PP the level of detail to be used must also be agreed.
Documentation exchange schedule	Often a contentious discussion as the DCS supplier may need the input ranges before the boiler supplier has received that information from his equipment suppliers.
Database management	Will the I/O list, instrument schedules be presented in Excel, Access or a proprietary format such as SPI?
FAT procedures	Unlikely to be agreed at the first meeting.
Interface voltages and supplies	If digital inputs require 24 or 48 V supply does it come from the DCS?
Galvanic isolation	How are analogue signals isolated when exchanged between the DCS and other systems?

10.10 Demonstration of SIL achieved

Before the SIS design is manufactured it is necessary to demonstrate that the design can achieve the required SIL for each SIF. One way to achieve this is by using fault trees to calculate the probability of failure of each SIF. The fault tree can include common cause failure modes. A popular cost-effective suite of tools is provided by Technis and comprises TTree to create the fault trees, Faradip to provide typical failure data and Beta Plus to calculate common cause failure probability. Another option is to use SILability by xSeriCon, which uses a well-known set of equations to derive the expected failure probability of a complete SIF based on input parameters supplied by the user. Major companies such as ABB and Exida also have here own suite of packages. One can conclude that from the need for specialised tools this is not normally done by the boiler C&I engineer but by the SIS supplier or a functional safety consultant. Hence only the briefest description is included here.

10.11 Implosion studies

A simple boiler with FD and ID fans does not need an implosion study. However, as flue gas clean up equipment such as SCR, flue gas desulphurisation, electro static precipitators or bag house are added they increase the pressure drop from furnace to the ID fans, which in turn leads to the ID fan being upgraded or a booster fan added. An 800 MWe unit with FGD, for example, may have two 14 MW ID/booster fans with considerable suction. They ID fans pull >15 kPa (60 in.)

exceeding the 8.7 kPa (35 in.) limit set by NFPA 85 and potentially can cause a furnace implosion. Where this flue gas clean-up equipment is added as a refurbishment, there is also a risk of a furnace or duct implosion. In these circumstances, it is normal practice to carry out an implosion study.

This is carried out by a specialist company. More than one company offers this service, but the following description is based on the approach of an American company TRAX LLC ENERGY based in Lynchburg, Virginia. TRAX has supplied 200-plus fossil fuel power plant simulators. They use a subset of these simulation programmes to model the furnace during upset conditions. TRAX has provided numerous furnace implosion studies. The simulations are run in ProTRAX on a PC. The software models the physical properties of the furnace, including dimensions, gas path pressure drops, fan heads, fan rundown times and maximum fan flows.

ProTRAX is a modular, dynamic simulation code. All modules are based on first principles of mass, energy and momentum conservation. No transfer functions are used, and the model is refined from plant data.

The proposed control system is modelled using a complete set of SAMA modules, which can be used to simulate controls, are part of the ProTRAX software. In addition to actually running the programme there is additional engineering to fully explain the causes and risks and potential solutions. The programme runs in real time or may be slowed down. The engineer may desire to analyse a phenomenon as it happens.

For a typical project TRAX will collect physical design control system intent information from the contractor and in the case of a retrofit go to site to collect data from the historian or PI system. They then carry out a series of computer simulations. Typically, five or six simulations are included but can range between 2 and 30. Additional simulations may be provided but are costly. To reduce the number of simulations some users want to consider just the worse-case scenarios. These include loss of all power, loss of one FD or ID fan, master fuel trip (MFT) or boiler trip all while at full load. However, low load is worst case for events that involve excessive furnace exhaust flow.

While this helps the boiler designer it really helps the control engineer as the simulations are accurate enough to predict controller settings. For example, following an MFT, the study will suggest how big a kick to apply to the ID fan vanes and if there is a variable speed ID fan, how much to kick the damper and how much to change the speed. The suggested settings are not normally changed during commissioning and help to prevent pressure overshoot or excessive negative pressure excursion.

One of the risks of a using a variable speed ID fan with a fixed speed FD is that it might respond slower than the FD. This conflicts with NFPA 85 recommended practices.

However, variable speed ID fans with fixed speed FD fans are common and an implosion study may be used to demonstrate that this is safe. Studies have also shown that a small over pressure may be vented via the water seal of the submerged chain conveyor.

The results can be so good that they allow counter-intuitive logic to be included. For example, on loss of one FD fan it is normal to trip some mills. The study may recommend tripping only one mill and delaying the subsequent trips to avoid excessive pressure dips.

The discussions at the report presentation are often insightful and result in additional model testing and/or modification to control measures.

10.12 Factory acceptance tests

The FAT for a complete boiler control system is in many parts. These can run in parallel or as separate tests. They include hardware testing, application software testing and graphics testing.

The FAT for the boiler SIS is normally held as part of the FAT for the control system, with provision for additional testing and test documentation, as required, to check the logic solver hardware and software are integrated, that fault tolerance is correctly functioning and that the functionality meets the specified safety requirements.

10.12.1 Hardware testing

During these tests, the client ensures that he has the right equipment in terms of quantity of processors I/O cards, cabinets, termination marshalling visual display unit (VDU) large screen displays, graphic display units, desks, archive and engineer work stations. The depth of testing is determined during the contract negotiations and early specifications and might include special tests coping with power failure, or operation at elevated ambient temperature.

Proper integration of third part equipment such as electronic drum level monitor (EDLM), CCTV displays water cannon, acoustic leak, etc., and communications to other systems if available at the FAT.

Review of earthing connectors, internal lights and flexible cables where any lamps are door mounted. Installation of client-specified fire detectors or fire-fighting equipment. Pay attention to panel plinth dimension as floor tiles may have been cut ready for the cabinet.

The supplier's hardware test should include a point-to-point test on each input and output. The supplier will normally have a database to keep track of input terminations, input card, human–machine interface displays, etc., that can be used to record these checks.

10.12.2 Software testing

There are several opportunities to do some software testing.

10.12.2.1 Informal review

Ideally an informal review should occur, for example, when the first set of mill logics or a steam temperature loop has been configured. This can be a laptop review or even review a print out and a screen dump of the graphics. The intent is to set the scene for the following reviews as well as to ensure that the software listing or block drawings can be understood. For example, the DCS blocks may be automatically allocated a block number that reflects the order in which they are executed. This cannot be easily changed but the DCS engineer can add dumb block numbers that reflect those used on the original SAMA diagrams.

This is an opportunity to contribute to the graphics. Avoid the detail design but ensure that tag numbers are correct and fully test one controller faceplate to ensure that auto/manual transfer, set point, alarms, manual operation all work as expected and specified.

If the DCS supplier is not able to do this, similar logic from a previous project could be reviewed to ensure that auto/manual displays are acceptable and that the logic for controller tracking, and any special functionality such as inhibit raise/lower, maximum and minimum limits function as expected. For example, a controller output can be locked, depending upon the DCS, by a 'freeze' command or by simultaneously applying an inhibit raise and inhibit lower signal or by connecting the controller output to both the maximum and minimum limits.

10.12.2.2 Pre-FAT

In many ways, this is the most important test for the boiler controls engineer. The intent is to finish up with at least one example of each loop working correctly. For example, one air-heater, one ID fan, one FD fan, one mill start-up and shut down is demonstrated and debugged. The drum level and furnace pressure control are to be fully tested, together with the steam temperature control for one steam leg and one set of combustion logic. Ideally the boiler engineer who created the logic drawings should attend the FAT and be prepared to justify his design or agree to the DCS supplier's alternative but equivalent solution. (See earlier text for three ways to freeze a controller output.)

This approach only works if the test engineer understands boilers well enough to actually know what the logic is intended to do.

At this stage, the worst possible test is one that is fully detailed and requires a test of the full functionality of every loop, each button on each auto manual station, each alarm, etc. It takes forever and distracts from the real purpose of a pre-FAT. Other than agreeing the order in which the loops will be tested and how the inputs will be simulated this should be a free-format test. The aim is not to show how good the software is, but to consider realistic operational events and tests that find faults. One approach is to consider the actions during a plant start-up. This approach tests the concepts and implementation.

Inputs used to be simulated by a panel on wheels that had resistors to simulate analogue inputs, gauges for analogue outputs, switches for digital inputs and LEDs for digital outputs. They took a long time to connect and hence the order of testing had to be known in advance. Some DCSs allow their I/O cards to be switched to engineer mode, where each input can be forced from an engineer's console and every output viewed on a screen. Many suppliers provide simple feedbacks with a valve output being fed back to a flow input. Others offer a more comprehensive process simulation package, including automatic valve and drive I/O loopbacks. If it works it is great but there is a risk of delaying the FAT while correcting the simulation code.

The pre-FAT has two purposes; it checks that the DCS implementation meets the design, and it gives the tester the opportunity to check the design. There will be instances where they have chosen the wrong interpretation of an ambiguous design. We will consider each of these in turn.

1. DCS supplier's errors

   It is usual to find several blocks not correctly configured at this stage. Often because of copy and paste errors. For example, using a median select instead of an average block. There will also be other cases where the SAMA logic such as a ramp function does not directly translate into an equivalent block. A subtler error is where one block can perform many functions. Even a simple ADD or SUMMER gate is open to misinterpretation unless close attention is given to the detail (Figure 10.1).

   This block can add the two inputs I1 and I2 as we might expect, but if I2 is set as –1 then it is subtracted from I1. Effectively the signals connected to 11 and I2 are multiplied by the preset values. Hence it can also be a simple multiplier or perform quite complicated calculations.
2. Designer faults

   Examples of designer faults include engineering unit or range mismatches. The boiler process engineers design in kg/s, the field transmitters and DCS are set up for T/H.

   The boiler process engineers design in absolute pressure, the field transmitters and DCS are set up for gauge pressure. These conversions should be part of the basic design. Complicated equations such as drum-level density compensation, a steam enthalpy calculation or a flow correction should be checked against a worked example.

   Novel control philosophies should be carefully reviewed to check the concept and implementation.
3. Ambiguous design

   Examples of ambiguous design include errors associated with mass or volumetric correction for flow compensation, or the requirement for the inclusion of a square root on a flow signal. These errors may arise when the DCS configuration engineer is not made aware of the field devices the signal originates from.

Design enthalpy may be in MJ/tonne while their calculation is in kcal/kg. The ambiguity can be minimised by showing engineering units on the logic drawings.

An air flow compensation block could be set up for mass flow if used for combustion or volumetric flow if used for stall correction. The logic drawing must identify the correct requirement.

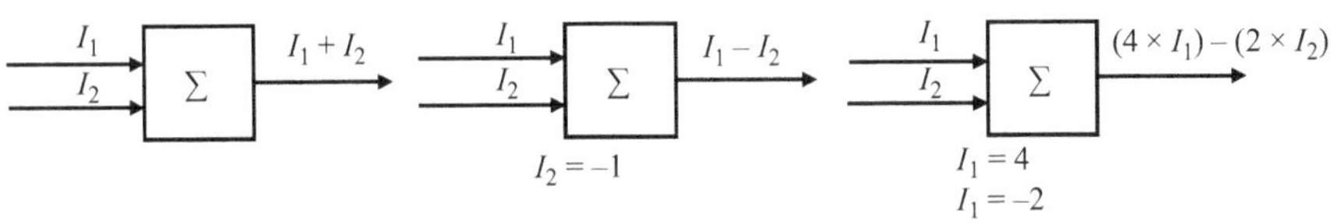

Figure 10.1 Various uses of an ADD gate

Some areas that always need checking are track and reset logic. See Chapter 6 for explanation of these terms.

Practical guidelines

Do the test in two stages. The first is to check as many blocks as possible and note any changes required. Let the supplier work the night shift to fix them and retest the next day.

It is best to have two boiler design engineers supporting the pre-FAT one looking after sequences and BMS and the other the modulating controls. They should be in the same room but each with a dedicated console. Note if a SIL-rated BMS is provided this may be tested separately. Where the DCS and BMS share signals then the signal exchange should also be tested.

The QA guide is 'An hour spent now saves eight hours on site' which might be acceptable if there is the luxury of further simulation pretesting on site. But if the only site testing is during the actual commissioning then it may take much longer to resolve and any delay could lead to liquidated damages. At this stage, the designer and the DCS supplier are on the same team trying to resolve problems before they are seen by the end user.

Sometimes the schedule overruns and it is not possible to do a pre-FAT. One alternative is that while the system is in transit or being installed do a paper review of the code against the design intent. This requires a DCS engineer to demonstrate the software using an emulator. At least 80% of the software may be checked in this way. Then there will be, say, 10% of the software where the actual configuration is not as shown on the original logic drawings but achieves the same functionality. Leaving 10% of the software where the DCS engineer did not understand the requirement and it's completely wrong!

10.12.2.3 FAT tests

FAT logistics

As the name implies FAT is a set of tests conducted before shipping the DCS to site to ensure the equipment and designed application software are acceptable.

Generally FAT tests are cooperative and all parties want fully signed off drawings. If there is any doubt about this, an obligation can be written into the contract. For example, in addition to the demonstrator, ask for a full-time configuration engineer to be available during the FAT or during the night shift to make all necessary changes.

For a power station boiler, there are likely to be several tests happening simultaneously. Additional integration tests are required for BOP systems that connect to the DCS via a communication network: the SIS/BMS/ sequence, the modulating controls and the graphics. The latter should be driven by the end user and reviewed by the boiler supplier to ensure that all the information needed to 'drive the boiler' is present.

The site commissioning engineer may be unfamiliar with the detailed logic, so it is also good if he or she can attend the FAT. The commissioning engineer may be a boiler specialist rather than a control specialist but it is important and useful that he understands the intent of the logic.

Up until the FAT the boiler engineer and turbine engineers may have used slightly different SAMA logic block depictions. However, at the end of the FAT all logic is typically available in DCS logic drawings. These may then become the master drawing set. This has the advantage that all parties start with a completely accurate representation of what exists and they do not have to update and reissue their drawings.

The disadvantage to this is that any mistakes made are forgotten and will be repeated on another job unless time is taken to update standard drawings. Any future modifications are made by marking up the DCS logic drawings. This may require some training so that all designers understand the configuration well enough, to be able to do this! Another disadvantage is that not all DCSs have easy-to-read SAMA-style logic configuration drawings. A set of say 100 SAMA functionals may translate to, say, 2,000 DCS sheets of generic-looking blocks with separate documentation for parameters, calculations and other scripts.

At the FAT stage, the emphasis switches to demonstrating that it works and in some ways a less rigorous but more time-consuming test is applied. Typically, a FAT will take at least 2 weeks but can take much longer depending upon the scope of simulation and testing required and the thoroughness of the PRE-FAT and the criticality of the plant.

10.12.2.4 Site functional tests (FTs)

Typically, the site FT will highlight field hardware and field interface problems. It might be the first time that some communication protocols to other people's PLCs are tested.

If the FAT was for the boiler only then the site FT will check overall boiler turbine integration including interfaces. Each loop is cold commissioned, field transmitters pressurised to produce a changing 4–20 mA signal enabling the field wiring marshalling cubicle connections, I/O and graphic displays to all be tested. And as the DCS is used to stroke actuators the auto/manual logic and field wiring will be tested. Field and interfacing faults will significantly delay the software functional testing.

10.12.3 Graphics

The most important check is to ensure that the database has been properly configured so that the correct field input goes to the correct graphic displays. Perhaps best achieved by reviewing the database itself and then making sufficient tests give confidence in the database. Best practice is for the control room operators to be involved in designing and testing the graphics. Refer Chapter 12 for more details. An experienced operator should also review the screens as often a change in the fuel loop may have an immediate effect on the air loop and so both must be on the same screen. Similarly, a three-element drum-level control system can only be tested and controlled if steam flow, feed flow and drum level are all displayed on the same screen as well as the corresponding auto manual stations for the feed valves and feed pumps.

10.13 Audits

Most EPC contractors will hold a series of engineering audits during the design review. Compliance with IEC 61511 also requires one or more functional safety assessments (FSA). The standard identifies

FSA 1 Following the LOPA or safety requirement specification.
FSA 2 following detailed design of the SIS.
FSA 3 Before the process goes 'live'.
FSA 4/5 later in the life cycle (by the owner/plant operator).

As a minimum FSA 3 is required but this is late in the design process. Potential changes to the design could be avoided by assessments FSA 1 and 2.

10.14 Final checks

An FSA and SIS validation are required by IEC 61511 before the unit is placed into service.

10.15 Non-C&I-related safety reviews

Just to complete the picture there are other safety-related activities taking place, where C&I does not dominate the review. These include:

Compliance with control of major accidents and hazards (COMAH).
COMAH 2015 is enforced in the UK by the COMAH competent authority (CA). This comprises five public bodies including the environment agency and the Health and Safety Executive.
The Construction (Design and Management) Regulations 2015 (CDM 2015)
It describes the laws that apply to the construction process.

There is a raft of regulations that cover working at heights, maximum loads for cranes, scaffolding, etc. These should all be listed in the basis of safety document. Clearly these are outside the scope of the boiler C&I engineer.

10.16 Operations and maintenance personnel training

Training is an obvious requirement to ensure that competent operators control the plant. Typically for a new power station boiler there will be a mix of off-site and on-site courses. The off-site boiler course might take 3 weeks, is often held at the boiler supplier's offices and might include some site visits. Within this period 1 day to 1 week is set aside for the control and instrumentation topics. The operator learns that steam temperature control is based on a two-loop (cascade) system but probably cannot actually follow the logic in detail. His training is supplemented by on-the-job training during commissioning. Site DCS engineers and technicians may have more comprehensive training at the DCS supplier's factory.

While the basic need for training is obvious it becomes important on SIS where knowledge of safety life cycle is important to maintain the validity of the designed SIF. This type of training is in addition to the basic boiler training. Any such training should be formally recorded.

10.17 Ongoing site activities and reviews

The preceding text covers the activities that typical boiler control engineer might be involved in. It assumes that he/she is normally office based, but will go to liaison meetings with the client and DCS supplier. Attend FATs and maybe assist at site for short durations. On the assumption that he/she is then dedicated to the next project, ongoing site works and support are carried out by commissioning engineers and in the case of a SIS a team led by a functional safety engineer will validate the designed SIFs at site.

The plant systems and SIFs need to be reviewed from time to time. Specific recommendations are given after a significant change or after any safety-related modification but with additional reviews, say, every 5 years.

Chapter 11

Improving plant automation and operational flexibility

Don Parker[1]

In this chapter we shall describe modern practices to improve the operational flexibility of conventional power plant, including characteristics of highly automated systems designed to improve start-up times and reduce the operator's role in direct plant control, in both normal and abnormal situations. We shall also examine the design outlines of modern boiler/turbine coordinated controls to provide fast ramping capability and wider, more stable frequency regulation performance.

The processing capability of modern digital control systems is enabling power plant to be controlled and operated in more flexible and economic ways, without adversely impacting on plant life or reliability. Indeed, many examples exist where control system upgrades have significantly improved power plant availability and reduced forced shutdowns, while extending the normal operating range and allowing faster ramping. Newly constructed plant, too, will typically specify higher levels of sequence automation and higher ramp rates than in the past, to support ongoing economic viability.

Of course, it is not just the digital equipment itself or the system software that delivers these economic advantages, but the more advanced applications that can now be deployed – ranging from extensive sequence automation and advanced modulating controls to context-sensitive alarm and protection systems.

In some countries, such as Australia, automation levels have been raised to provide 'single push-button' start-up of coal-fired units and multi-unit operation (i.e. one operator for two or more units) as well as unit operators who can leave the control room, working in an 'unattended' mode that utilises pocket-pager alarming and portable digital monitoring and control devices.

11.1 Features of highly automated power plant

Each aspect of power plant control, protection and monitoring has its own range of features that might be considered highly automated. The following are therefore general indicators of high automation within those ranges.

[1]Provecta Process Automation LLC, USA

In highly automated plant:

- The control systems provide start/stop sequences, modulating controls and plant protection designs with application scope and features that exceed those based on traditional design principles.
- The controls design minimises the requirement for operator actions or interventions under a range of situations including unit start-up and shutdown, normal load ramping and contingency situations.
- Hierarchical system sequences are provided that enable concurrent starting of auxiliary equipment such as boiler fans, feed pumps and turbine systems to improve unit start times.
- The controls provide plant protection layers that ensure equipment is not damaged if the operator is not immediately available (e.g. if the operator is attending to another unit).
- Unit output is maximised in abnormal situations, for example, by increasing instrument redundancy to retain automatic control upon failure of critical instruments.

11.2 Features of operationally flexible plant

What does it mean to be operationally 'flexible'? The description applies to a range of aspects in power plant operation and generally relates to some or all of the following:

- Faster starting – achieved by a combination of high-level sequences and wide-range modulating controls.
- More consistent start times and start-up load profiles – minimising market non-compliance costs.
- Faster ramping – through improved boiler control loop performance and boiler–turbine coordinated control design.
- Wider range of normal operation, with adaptive control schemes to improve wide-range stability, and initiatives such as automated HP heater cutting for overload in Automatic Generation Control (AGC) [1], and automated feed-pump and mill start-up/shutdown scheduling.
- Capability to handle a range of fuel sources, such as with coal/gas co-firing, or a wide range of coal types.
- Providing extended ancillary services to the electricity market (fast-responding AGC and contingency frequency response services).

11.3 Benefits of highly automated plant

Many of the economic benefits of highly automated plant can be inferred from the features listed earlier: faster starting units can take advantage of market conditions to increase revenue; costs associated with start-up fuels such as oil or gas are minimised, particularly during the period before synchronisation of the generator to the grid, or while a steam bypass system is still in service; fuel flexibility can deliver huge savings as costs of each source vary over time.

But other benefits can also be found. These may include reduction in operator stress levels that can affect both employee health and plant reliability. A boiler that requires constant monitoring and frequent intervention is more prone to trip offline or sustain damage. The more that operators can be spared from supervising a mill starting or alarms and disturbances during load ramps, the more they can concentrate on fault diagnosis and unit optimisation.

As well as high-level sequences allowing for parallel operations of plant systems, each lower-level automatic sequence can also improve start times by checking the availability of all equipment (valves, motors, instruments, etc.) required before it starts, and advise the operator. Moreover, sequences move on immediately through each step when plant conditions are met, not waiting for the operator to return to the task and give the next command. Hold events can be flagged with a detailed fault description, providing for a faster correction.

Improved plant reliability is a key economic benefit of highly automated and protected plant. An important element in its success is the introduction of self-diagnosing process transmitters in place of switches for interlocks and trip inputs and duplicating the instruments when the consequence of failure warrants it. Extensive triplication of major process instruments, with median signal selection, ensures erroneous signals are ignored and alarmed as outliers. These more reliable instrument configurations have provided two main benefits:

1. Modulating control loops operate on automatic for a greater portion of the time.
2. More reliable plant protection not only reduces both spurious trips and failure to protect, but has allowed for an increased number of process conditions being protected against, to reduce the possibility of equipment damage and extended outages.

11.4 Design principles for high-level automation

11.4.1 Hierarchical sequence design

Sequences should be constructed in 'layers'. To start an ID fan, for example, the lowest control level may have an automated start of a main/standby lube oil pump pair. The ID fan sequence would place this subsystem to auto, then may start the oil cooler, position the control vanes and dampers, start the main motor when conditions are ready, open the inlet damper and place the vane actuator to auto.

At one level higher, the ID fan sequence itself may be initiated by an air and gas system major (or master) sequence that starts all ID and FD fans in order, first ensuring a clear gas path exists, then starting the first ID fan, setting furnace pressure to auto, starting the first FD fan, raising air flow to minimum and selecting auto air flow control. The second pair of FD and ID fans will then be started at the appropriate time and balanced with the first set. To ensure the sequence will not encounter any hold condition that is observable before the sequence starts, (e.g. instrument signal quality and motor drive health) equipment availability status is added to the start release condition of lower-level sequences and passed up to the major sequences. A single command will then start all FD and ID fans at the appropriate time and place both furnace pressure and air flow to auto.

Highly integrated sequence development for starting major systems or even a complete unit start-up requires a top-down design using this control hierarchy concept to ensure all process logic is properly and consistently managed. Figure 11.1 illustrates a typical sequence hierarchy. However, a bottom-up approach to fault analysis must also be included to ensure the most common fault scenarios have been catered for such that sequence halting, alarming and shutdown all act consistently. Philosophies such as whether a sequence can place a drive or sub-sequence to auto must be decided and made consistent across all sequences.

The hierarchical approach to sequence design provides flexibility in start methods, enabling the operator to choose some or all systems to start by sequence, introducing hold points and even breaking into and then restarting sequences. Logic to enable a sequence to step quickly through a partially completed start and then synchronise to the correct step is not simple to design (and not always achievable) but applying the philosophy where possible helps ensure sequences can be used to maximum advantage. Protection commands at the drive level must also be input to the sequences as override shutdown commands to ensure the sequence status matches the plant condition.

These principles enable sequences to be designed that not only are relatively simple to follow and diagnose operationally but function more reliably, and so will be more frequently used by operators.

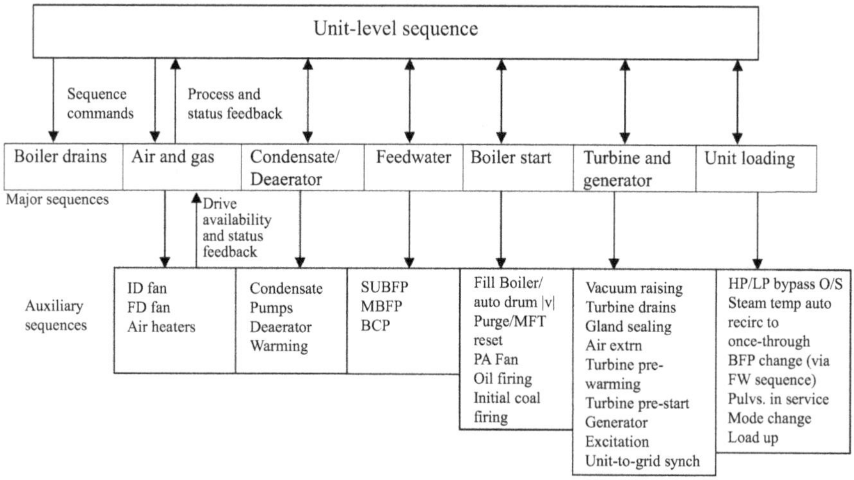

Figure 11.1 Typical hierarchical unit sequence outline

11.4.2 Extensive plant self-protection and situation recovery control actions

Control and instrumentation requirements for the protection of boilers have long been established both in recognised practice and codes such as NFPA 85, and more recently to meet functional safety design principles, as described in Chapter 10.

Plant protection design can also extend to ancillary equipment and systems, providing, for example, fan trips on high-bearing vibration or temperature. The extent of protection should take into account the availability of the operator to quickly attend to such urgent plant conditions and initiate the necessary equipment trip or load reduction. The less available the operator may be, the higher the required level of automatic equipment protection. The design process thus extends beyond considering the probability of equipment failure to include the maximum time an operator may be unavailable to attend to a particular situation. Taking such a risk-based approach in this wider operational context to achieve an optimal protection design is presented in some detail in [2].

Not all protective actions require tripping of auxiliary equipment or the unit. Other actions to recover safely from a contingency may include:

- load holds;
- reducing the rate of change of load;
- additional unit runbacks (e.g. upon loss of a low pressure (LP) feed heater or condenser circulating water pump), rundowns and turbine unloadings;
- control mode changes and
- automatic scheduled starting of additional plant (e.g. feed pump).

During such override control actions, appropriate automated advice to market traders may be necessary to avoid output non-compliance problems.

11.4.3 Improving modulating controls design

11.4.3.1 Applying a 'remain on auto' principle

Improving the performance of modulating controls enables processes to be kept within alarm limits and with minimal operator intervention. In the past, control loops were designed to trip to manual on many process and field device conditions. With reduced operator availability and more frequent changing of load, design principles should be applied to keep loops on auto wherever possible. This may include:

- Increasing instrument redundancy (refer Section 11.3).
- Removing tripping to manual on deviation where dual instruments are installed, but selecting the 'safe side' signal rather than the average of the signals. For example, the *maximum* measured value of two superheater outlet thermocouples would be selected (for creep damage protection), but the *minimum* of two superheater inlet temperatures (for steam saturation protection).
- Eliminating actuator position feedback as a control input to the distributed control system (DCS).
- Duplicating critical DCS input and output (I/O) hardware.

11.4.3.2 Advanced control strategies

'Advanced process control' (APC) applies to a range of modern control strategies that provide superior performance to the conventional stand-alone Proportional-Integral-Derivative (PID) controller. One of the limitations of the PID controller is that it has very weak predictive capability: the integrator accumulates past deviations over time to eliminate offsets and the proportional component (in a velocity-based algorithm) considers the change between the current deviation and previous scan's value. While the derivative term acts in anticipation of the future effect of measured process changes, its 'model' is a simple exponential response. In fact, the derivative component is often disregarded in tuning a PID controller due to a generally poor understanding how to set it and concerns over amplification of process noise at the controller output.

Advanced control designs generally include an explicit or inferred model of the process to better predict and hence compensate for disturbances. Common industrial APC classifications include:

- enhanced PID control (such as adapted dynamic feedforward; two-degree-of-freedom (2-DOF) structure; Smith predictor; parameter scheduling and auto-tuning);
- process optimisation (making optimal setpoint adjustments);
- multi-variable, such as model-predictive control (MPC) and state feedback;
- fuzzy logic and
- soft sensors (inferred measurements) for state-based controllers.

In many respects, a steam generator is an ideal candidate for advanced control applications. The major processes are highly interactive, noisy and non-linear. The responses can vary with load, fuel properties, mill wear and furnace cleanliness. There is a very wide range of response times, from many minutes for steam pressure, temperature and mill grinding processes to just a few seconds for the steam flow reaction to a turbine valve's movement.

Multivariable high-order model-based designs have thus been the subject of testing and development in power plant over many years and have been applied to some of the more challenging control loops. The most successful of these has been applied to steam temperature control using MPC, state-based/observer controllers and the 'two-loop' design which is described in detail in Chapter 7. In the past, a barrier to take-up of the more computationally intensive approaches has been the requirement to perform the calculations in a dedicated PC or on a server, adding to the cost and risking equipment and communications failure. All major DCS suppliers now have APC suites available for deployment in their control processors, simply treated as additional control blocks. Tools for system identification and tuning guidance usually come with the products.

Case studies in successful power plant applications such as steam temperature have been reported by DCS suppliers including Siemens, Emerson and ABB, as well as plant owners and research authorities [3,4]. The list will certainly grow over time. Yet the examples of boiler applications are still rather limited.

The main reason is that the PID controller holds several key advantages, including the following:

- Many control systems provide seamless linking from PID blocks to HMI graphics with automated operator faceplate creation.
- Individual drives can be operated on manual without impeding the control performance of others – not the case with multivariable control.
- Tuning is generally straightforward and training for all levels of design and adjustment is readily available.
- The controller design is robust, capable of performing through a range of process variations and noise levels.

A popular approach in applying APC is to utilise PID controllers to control a secondary loop and drive the field devices, with the advanced algorithms providing or modifying the setpoints of the PID controllers. Some APC designs go further, providing both a 'standard' and 'advanced' mode that assist with initial commissioning and offer a fallback option in case of any problems encountered.

11.4.3.3 Enhanced PID control

PID-based control design has itself undergone many enhancements in power plant control. These include extensive use of a feedforward/feedback structure and adaptive or load-scheduled controller parameters, employing steam properties calculations and response models to improve ramping, disturbance responsiveness and stability. Many controllers now provide inherent signal filtering, saturation protection through integrated tracking and non-linear deviation regions to reduce cycling around the setpoint.

A useful variation of the PID feedforward/feedback design for load ramping applications is the 2-DOF controller, depicted in Figure 11.2. In this arrangement, a feedforward is derived from a process or demand signal that can be modified dynamically to minimise disturbances to the controlled process during a load ramp. The remaining disturbance or trajectory of the controlled process is modelled and applied to the setpoint as a filter to minimise any unnecessary reaction of the feedback controller. The controller can then be tuned just for disturbance rejection,

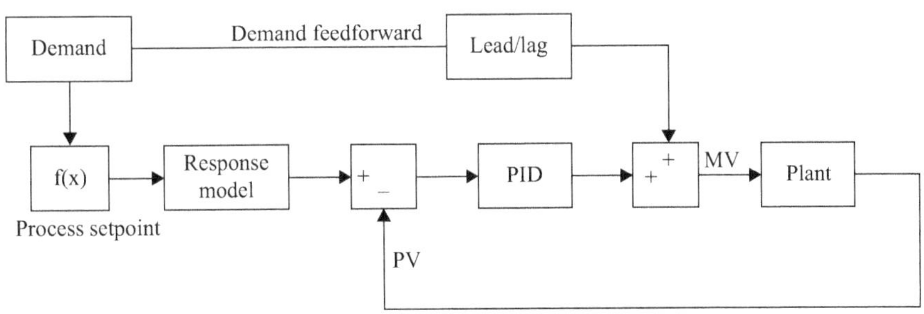

Figure 11.2 Two-degree-of-freedom feedforward/feedback control design

that is, as a regulator, and the response model and feedforward dynamic tuned as a pair for ramping. Two degrees of freedom thus exist to adjust the controls both for disturbances and for ramping.

The design is often applied to several loops in modern power plant controls including, as the controlled process: steam pressure; MW output; flue gas O_2; drum level and primary air bus pressure. The dynamically modified feedforward variables are, respectively, unit load demand, throttle valve position demand; air flow demand; steam flow and maximum mill feeder demand. Setpoint response models range from a simple first-order delay for MW to a third-order delay for steam pressure setpoint and complex dynamics for flue gas O_2 and drum level [5].

By applying model-based feedforward and setpoint modification dynamics, the standard PID design is thus expanded to include forward-looking or predictive elements. The 2-DOF design is an essential component in modern coordinated control designs, as will be seen in the following sections.

11.5 Boiler–turbine coordination: control design for operational flexibility

As outlined in Chapter 4, boiler operation is determined by the demand for steam production. For most large conventional power plants, this steam demand is determined directly by the requirement for electrical power generation. With the increasing volatility of system generation due to the impact of intermittent and often unregulated renewable energy, many power plants are required to follow load dispatch signals that can vary from minute to minute and ramp over a wide operating range.

It is therefore beneficial to examine in some detail how the boiler and turbine controls are designed and adjusted to enable accurate following of these varying load demand signals while achieving stable, efficient operation with minimum stress to boiler and turbine components.

In this section we consider more closely the design of the coordinated unit control modes employed for remote dispatch or AGC operation, their dynamic characteristics and those of the subsystems they control, together with the automatic limiting functions included to ensure safe operation when various contingencies are encountered.

11.5.1 Understanding boiler and turbine steam dynamic responses

Boilers and turbines both have direct influences on steam flow and pressure. The nature of the steam response dynamics, and hence the appropriate control arrangements for stable, coordinated operation, can be best understood by examining the physical principles behind these influences.

11.5.1.1 Stepping the throttle valves

If a boiler's fuel flow is held constant and the turbine's throttle valves are stepped open, the steam pressure will fall as the flow to the turbine increases. For the first few seconds the pressure will fall very quickly as the steam available in the piping from the boiler is consumed. But the lower pressure brings about a supporting effect from the stored energy in the boiler, increasing the steam production as the water and metal temperature in the upper part of the furnace walls is now above the saturation temperature for the new pressure. This phenomenon was also described in more detail in Chapter 6 in the context of the 'shrink' and 'swell' effect on drum level. The increased boiler steam flow slows the reduction in pressure along an approximately exponential trajectory until it settles at a new equilibrium. At this point the steam flow, which initially rose when the throttle valves were opened, will have fallen in proportion to the steam pressure until it is back to the original value. A trend of the pressure response from a throttle valve step-down test on a small drum boiler is shown in Figure 11.3.

This 'boiler energy storage' effect is most pronounced in large drum boilers, where the pressure change is relatively slow, with a time constant of around 120–150 seconds. Even supercritical spiral-wound once-through boilers, which are commonly considered to have very little steam storage capacity, exhibit similar characteristics, with storage time constants of 80–110 seconds.

11.5.1.2 Stepping the fuel

Alternatively, if a boiler's fuel flow (and associated air and feedwater flows) is stepped upwards with the throttle valves held constant, both steam flow and pressure will rise in proportion to each other, but relatively slowly. The reaction time depends upon several factors: the type of fuel being burned and, if coal, the grinding and transport processes employed, as described in Chapter 5. A trend of the pressure response from a fuel step-up test on a drum boiler with vertical spindle coal mills is shown in Figure 11.3.

For gas firing systems, the main delay is in the furnace combustion and heat transfer through the water walls, provided the gas flow control system is tuned for fast response. For pulverised coal boilers, it takes time to grind the coal to the required fineness. Therefore, since mills are generally fixed speed, the coal inventory within each mill will increase as the throughput increases. This has the effect of delaying the pulverised coal being delivered from the mill which, together with a longer transport delay to the burners than gas and a different flame

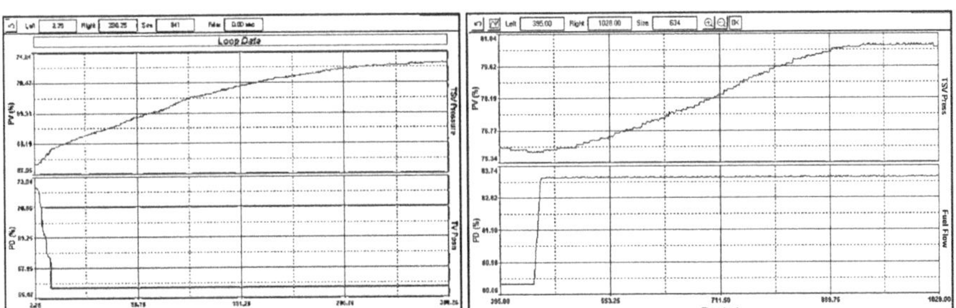

Figure 11.3 Pressure responses of a coal-fired drum boiler to a throttle valve step down (left) and fuel step up (right). Source: Provecta Process Automation. Used with permission

characteristic, results in a significantly longer delay for coal-fired boilers. The response to flow and pressure will have an initial delay where there is no visible response, followed by a rising rate of change, an inflexion point and then slowly settling to the to the new output to match the load. This response is typically of third order (equivalent to three lags in series) and represents a combination of the boiler/fuel inertial response and the boiler steam storage response time.

In order to design a coordinated control system, the boiler steam storage and inertial delay responses must be isolated. This can be done by either

1. conducting the aforementioned fuel step test, performing a third-order system identification on the steam flow and removing the measured boiler storage time component, or
2. conducting a load ramp in turbine following (TF) mode, with constant pressure setpoint and measuring the time delay between the ramping fuel demand and the measured steam flow. This measured delay, typically between 120 and 180 seconds, can be treated as a second-order response with two equal time constants, each of half the measured delay.

11.5.1.3 Response modelling

The measured responses may vary with steam flow and with the loading of each mill. Once determined for a range of operating conditions, the required MW and pressure response can be incorporated into the control system to dynamically shape anticipatory signals to the fuel demand.

Simple modelling of the boiler/turbine process can enable a good approximation of the control setup to be tested before applying on the plant. In some cases, model-based setpoints are developed within the control system to reflect the expected outcome to steam pressure and flow, thus reducing process deviations presented to the controllers during ramps and improving unit stability.

Figure 11.4 shows how to arrange a very simple, linear model in a visual modelling tool such as Simulink® by applying the relationships between throttle valve position, steam flow and pressure and the fact that an imbalance between heat input from fuel and steam flow admitted to the turbine will generate an integrating response.

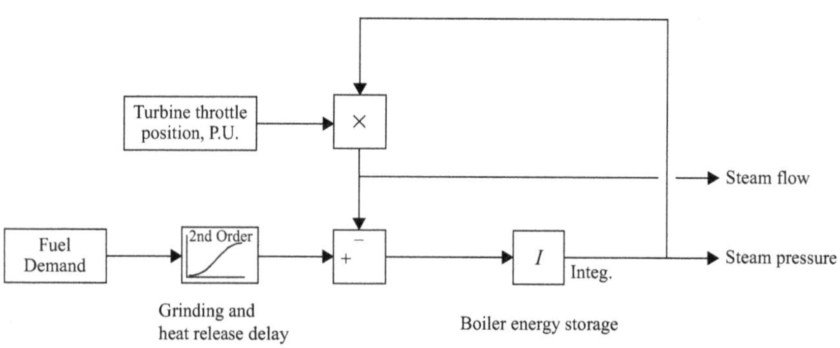

Figure 11.4 Simplified boiler–turbine heat input and steam mode

11.5.2 Coordinated modes for MW control

Boiler following (BF) and TF control modes, as described in Chapter 4, do not provide for accurate electrical load control and are mainly used to control steam pressure under abnormal situations such as a forced turbine load reduction or loss of one FD fan.

Coordinated boiler–turbine control modes are required to meet load-following frequency response requirements for commercial operation. These modes can have a range of structures but largely fall into two groups, labelled as Coordinated TF (C-TF) and Coordinated BF (C-BF). The titles are a little misleading however because, while they correctly refer to which component is 'following' by controlling pressure, the boiler and turbine in both cases are also driven directly with feedforward signals from the unit load demand.

Of these, C-BF is the default mode for accurate MW control in AGC and is often the only option provided; it is then typically referred to as 'Unit Coordinated' or UC mode. When the control room operator is setting the target MW load, the mode is 'Local Coordinated' or 'Local UC'; when the dispatch centre is setting the target, the mode is described as 'Remote UC', or simply AGC mode.

11.5.3 Coordinated Turbine Following

In this mode, the boiler demand to fuel, air and feedwater is modulated to control generated electrical load as shown in Figure 11.5.

Referring to the diagram, the turbine controls steam pressure to a setpoint as a function of unit load demand (1) or the valves may be kept fully open to allow the steam pressure to 'slide' naturally with steam flow. Because fuel input to a coal mill requires grinding, transport and combustion before transferring its heat through the water-walls to generate steam, using fuel to control electrical load is a delayed process. This design however has good inherent stability because the pressure response is 'self-regulating'. That is, if fuel were stepped up on manual with the turbine controlling pressure, the megawatt output would come to a new stable point.

The load-following capability of C-TF can be improved by delaying the pressure setpoint and MW demand signals (2) and carefully calibrating the feedforward signals to the turbine and boiler demands (3, 4) at all loads and valve positions. Adding transient 'overfiring' elements as described later in this chapter improve the unit's responsiveness during ramps. In situations where a longer MW following delay (say 60 seconds) and perhaps $\pm 1.5\%$ allowable output variation can be negotiated, C-TF can be used for ramping in AGC. However, it is more typically used for start-up on coal-fired plant when coal mills are coming into service, as the mode keeps MW within the bid load tolerance with slower but stable automatic fuel adjustments.

11.5.4 Unit Coordinated (also referred to as C-BF) mode

An outline of UC mode is provided in Figure 11.6. Because designs similar to this are the most common used for AGC operation, it is the default mode being described in the rest of this chapter.

UC mode offers accurate, fast megawatt control by modulation of the turbine valves to control MW. However, the pressure response in this mode is 'integrating' since any mismatch between fuel input and the steam flow drawn by the turbine results

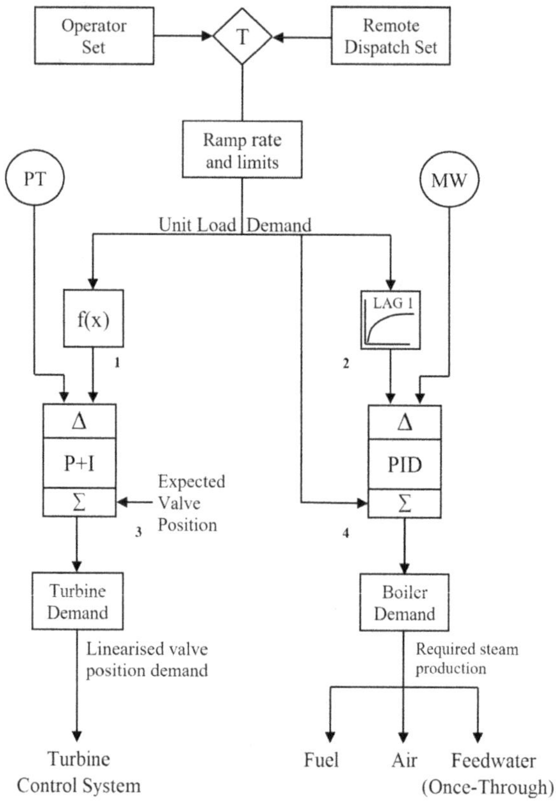

Figure 11.5 Coordinated Turbine Following mode

in pressure rising or falling until the mismatch is removed. Thus, if fuel flow were stepped up while the turbine controlled MW to a set value, the steam flow being produced by the boiler and the steam being drawn to the turbine would not balance, resulting in an ever-increasing steam pressure (or at least until the boiler's safety valves operated!). For this reason, the control logic will generally prevent a unit from being operated with the fuel system on manual while the turbine is controlling MW output.

A process with an integrating response has at least a 90° phase delay in its feedback at all frequencies. This makes it highly prone to instability if the process is also of high order and/or has a dead time (as in the case of steam production in coal-fired boilers). Coordinated boiler-follow mode therefore requires careful tuning to achieve fast, stable load control.

11.5.5 Components of the Unit Coordinated controls

11.5.5.1 Setpoint delays and overfiring

The part of a power plant's control system that develops the demand signals to the turbine and the boiler's fuel, air and feed water systems is often referred to as the 'Unit Master' area. The Unit Master typically includes both the steam pressure

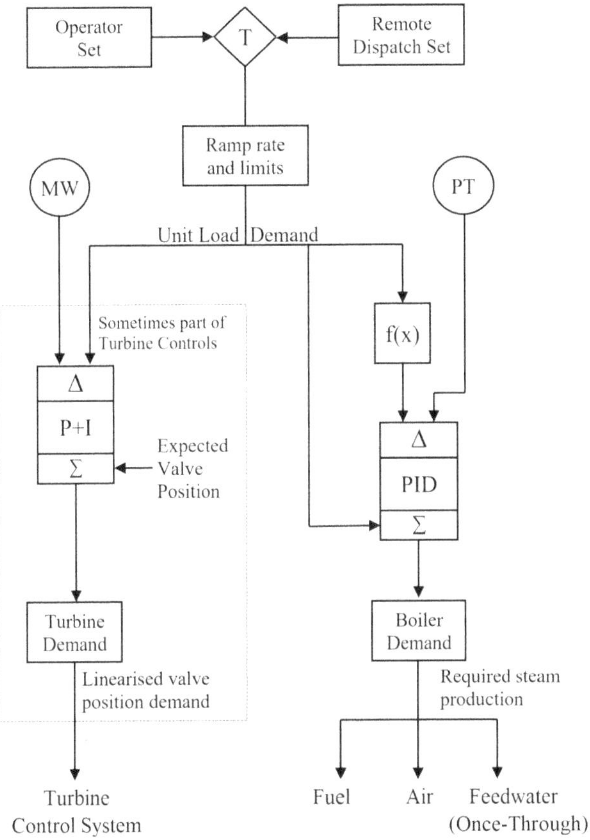

Figure 11.6 Unit Coordinated mode outline

and megawatt controllers, and will perform the interlocking, tracking and switching required to ensure smooth transfer between unit control modes. UC mode is the top level within the Unit Master. Since the mode's design is based around a feedforward/feedback control structure, stable and fast-ramping operation is achieved in a similar manner as for C-TF mode by calibrating the feedforward signals and adding dynamic (transient) components to account for boiler inertial delay and steam storage capacity. The Unit Master also continually determines the limits of the plant's production capability and generates appropriate actions (such as block increase, and runback) to keep the unit secure.

The main dynamic component in the Unit Master is a complex transient signal added to the boiler demand feedforward. This 'overfiring' signal is derived from two sources:

1. Rate of change of load demand – to overcome the boiler delay.
2. Rate of change of pressure setpoint – to compensate for the variation in stored energy in the boiler water and steam with varying pressure.

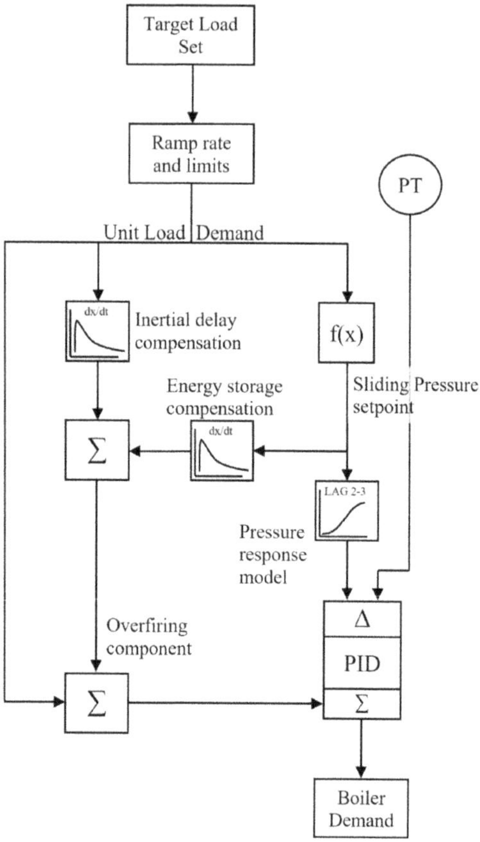

Figure 11.7 Overfiring components for improved ramping capability

Take, for example, a boiler with a known fuel-to-steam delay of 150 seconds. If the unit MW output is allowed to be delayed by, for example, 40 seconds and the known delay from a change in steam flow to MW response is, say, 8 seconds (allowing for the delayed effect of reheater volume and IP/LP turbine expansion on MW generation), then the fuel demand must be brought forward by a total of 118 seconds. This is achieved by adding a filtered derivative (or 'kicker') function as shown in Figure 11.7 to the boiler demand feedforward.

Similarly, the overfiring signal for the steam pressure is also a filtered derivative function based on the rate of change of the base pressure setpoint, as also depicted in Figure 11.7. The sliding pressure setpoint itself is often delayed by a second- or third-order lag to properly reflect the time required for the pressure-based overfiring signal to affect steam production and to reduce overfiring. Any adjustment to these lags, for example, to increase the minimum MW reserve margin for load ramping, will require an equivalent adjustment to the pressure-based overfiring signal.

Figure 11.8 presents trends from a simulation of a load ramp in sliding pressure. The overfiring components are traced at the bottom of the figure. Figure 11.9

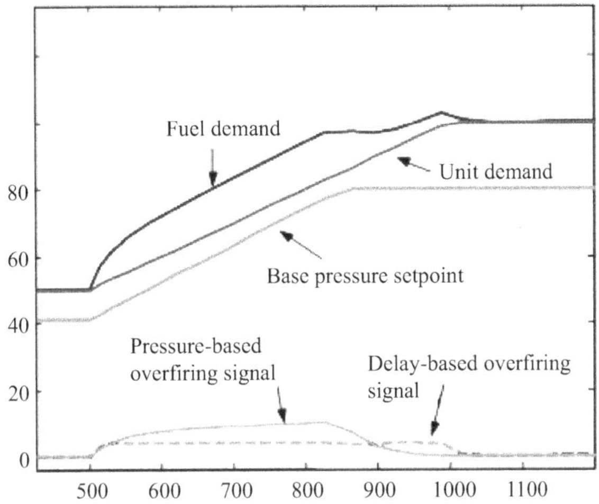

Figure 11.8 Simulated load ramp showing boiler demand overfiring components

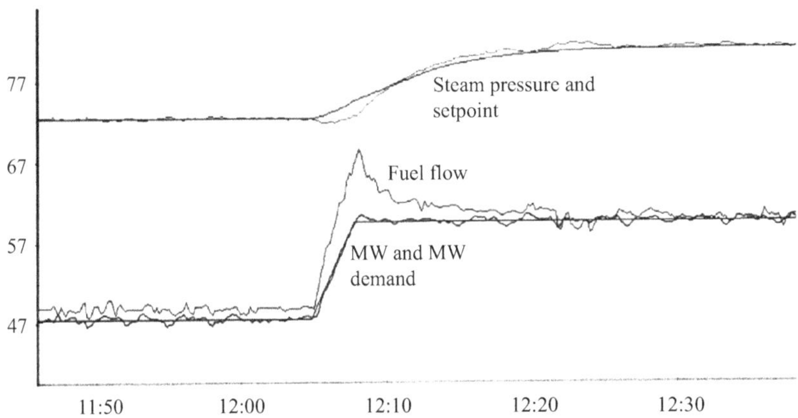

Figure 11.9 Field data trend from a gas-fired plant ramping in sliding pressure at 3%/min. Source: Provecta Process Automation. Used with permission

shows actual plant trends from a gas-fired unit during a fast ramp in sliding pressure. Note that there is an initial fall in steam pressure since no MW demand delay is allowed on this unit. The compromises associated with setting the quantity of overfiring for fast ramping while retaining turbine reserve margin for frequency response and minimising steam temperature and pressure deviations are often the limiting factors determining the load ramp rates and controllable operating range of a unit. Some control systems vendors provide more complex 'model-based' setpoint development and overfiring designs that can shape more accurately the steam pressure and MW responses. Such designs can improve ramping capability, unit stability and the working range for AGC operation, while maintaining both boiler and turbine demand reserve margins for frequency response.

The use of compensating dynamic feedforwards in combination with approximate response delay models applied to the MW and pressure setpoints form the '2-DOF' control configuration described earlier that enables ramping and disturbance responses to be independently tuned.

11.5.5.2 Throttle valve demand feedforward

The turbine's MW control loop is very fast. If the throttle valves are stepped, typically about one-third of the MW response is delivered in the first few seconds as the steam passes through the HP turbine, with the remainder achieved over about 20 seconds as it passes through the reheater and expands in the IP and LP sections of the turbine.

In many cases a simple P+I controller, tuned with a fast response, is sufficient to control unit MW. However, such a design requires a deviation to exist before the controller can respond, and MW control accuracy is becoming a more important in AGC operation as the grid's unregulated generation increases. To improve MW-following performance during load ramps and frequency events, a feedforward can be added to the turbine demand that calculates the expected position for given MW setpoint and pressure conditions.

The linearised relationship between turbine valve position, steam flow and pressure can be seen from Figure 11.4 as

$$TV\ position \propto steam\ flow/steam\ pressure$$

Since steam flow closely approximates a linear function of MW output (for constant steam temperature), the expected throttle valve position to achieve the required MW can be developed as

$$TV\ position\ demand = K * MW\ demand/steam\ pressure\ setpoint$$

where K is a scaling adjustment factor.

A dynamic adjustment function such as a lead-lag network can be added to drive the turbine valve in the desired trajectory without requiring a megawatt deviation to develop before responding.

11.5.6 *Frequency response*

The turbine governor control system has a 'droop' characteristic, which causes the throttle valves to open or close in proportion to the deviation of the turbine speed from the base synchronous speed. This means that if the grid frequency falls the throttle valves will open and generate a higher megawatt output. However, if there is no change in the boiler firing rate, steam pressure will fall and eventually the generated power will return to its original value.

To provide a sustained output change during a frequency deviation, the Unit Master control uses this deviation to influence both the turbine and boiler demand signals.

For the turbine, the required proportionate MW response is applied to the MW setpoint. Initially the local droop reaction of the throttle valves will generate a sharp MW response and the controller will see very little deviation. But as the steam pressure is affected by the valve movement, the MW will begin to recover and so the controller will respond to move the valves further, leading potentially to a larger pressure deviation.

For the boiler, a combination of the increased unit load demand and falling steam pressure produces a strong fuel reaction to assist in restoring steam pressure. To prevent pressure overshoot after large frequency events, some designs include a dynamic adjustment to the pressure setpoint to reduce the deviation applied to the pressure controller and so avoid an excessive correction to the fuel system.

One issue with frequency response for sliding pressure boilers is that the MW output change, as a proportion of turbine valve movement, reduces with steam pressure and hence with unit load. Since many connection agreements require a fixed MW droop response regardless of load, the MW controller must adjust the turbine demand further at low load. Care must therefore be taken in tuning of the MW controller, usually with an adaptive gain function, to ensure response compliance over the full load range.

11.5.7 Capability limits and runbacks

Boilers must always be operated within safe working limits – if mills are run above their design throughput they can 'choke' and trip from low air flow. Fans can reach a stall condition and feed pumps can be overloaded. The capacity of each item of major auxiliary plant is set as a proportion of the total unit output. For example, an FD fan may be capable of 60% MCR rating; a mill may be capable of 20%. The lowest capability of all the sets of auxiliary drives establishes the maximum output of the unit. Similarly, a minimum capacity is established.

If the unit is ramping and a capability limit is reached, the capability value replaces the target load to prevent further change in that direction. Other process inputs such as fuel or steam pressure deviation can also generate commands, called blocking signals, to prevent load demand from increasing or decreasing by setting the ramp rate to zero while the condition exists.

If a major drive trips, the unit must be quickly unloaded to avoid a complete shutdown. This event produces several actions in very fast succession. While the details will be different across power plants, similar principles apply:

- A 'Runback' or 'Unloading' condition is detected.
- The capability value drops to the new limit.
- The target load falls to match the new capability.
- The load ramp rate changes, in some cases up to 200%/minute.
- Unit control mode changes, typically to TF since reducing air, fuel and feed-water quickly takes precedence over MW control. If the runback is conducted in UC mode, the boiler overfiring components must be removed.

An outline of the capability and runback control design is shown in Figure 11.10.

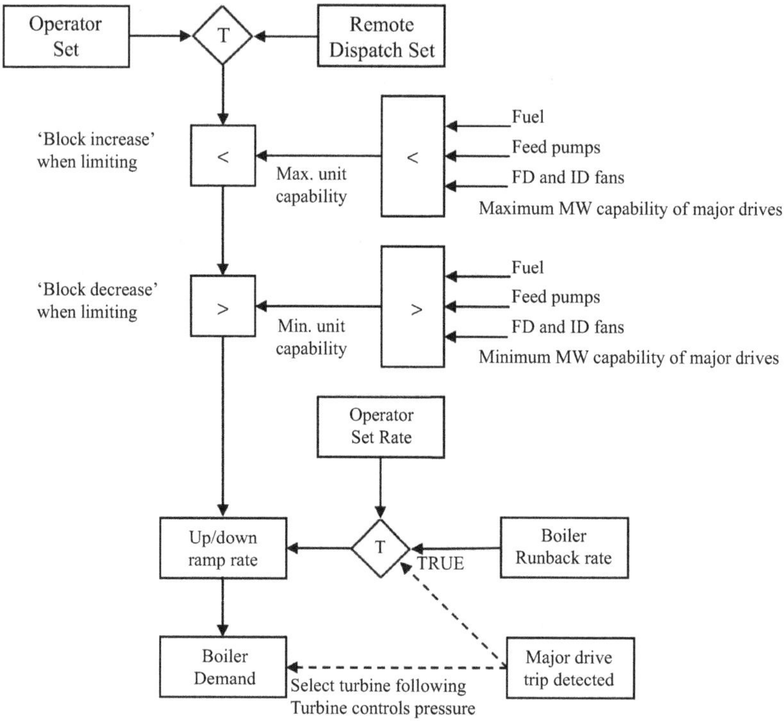

Figure 11.10 Typical unit capability and runback control design

11.6 Summary

As we have seen, the various impacts of the boiler subsystems and the turbine on steam flow and pressure must be taken into account in the Unit Master design to ensure predictable, stable operation is achieved, particularly for units that must follow a frequently changing load demand.

References

[1] Siemens A G, 'SPPA-P3000 Solutions for Process optimization'. 2006 (p. 4). www.energy.siemens.com. [pdf] Accessed March 2018.

[2] Ronan N, and Parker D. 'Assessing operational risks associated with remote and multi-unit supervision initiatives: an Australian case study'. *Proceedings, 60th Annual ISA POWID Symposium, 2017*. International Society of Automation, USA, 2018.

[3] Moelbak T, and Mortensen J. 'Steam temperature control', in Flynn D. (ed.) *Thermal Power Plant Simulation and Control* Ch. 5. Institution of Electrical Engineers, London, 2003. ISBN 0-85296-419-6.

[4] Taft C, Sorge J, Lupu M, and Kephart R. 'Application of model predictive control (MPC) to improve steam temperature control on a pulverized-coal unit'. *Proceedings, 59th Annual ISA POWID Symposium* 2016. *International Society of Automation, USA, 2018. ISA* Volume 514 ISBN: 978-1-941546-97-0.

[5] Parker D. 'Two-degree-of-freedom controller applications in fossil-fueled power plant'. *Proceedings, 59th Annual ISA POWID Symposium, 2016.* International Society of Automation, USA, 2018. ISA Volume 514 ISBN: 978-1-941546-97-0.

Chapter 12
Upgrading and refurbishing systems

David Lindsley[1] *and Don Parker*[2]

Even the newest and most advanced control systems eventually become dated. Up to about the middle of the twentieth century, control and instrumentation (C&I) systems were mainly pneumatic, the pace of development was slow, and it was generally possible to keep a system in good functional order, however old it may have been, simply by repairing the equipment or, if the worst came to the worst, buying replacement components from the original manufacturer. There was a quite well-founded expectation that the manufacturer would remain in business and parts would continue to be available for the life of the power station. In addition, plants that were originally designed for an operating life of around 20–25 years were not expected to require major changes to their control systems.

Long-term support continued to be expected and offered by vendors as (analogue) electronic controls were introduced. These systems were generally designed around high-reliability, replaceable, slot-in cards, so spares of each card type could be kept as a good risk management strategy. There were some operational enhancements over time: addition of upgraded historical data logging facilities; replacement of paper-based chart recorders; corporate PCs for operators, communication system upgrades, etc. But once again, there did not seem to be a reason to expect a large-scale control system replacement during the station's operating life.

The advent of computer-based control systems cut much of this extended support short as vendors jumped across to the nascent digital technologies. This brought about a sea-change in the maintenance of control equipment and systems. The frequency of updates to vendor control systems or the associated third-party operating systems increased, and the support periods for legacy systems plummeted.

12.1 Drivers for change

The most common and essential reason to change a control system stems directly from the reduced support periods for the older systems. Instances have occurred where stations have attempted to continue operation well beyond the manufacturer's

[1]Retired
[2]Provecta Process Automation LLC, USA

support cut-off by scouring the world to source additional spares, purchasing used electronic modules or distributed control system (DCS) processors (often with unknown operating and storage histories) or arranging for costly customised equivalent circuitry to be manufactured. The approach can lead to increased equipment failure with sometimes severe consequences such as unit trips or even plant damage. A full replacement is eventually seen as the only viable solution.

For electronic-based systems, two other factors emerged that increased the pressure to upgrade. The first of these was a consequence of life extension projects. In many parts of the world it was found to be very economical to undertake mechanical refurbishments to power stations – often with the aim of doubling the original service life. While many of the electronic systems had survived their design life well, the possibility of a 40-plus-year stretch for the controls was unthinkable. Life extension projects also introduced plant changes and improvements that the existing control system could not easily accommodate, either due to capacity or programmability constraints.

The second factor was a growing need for the electronic control systems to be modified to improve operational performance. This was not simply to improve controller performance – the original systems were typically carefully designed and tuned, and provided good control for the original operating requirements. Any performance degradation generally related to field equipment issues rather than the control hardware. But the introduction of more volatile automatic generation control (AGC) demand profiles, frequency response ancillary services, operation at lower load, fuel source changes and demand for more automated starts were among the drivers to move across to programmable systems. The capability of digital systems themselves for monitoring and diagnostics also proved attractive: remote monitoring for engineering and corporate functions, improved plant status logging and integration with plant safety isolations and maintenance systems all gave weight to an argument for change.

Over the period 1995–2015, most electronic control systems had been replaced with DCS equipment. But that was not the end of the story – now the systems are all fully digital, the components are on an upgrade cycle comparable to the equipment inside your home PC, tablet and smartphone!

12.2 The DCS upgrade cycle

Introduction of the first DCS equipment led to a fall in support periods from 20 or more years to around 10–15 years. This has edged progressively closer to an 8–12-year support window for the control processors and input and output (I/O) equipment and as low as 5 years for industry standard components such as data servers, workstations and networking hardware.

It is understandable that the industry-standard parts of a DCS have relatively short lives: they are essentially high-reliability versions of domestic and office computer and networking equipment, and so still need to move in step with software, storage, processor and communications upgrades of their hardware and software technology suppliers. These systems can, in just a few years, become incompatible with the end user's corporate policies on spares availability, remote access, network security and connectivity.

The basic reason for shorter support periods of the specialised digital control equipment is simply that the DCS manufacturer can no longer source all the microelectronic components – including microprocessors, real-time processing units, memory, and connectivity controllers such as UARTS – over very long time frames. With new components comes the need to manufacture new circuit boards, and program new system software. There is a need to keep abreast of what is soon to be released by watching (or, if possible, influencing) design standards and models and to utilise computer-aided design and simulation platforms to bring the new products as quickly as possible to market. Even then, new components may be well into their product cycle before any new system is actually installed and controlling a power plant process.

For the C&I manufacturer, this constant evolution of systems and technologies is both an engineering challenge and a market opportunity: the ability to provide more compact hardware, with a more open system than other vendors, (e.g. by widening support of fieldbus communication and other interface protocols), introducing support for thin client browsers or now requiring fewer control processing units (known variously as drops, stations, field controllers or automation controllers) makes for a growing market share. Thus, an expensive game of 'technical leap frog' ensues.

Control system vendors are naturally keen not to lose customers through these upgrades. An attractive option that is now being widely adopted is that of 'migration', where none of the field cabling, I/O modules or associated termination and controller cabinets are replaced. New controllers are installed, along with the system servers, historian and the human–machine interface (HMI) – sometimes referred to as a 'front-end' replacement (as opposed to the term 'top end' which we later use to refer to a replacement where the process controllers are not included). Often the new servers can be installed and some of the new screens placed into operation before the full controller cut-over, giving operators a chance to become accustomed to any changes. Only a short outage may be required to replace and commission the controllers, which are capable of either running the old control logic project files or have a translation process to make them compatible. Either way, the control logic does not need to be redrawn. Similarly, historical data records and I/O and alarm databases are all kept compatible with the upgraded system.

As one can imagine, the original manufacturer would be the first company expected to be able to provide a migration pathway; but today it would not be surprising to find several DCS suppliers offering comprehensive migration solutions, including application software translation, to legacy systems of other manufacturers.

While migration will have the least impact to operators and maintenance staff, it is important to test certain aspects of any new equipment, whether that be supplied by the OEM or not: Do the new processors start with the same logic memory states? Do customised application modules transfer across with all the same internal settings? Has the order of processing changed for a configured sheet, which might affect logic operations?

So, for DCS systems, while the interval between upgrades has tended to decrease due to the technology life cycle, the relative cost of the upgrade has also decreased, because the technology permits backward compatibility and competitive pressure has required its development.

12.3 The impact of change

Replacing a control system will have an impact across a wide cross section of a power station's people and many of the organisations' work systems. Some areas of greatest impact are described next.

12.3.1 Operators

System refurbishment can produce major changes to the operator interfaces. Such changes are greatest when a system based on the use of a hardwired operating panel (or 'hard desk') is replaced with a screen-based operator console (or 'soft desk'). The screen-based console is also referred to as the 'Human-Machine Interface' (HMI).

12.3.1.1 Loss of operational redundancy; increase in instrument and controller redundancy

In the hard desk environment, every actuator is controlled via its own discrete hand/auto station (alternatively called an auto/manual station). These stations enabled the operator to manually position an actuator directly via a discrete control circuit which was separate from the electronics of the automatic controller. Such a system provided a high degree of operational redundancy. Hard desks also provided discrete indicators showing the important plant parameters. Sometimes these indicators were driven by transmitters which were completely separate from those feeding the electronic controller, again providing redundancy in case of the control system's failure.

With a screen-based console, only the most essential of command signals for plant safety, for example, boiler and turbine trips, will have 'hard-wired' connections that bypass the DCS. It was therefore not until highly reliable digital controllers coupled with sophisticated fault-detection systems that the more conservative engineers began to accept that it was safe to entrust the control of large-scale hazardous processes to a soft desk concept. If the DCS suffers a major failure, the operator will generally no longer be able to keep the plant running using independent indicators and manual actuator controls. The situation however is unlikely, as control equipment and instrumentation reliability has progressively improved and the risk of failure has continued to fall. Furthermore, the system, particularly the processors handling unit protection (the 'safety systems'), can be designed to generate fail-safe responses under loss of processing capability or certain essential instrument signals, and trip the plant under predetermined circumstances.

With the introduction of a DCS comes the opportunity to improve both redundancy (i.e. fault tolerance) and fault detection of field instruments since they are all now connected to the same system. In cases where, say, two transmitters were previously used as a redundant pair for control, and a third for indication (sometimes also looped to the unit computer), the three could now be combined to improve fault detection by using median selection with deviation and quality monitoring for both control and historian data collection. A pressure switch that may have driven an alarm facia or was part of a start interlock could now be replaced with a transmitter that can also provide an analogue process value and be health-monitored by the DCS's diagnostic functions.

12.3.1.2 A new view of the process

In moving from hard desk to screen-based console, operators are confronted with a completely new way of viewing the process: in the former, most control stations, trends, indications and alarms were accessible to the operator in 'parallel'. It was not uncommon for an operator to be driving, say, two actuators from different systems at the same time, while watching trends, viewing a set of indications with a sweep of the eye and noticing two new alarm facias that have just lit.

Now, the interface is largely 'serial': what is visible depends upon the screens that are selected. Often a screen hierarchy is provided, with the upper levels providing an overview with key indications, trends of major variables and control faceplate access points. The coordinated control overview is usually set out in this way. On a lower-level process-flow graphic, system-specific trends may also be embedded or referenced with access points to open them in separate windows. Control faceplates are typically hidden until the valve, motor or pump, etc., icon is selected. Process values shown on the graphics will be highlighted in the relevant priority colour if they reach an alarm level. A greater reliance will be placed upon alarming and automated responses to contingencies in the screen-based environment.

Some projects have sought to resolve this change in view by providing a large number of operator screens – as many as 20 per unit – to try to replicate the panel overview effect. A more successful approach has been to ensure the graphics are well designed and accessible, that operators make extensive use of multi-trace trends, and the alarm system is designed to focus operator attention in accordance with process requirements and has a sufficiently low alarm frequency such as to be always manageable. Increasing the level of sequence automation and ensuring all modulating control loops are generally operating on auto without operator intervention are also important success factors. In those cases, 6–8 screens per unit (including two large displays, but excluding corporate computers and screens) are often found to be ideal.

12.3.1.3 DCS to DCS upgrades: display graphics and other changes

Most upgrades in the future will be from one screen-based system to another. Even these can have a large impact on operators, particularly if the graphics have been revised, for example to meet new 'high performance' screen layout standards such as ANSI/ISA-101.01-2015 [1] that covers the philosophy, design, implementation, operation and maintenance of HMIs for process automation systems. The new concepts steer away from P&ID-based graphics towards low-colour/contrast indications of process variables as bar graphs or other visualisations and embedded trends, particularly for the overview-level monitoring screens. Values represented on a scale can set them in the context of the normal range and alarm limits; trends are essential to show recent history, instability and trajectories towards alarm conditions. Bright colours are used mainly to draw attention to alarm events and processes in alarm. Since this is a general process industry standard, application to graphics for power plant, given the highly interactive nature of its processes, should be reviewed by those experienced in the field. As power plant is operated (and thought of by the operator) in systems, there remains a place for system-based graphics – at a lower tier in the new graphical hierarchy – that (now more clearly) represent the flow of material and the measurements, controllable valves, etc., shown in their proper locations in those processes. These screens provide unambiguous access to process values and control stations for more complex but lower-level operating activities such as preparation for a feed pump changeover. An example of the new design graphics is provided in Figure 12.1.

Arrangements should be made for adequate training in all aspects of the new system, including any changes to security access levels, graphics symbols, control station functions and display of underlying control logic for tracing the status of signals and interlocks and interaction with the plant safety isolation systems.

12.3.2 Maintenance

As with any system change, maintenance functions will also change, and re-training of maintenance groups, as well as operators, will be necessary. The most dramatic control system impact occurs the first time a screen-based system is installed. The alarms being presented to the operator can increase 10-fold as each individual field device and control system component can generate many more individual alarms

Figure 12.1 An application of new graphics standards. © Emerson Process Management. Used with permission

than before. In some cases, the work of logging equipment faults and raising defect notices can be taken from the operator directly to the maintenance groups, as remote access to all alarms becomes available.

Maintenance of the equipment itself also changes significantly. Many components will be self-diagnosing, and generally too complex for on-site repair. Redundant processor failures will involve in-service replacement and returning to the supplier for memory dump interrogation and review.

Where the upgrade is DCS-to-DCS, it is valuable to consider how any new features may improve plant fault diagnosis and maintenance coordination efficiency. An integrated alarm database for instance, complete with setpoints, test procedures and test records can streamline alarm investigations and routine alarm maintenance.

12.3.3 Cyber security

As digital control systems make greater connections to corporate IT systems and via the Internet for remote monitoring and machine data analytics, they become more potentially vulnerable to cyber attack. Being part of a nation's critical infrastructure, large power plant control systems often come under government regulation requiring strict compliance to certain hardware and software standards, along with management and access procedures. This will include regulation of procedures for virus checking, patch management, version updates and system upgrades.

Care must be taken in such matters as how USB memory 'thumb drives' are used, and the set-up of smart field devices and portable control tablets with IP addresses and wireless access [part of the growing Industrial Internet of Things (IIOT)] which might provide 'back doors' to the system.

12.3.4 Corporate expectations

Once a modern DCS is installed with a fast historian facility and remote monitoring capability, there will be great interest in providing data to various corporate bodies. There may be requirements to provide alarm summaries and frequency analysis, plant efficiency figures and cost of losses, real-time operating data and cloud connections for data sharing with third-party monitoring and diagnostic (M&D) service providers.

To meet these expectations employees with a new set of skills in industrial IT may need to be engaged, trained and dedicated full time to the task. In some cases, the task is handed to the corporate IT support team. This has varying success, depending on the size of the organisation and whether the team has a strong working knowledge of the control system's architecture, set-up and functions. A control system may look like another IT facility from the outside [essentially a set of servers with multiple firewalls and possibly a demilitarised zone (DMZ)], but it is fundamentally different because of its focus on reliable real-time execution with a secondary focus on data storage and data security. Behind the servers and firewalls, there are specialised network connections to devices performing mission-critical functions which must not be compromised.

12.4 Refurbishment case studies

Refurbishment has taken several forms over the years. Some of the most common phases include

- Upgrade of field devices: replacement of switches with transmitters; introduction of 'smart' transmitters that can be interrogated with digital communicators; replacement of pneumatic positioners with electro-pneumatic or digital versions.
- Changes to control room components: replacement of unit computer CRT displays with flat-screen monitors; replacement of paper trend recorders with slot-in screen-based replacements; addition of historical data storage and retrieval systems; replacement of hard-wired hand/auto stations with digital equivalent faceplates.
- Hybrid upgrade: upgrade of some systems to digital technology, for example, introduction of a digital burner management system (BMS) or turbine control system.
- Upgrade only to the HMI and associated servers.
- Integration of independent systems: extension of DCS controllers and replacement of independent systems such as feed pump or turbine control units.
- Full DCS to DCS replacement.
- DCS version upgrades (migration).

In the following sections we examine several examples.

12.4.1 Multi-phase to DCS

The first case is a four-unit drum-boiler station, fired with either fuel oil or gas, constructed in the mid- to late 1970s that progressively had some control system and control room parts replaced and facilities added, before finally completing a full DCS replacement in 2010–2012.

In the two photographs of Figure 12.2, the original hard desk is shown (a), and also a photo taken just prior to commencement of the full DCS replacement project. In Figure 12.2(b), the unit computer interface had been replaced with a proprietary system, installed as three-unit computer screens in the vertical section of the panel. Paper trend recorders had been replaced by two dedicated trend screens also seen in the vertical panel. Between those two screens are three single-loop digital controllers that had been added.

Also added was a new data historian facility for both operator use and remote data logging, trending and analysis. With this change a desk was added, seen to the very right of the lower picture, for the new facilities with double-stacked screens, as well as a corporate PC and market trading screens. The station has always operated with one operator controlling two units.

In the more recent total control system replacement project, a Yokogawa DCS replaced all separate systems including the BMS, turbine protection system, modulating and logic controls, alarm system, unit computer and historian.

The new control room, shown in Figure 12.3, also brought in four units from an adjacent station on the same site, resulting in an eight-unit control room. Single push-button unit start-up sequence automation was added, as well as enhanced

(a)

(b)

Figure 12.2 *(a) Original panel for one of the four units and (b) panel for two units following partial upgrade and addition of monitoring screens.* *© AGL Energy Ltd. Used with permission*

Figure 12.3 *New control room for the four-unit gas-fired plant. Control consoles for four other units of an adjacent station were also located in the same room. Note the 'Yarway' drum level indicators (circled).* *© AGL Energy Ltd. Used with permission*

Figure 12.4 Original DCS HMI. © Millmerran Operating Company. Used with permission

control strategies to improve ramping, frequency response and AGC following capability. Nominally three screens are assigned per unit plus one for common plant, but all are freely assignable to any unit.

Improved steam temperature control was also required to assist in extending the operating life of boiler components. To provide a wider range of AGC operation with minimal operator assistance, automatic start/stopping of a boiler feed pump and scheduling of gas burners was introduced.

12.4.2 DCS HMI upgrade

The second case study is of a two-unit supercritical station commissioned in 2003 with an ABB-Bailey DCS and fully screen-based HMI, shown in Figure 12.4. Soon after commissioning a third-party historical data logging facility was added. The DCS however was past half-way through its life cycle by the time both units were in commercial operation.

This was not an issue at the time for the controllers, but the UNIX-based servers for the HMI, alarm system and other services ran out of support just a few years later and essential components such as hard drives and workstation controller boards and had no directly compatible replacements. It was decided to undertake a 'top end' replacement, but to retain the controllers which had a compatible future upgrade pathway available when required.

The task involved installing a new operator console in the same control room and translating all existing graphics, control faceplates and alarms to the new system's format but with minimal change from the operator's viewpoint. A compatible ABB system was selected as the migration platform. The new HMI (Figure 12.5) was factory-tested and then set up in its new location to operate in parallel with the old screens, for familiarisation prior to full cut-over of operation.

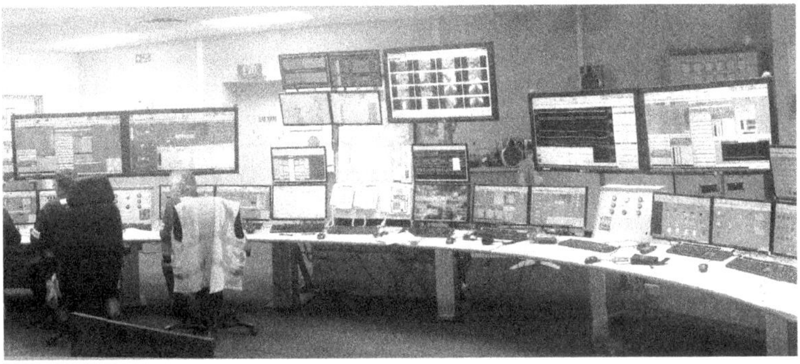

Figure 12.5 Replacement HMI for both units and common (station) plant, located in a different area of the same control room. © Millmerran Operating Company. Used with permission

Opportunity was also taken to update other aspects of the room including lighting and sound attenuation, a central supervision and assistant's desk, training area and file storage facilities. One issue with the graphics translation related to a change in the aspect ratio: if a 4:3 screen graphic is now to be viewed on a 16:9 or 16:10 screen, how should it be displayed? Some vendors have simply stretched the graphics, distorting the icons and text, and blurring the image due to aliasing. In this case use was made of the vacant band on one side to contain a control faceplate or to reserve as a location for control faceplates to 'pop up'; some graphics were also redrawn.

12.4.3 Complete DCS replacement

A four-unit station that had become fully operational in 1996 was built with a Siemens Teleperm ME digital control system that was designed with high levels of automation and reduced attendance operation: operators could leave the control room with pocket-pager alarm facility for recall. High levels of instrument redundancy were installed, and extensive plant tripping functions provided both high plant reliability and safety. The original operator interface had a combination of screen-based functions and a hard-wired panel with a small set of control stations considered necessary for ease of operation.

In the early 2000s the operator interface, historian and alarming systems were replaced with the UNIX-based Siemens T2000 DCS, also known as Teleperm XP. The original Teleperm ME controllers, which enjoyed extended-life OEM support, were retained. Control of all four units was brought into a single room, where a screen-based system, including large screen displays, was installed. All controllers and field I/O were retained, as were a subset of the hard-wired control faceplates considered essential for certain operations requiring multiple control station activities in quick succession.

As it had been decided to extend the operating life to around 40 years, in 2016 the first unit's complete control system was replaced with a Siemens T3000 DCS.

Several independent systems, including the turbine controls, were also replaced and integrated into the DCS. Many of the original control principles remained, with some enhancements to tripping functions, alarming, sequence design and updating modulating control concepts to improve unit flexibility. The new system also continues to support the capability of operators to leave the control room with portable monitoring and alarm equipment.

In this case the original DCS configuration could not be converted automatically as a complete generation of controller technology had been ‘stepped over’.

12.4.4 ‘Transparent’ control system migration

In this last case, a two-unit supercritical plant, which was commissioned in 2011 with an Emerson control system, underwent a ‘front-end’ migration by the OEM as part of a corporate technology policy implementation. The project involved replacement of all control processors, system servers and networking equipment, and translation of system databases, control application software and operator graphics to the new system. Some enhanced features, such as a new trending application, came with the upgraded software. The I/O and field devices were retained.

The system was factory tested to check the correct operation of the logic and operator graphics in the new processors, and the change-over was aligned with planned maintenance outages to avoid any unit downtime requirements.

The change provided a significant extension to the future equipment support cut-off date, enhanced engineering, control and interfacing functions, met changing cyber security requirements and yet was largely transparent to the operator.

12.5 Preparing for change

The steps in preparation for a DCS change, from the first discussions to float the idea, through to having a complete technical specification published, are complex and many. This section provides a taste of what is involved and highlights some important aspects along the way.

12.5.1 Making the business case

A control system replacement is a costly exercise requiring plant downtime and carrying risk to plant reliability (at least initially) afterwards. Replacement projects tend not to be embraced by management with gusto, who may instead prefer to choose lower-cost options to continue operation with the current systems. But, as pointed out earlier, eventually replacement will be required. A cost–benefit analysis will be a necessary part of the case to justify the extent, cost and timing of the replacement project.

12.5.1.1 Analyse system fault data

As equipment ages, failures may have a growing impact on operations. Of particular interest will be events caused directly by control system failures, such as trips, loss of production and unit start-up delays. If trended over time, a pattern may emerge of growing failures, and so a projection of future lost production revenue can be estimated.

In some cases, current failure costs may be insignificant if spares have remained available. In that instance, the risks in retaining equipment after support for spare parts has stopped in future will need to be carefully assessed.

12.5.1.2 Review opportunities for improvement

Opportunities to improve production revenue or reduce operating costs should be reviewed by assessing suggestions from all stakeholders. The mechanisms for realising the opportunity can be either specified in detail or left to the prospective DCS vendor to put forward their experience to provide solutions. Examples include:

- Efficiency improvement: conversion to sliding pressure; apply advanced steam temperature control and raise the steam temperature setpoint.
- Market advantage: improve start-up times with higher-level sequence automation; raise ramp rates.
- Unit availability and forced outages: improve coal mill controllability; add further runback functions; improve unit stability; revise operator task burden by directing equipment fault alarms straight to maintenance crews.
- Fuel costs: change to cheaper coal source (requiring changes to mill, fuel, air and CV controls); switch ignition to natural gas.
- Operation costs: changes to operating roles, for example, enabling one of two operators to leave the control room for inspections, or reducing additional operations staff needed for start-ups by improving plant interlock diagnosis, adding instrument redundancy and centralising control of stand-alone plant, such as feedwater chemical injection and emission control systems.

12.5.1.3 Minimise project costs with schedule planning

The outage time required for a complete DCS replacement can vary significantly depending on the extent of change to field equipment, control and termination cabinets and the control room, but the period will generally fall between 2 and 4 months. The cost of lost production would therefore be very high unless installation is aligned with existing major overhauls.

Nonetheless, it is important not to rush the project to meet the first available suitable outage: quoted costs also escalate if a tight design and manufacture program is set by the owner; much better to plan well ahead and allow reasonable periods for project execution.

12.5.2 Technical preparations

12.5.2.1 Pre-project planning (PPP)

This stage of preparation goes by several names: in the United States it is often referred to as front-end loading (FEL), alluding to the task of examining detailed operational and design requirements very early in the project to more accurately establish the full range of the project scope. PPP has its roots in the construction industry, but applies equally to major technical renovation projects such as a DCS replacement [2].

PPP can be a large task but one which, if well executed, will reduce design, installation and commissioning time and recover the cost of the project much earlier. Because such projects do not occur frequently unless the owner has a very large fleet of generating units, it is often effective to engage specialist engineering services for the task. Some aspects of a PPP include:

- Project risk analysis and management
  A risk analysis with an associated set of strategies to mitigate those risks will help establish the responsibilities and actions required to execute the project effectively. Risks reviewed will be wide-ranging, including technical, schedule, operational, financial (such as project funding and returns), commercial, safety and environmental aspects. By way of example, an assessment of operational risks is provided in [3], in which a major change to operating arrangements was being considered. The project involved providing remote (off-site) control of a supercritical unit in combination with multi-unit operation of a four-unit drum-boiler station from a refurbished centralised control room.
- Establishing a detailed scope of the project
  The question of what to include in the project's scope of works must be carefully and fully answered at the PPP stage to avoid cost and schedule overruns, taking into account many factors such as by asking
  - What is the remaining operational life of independent control equipment that may be brought into the new DCS
  - Can existing control cabinets be used?
  - Will field cabling require additional termination facilities such as junction cubicles to bridge to new locations?
  - Do new instrument redundancy principles apply – if so, where will additional transmitters be required?
  - Do actuators, positioners, or, if electric, power control units need to be replaced?
  - What is the accuracy of existing documentation?
  - Is the control room to be refurbished or relocated? If so, is this to be included in the DCS replacement contract or to be completed by others?
  - Are there any regulatory requirements or safety standards requiring compliance?

Such an undertaking cannot be done as a 'desktop' review – it will require a clear understanding of the status of the existing control systems and, most importantly, the objectives of the upgrade (e.g. plant life extension, control improvement, plant availability improvement, operational efficiency improvement, etc.). The status of the existing control systems, instrumentation and interfaces usually requires extensive on-site surveys, discussions and documentation searches.

- Determining the time frame
  The periods required for each phase of the specification, design and implementation should be set out in an overall project schedule with sufficient detail to generate accurate contract milestone dates. The periods required will depend heavily upon the extent of the systems to be replaced, whether the control logic replacement is to be 'like-for-like' or with substantial enhancement, equipment delivery times and the dates of available maintenance outages for installation.

12.5.2.2 Benchmarking performance

Benchmarking a control system's performance is the process of recording plant parameters and controlled process deviations through a variety of tests such as:

- steady maximum/minimum loads and during narrow and wide-range load ramps;
- AGC MW-following through various profiles;
- auxiliary plant trips and runbacks and
- response to simulated frequency disturbances.

A benchmarking procedure can also include actuator step tests to gather plant response data for vendor reference and identify plant issues such as actuator hysteresis, which can then be addressed during the control system upgrade.

A tabulation of the results recording maximum deviations and times for processes to settle can provide a helpful summary for preparing the performance requirements for the technical specification. It can also identify particular areas, for example, steam temperature and pressure control, where the vendor may be requested to improve performance. It is wise to make a report of the benchmarking results available to prospective suppliers as it will help them understand the effort required to meet performance guarantees.

12.5.2.3 Documentation updates

When embarking on a replacement or refurbishment project, the task will be immeasurably assisted by the availability of comprehensive and accurate documentation of the systems as currently configured, together with detailed information on the reasons behind all changes to the original designs. Documents to review and update include:

- Any functional-level control drawings and control descriptions that may form part of the functional design specification, but not accurately reflect the current control system configuration.
- Loop diagrams: Ensure that loop information for new instruments or interfaces has been captured in the latest, registered copies of diagrams or the master loop-generating database.
- P&IDs: Add any new instruments and actuators to the drawings; replace local controllers with DCS control symbols where they have been modified for remote control.
- I/O database: It is crucial to have an up-to-date database of field instruments, drives and interfaces with other stand-alone control systems. This is an essential document because it defines the interface between the hardware and software design of the control systems. An accurate I/O database assists to decouple these two design aspects of a control system upgrade. The effort required can vary widely, depending whether a database exists and how it has been managed.

12.5.2.4 The control room and HMI features

In the past, changing from a hard desk to soft desk arrangement typically involved significant changes to the whole control room environment. Walls, ventilation, lighting and doorways may all have been modified.

Where the upgrade is from one screen-based system to another, the extent of change (unless the control room is being relocated) is not so great. Still, opportunity should be taken to reflect on how the whole operating environment might be improved. Some aspects to consider include whether a variable-height desk should be provided to enable standing operations; whether sound levels and lighting are adequate and if the opportunity exists to provide new break-out facilities for interaction with engineering and maintenance personnel. There will be budget constraints, but an ergonomically efficient control room can pay for itself through better decisions and actions to keep plant in service through potentially costly events. Two other examples of a modern desk layout were presented in Chapter 8.

The publication of a recent standard for HMI design, as mentioned in Section 12.3.1.3, should also be given consideration while the opportunity exists to secure resources for revising any graphics as part of the project.

12.5.2.5 Control enhancements

The opportunity will exist to make changes to the level of sequence automation, protection (inputs to initiate trips, runbacks, etc.), alarming and modulating controls. Influencing factors may include any intended changes to operator roles and attendance levels (for instance, having one operator control two units), market participation, operating load range or fuel sources (such as switching from oil to natural gas ignitors). These will all influence the extent of control enhancements. Some aspects to consider for improving flexibility of unit operation were discussed in Chapter 10.

12.6 Implementing the project

Where a control system is to be fully replaced, it is important to carefully follow the design, testing and commissioning processes to achieve a successful outcome. What follows are some pointers, based on experience, that will assist:

- If possible, appoint a 'champion' of the project to see it through to completion. The project will need to be clearly explained to all the stakeholders, and the responsibilities of different parties to provide information, safety isolations of equipment and access to different parts of the plant for equipment installation and then progressive testing will all need to be identified, recorded and coordinated. In some cases, the plant owner will provide technicians to assist in loop checking of field equipment and on-site witness testing of drive interlocks and trips. It is at this point that a definition of all terminal points for interfacing with field devices and other systems becomes a crucial piece of information.
- If the station owner is to be involved in the design of HMI graphics, it is often worthwhile to have operators participate in the activity as they are aware of the related signals needed to be visible when operating particular equipment. An example is to have fuel flow, drum level, main steam pressure and temperature, as preconfigured trends, on an overview graphic from which an operator would manually adjust the boiler master hand/auto station. But there are important considerations: graphics should follow project design guidelines to provide consistency and clarity, so training in the design principles is important. It is also

important to note that one piece of information should be available on all graphics, or displayed as a hard-wired indication: the boiler protection code National Fire Prevention Association 85 requires that drum-level indication be always visible to the operator. It is also useful to have other parameters located in the same area on related graphics; for example, include gross MW load on most screens, and display main steam pressure, total fuel and total air flow values as a minimum on all the boiler graphics.

- If an operator training simulator is to be supplied with the project, ensure that is it available well before the first piece of equipment is energised and that the operators involved with the commissioning activities are familiarised with all aspects of the new interface (this applies whether changing from hard to soft desk or whether both old and new systems are screen-based). Conducting the factory acceptance test using the new training simulator (if supplied) is the best way to ensure this.
- Prepare a detailed schedule showing all major milestones and the systems to be available for each. Each instrument, actuator and motor can be linked in a database with the plant test activities where they will first be required. As they are tested and signed off, 'what's left' lists for the testing milestones can be generated to set daily commissioning priorities.
- The current critical path should always be tracked to ensure it does not strike impediments.
- Prepare weekly look-ahead plans that are reviewed and updated daily.
- Ensure all performance tests are conducted and reviewed against the contract requirements. As noted in Chapter 9, this may often be inconvenient to production but is crucial to ensure the project meets all its aims.

12.7 Keeping track of system reliability and costs

Once the equipment has been acquired, the users must focus on how best to use it, and they must be prepared to ignore all information on new systems (which arc always better than the one that has just been bought!). The fact is that it *has* been bought, and now the users must gain the maximum advantage from the investment.

An important procedure is to keep careful records of all the markers along the road that stretches from initial acquisition to *its* ultimate replacement:

- Log each failure that occurs, recording the time and date of the event. This is useful information for the user and the vendor since it supports or disproves reliability claims.
- Keep a note of the time that elapses between placing an order and receiving the goods. This will help reinforce your arguments if delays in shipping spares are increasing, exposing the plant to increased levels of risk, causing reduced output or incurring increased operational costs.
- Monitor and record how long it takes to replace the failed item of equipment with a spare from stock.
- Record the cost of spare parts (I/O modules, control processors, power supplies, etc.).

- When recording details of a failure, make a note of any plant outage or reduction in output that can be directly attributed to the failure.
- It is also useful to try and relate the failure to any event that occurred at or just before the time at which the failure occurred, such as a severe thunderstorm, or the failure of any other electrical plant or machines. This can assist in post-mortem analyses which may point to a design defect such as inadequate screening or earthing, or poor design of power supplies.

12.8 Change management

It is almost inevitable that the design of a control system will alter between the initial conceptual stage, through commissioning and handover, and during prolonged use afterwards.

A change management procedure should be a requirement that is included in any technical specification, without mandating the exact mechanism. The DCS vendor should be required to present their system to describe, approve, modify, test and trace changes.

At the design stage, the changes may be to correct errors, but it is equally likely that they are required because of some misunderstanding of the plant's characteristics or functions. While the design is being functionally prepared, changes can be managed with drawing revisions, database updates, correspondence, etc. At some point, for example, when vendor factory testing begins, a 'design freeze' may be declared, after which point a detailed change management procedure will become active. Vendors are usually agreeable to this so that responsibility for costs for any later changes can be clearly allocated. In any case, it is good practice to manage all design changes once application logic preparation has reached an agreed milestone. A separate procedure and set of change records may be required once on-site commissioning has commenced.

Some fields to include in a change management record include:

- Reason for change
- Change details
- Impact on program
- Party to whom the cost is directed (i.e. whether the change is fixing an error, or a new feature done to the owner's direction)
- Approved by/date
- Test method or procedure reference
- Tested by/date (with space to sign for each unit)
- Witnessed by/date (with space to sign for each unit)
- All documentation updated (signed for each unit)
- For on-site changes:
    - Temporary or permanent?
    - If temporary, date restored
    - Unit outage required?

The change management logs also become a useful document set for future DCS upgrades or replacement since if the reasons for the changes are forgotten there is a possibility that when the time comes to refurbish the systems again, the errors or misunderstandings will be repeated. This is particularly the case if functional-level drawings were prepared at the design stage, but revised only to some intermediate point in the project rather than reflecting the final configuration.

12.9 Summary

The short life cycles for digital microelectronic components have made the need to upgrade control systems virtually inevitable. Most power plants older than 8 years or so have already been through at least one upgrade, with many changing electronic or first-generation DCS controls completely.

To ensure vendors retain their customer base, modern DCS upgrade paths are usually provided that enable controllers to be replaced without expensive replacement of power supplies, system backplanes or control network infrastructure which have compatibility with existing application configuration.

This chapter has discussed the preparation, implementation and management of both full and partial replacements and equipment upgrades.

12.10 Conclusion

We have now reached the end of this overview of a wide and complex subject. We hope that we have been able to throw some light on the technology and that the explanations may have lifted some of the veils of mystery that sometimes seem to obscure it. The fact remains that it is a complex matter and it is unwise to entrust the safety of a plant to people who do not understand either the control aspects or the plant operations.

It has been the privilege of the three of us to have worked with power plant throughout our careers, and we hope that some of what we have put down here will be useful and that it may encourage others to take up a very interesting and important subject.

References

[1] ANSI/ISA-101.01-2015. 'Human machine interfaces for process automation systems'. American National Standard, USA, 2015. ISBN 1941546463, 9781941546468.

[2] National Research Council. 'Starting Smart: Key Practices for Developing Scopes of Work for Facility Projects', Chapter 3. Washington, DC, USA, 2003: The National Academies Press. ISBN 978-0-309-08920-3 https://doi.org/10.17226/10870. Accessed March 2018

[3] Ronan N, and Parker D. 'Assessing operational risks associated with remote and multi-unit supervision initiatives: an Australian case study'. *Proceedings, 60th Annual ISA POWID Symposium*, Cleveland, OH, USA, 2017. International Society of Automation, USA, 2018.

Index

www.ingramcontent.com/pod-product-compliance
Lightning Source LLC
LaVergne TN
LVHW011912200926
848484LV00002B/599